London • New York

**Co-published by IWA Publishing**
Alliance House, 12 Caxton Street, London SW1H 0QS, UK
Tel. +44 (0) 20 7654 5500, Fax +44 (0) 20 7654 5555
publications@iwap.co.uk
www.iwapublishing.com
IWA Publishing ISBN: 9781780405902

# Climate Change and Water Resources

# Climate Change and Water Resources

Edited by
Dr. Sangam Shrestha
Prof. Mukand S. Babel
Dr. Vishnu Prasad Pandey

CRC Press
Taylor & Francis Group
Boca Raton London New York

CRC Press is an imprint of the
Taylor & Francis Group, an **informa** business

CRC Press
Taylor & Francis Group
6000 Broken Sound Parkway NW, Suite 300
Boca Raton, FL 33487-2742

© 2014 by Taylor & Francis Group, LLC
CRC Press is an imprint of Taylor & Francis Group, an Informa business

No claim to original U.S. Government works

Printed on acid-free paper
Version Date: 20140320

International Standard Book Number-13: 978-1-4665-9466-1 (Hardback)

This book contains information obtained from authentic and highly regarded sources. Reasonable efforts have been made to publish reliable data and information, but the author and publisher cannot assume responsibility for the validity of all materials or the consequences of their use. The authors and publishers have attempted to trace the copyright holders of all material reproduced in this publication and apologize to copyright holders if permission to publish in this form has not been obtained. If any copyright material has not been acknowledged please write and let us know so we may rectify in any future reprint.

Except as permitted under U.S. Copyright Law, no part of this book may be reprinted, reproduced, transmitted, or utilized in any form by any electronic, mechanical, or other means, now known or hereafter invented, including photocopying, microfilming, and recording, or in any information storage or retrieval system, without written permission from the publishers.

For permission to photocopy or use material electronically from this work, please access www.copyright.com (http://www.copyright.com/) or contact the Copyright Clearance Center, Inc. (CCC), 222 Rosewood Drive, Danvers, MA 01923, 978-750-8400. CCC is a not-for-profit organization that provides licenses and registration for a variety of users. For organizations that have been granted a photocopy license by the CCC, a separate system of payment has been arranged.

**Trademark Notice:** Product or corporate names may be trademarks or registered trademarks, and are used only for identification and explanation without intent to infringe.

**Library of Congress Cataloging-in-Publication Data**

Climate change and water resources / edited by Sangam Shrestha, Mukand S. Babel, and Vishnu Prasad Pandey.
       pages cm
    Includes bibliographical references and index.
    ISBN 978-1-4665-9466-1 (hardback)
    1. Water-supply--Effect of global warming on. 2. Water supply--Environmental aspects. 3. Climatic changes--Environmental aspects. I. Shrestha, Sangam, author, editor of compilation. II. Babel, Mukand S., author, editor of compilation. III. Pandey, Vishnu Prasad, author, editor of compilation.

TD353.C528 2014
628.1--dc23                                                                          2014003179

**Visit the Taylor & Francis Web site at**
**http://www.taylorandfrancis.com**

**and the CRC Press Web site at**
**http://www.crcpress.com**

# Contents

Foreword by Kuniyoshi Takeuchi ............................................................. vii
Foreword by Worsak Kanok-Nukulchai ..................................................... ix
Preface ........................................................................................... xi
Acknowledgments .............................................................................. xiii
Editors ............................................................................................ xv
Contributors ..................................................................................... xvii

**Chapter 1** Global Climate System, Energy Balance, and the Hydrological Cycle ..................................................................................... 1

*Sangam Shrestha and Prasamsa Singh*

**Chapter 2** Climate Variability and Change ............................................. 31

*Anthony S. Kiem*

**Chapter 3** Detection and Attribution of Climate Change ........................... 69

*H. Annamalai*

**Chapter 4** Uncertainty in Climate Change Studies ................................... 81

*Satish Bastola*

**Chapter 5** Climate Change Impacts on Water Resources and Selected Water Use Sectors ............................................................. 109

*Mukand S. Babel, Anshul Agarwal, and Victor R. Shinde*

**Chapter 6** Economics of Climate Change .............................................. 153

*Sujata Manandhar, Vishnu Prasad Pandey, Futaba Kazama, and So Kazama*

**Chapter 7** Climate Change Vulnerability Assessment ............................. 183

*Vishnu Prasad Pandey, Sujata Manandhar, and Futaba Kazama*

**Chapter 8** Climate Change Adaptation in Water .................................... 209

*S.V.R.K. Prabhakar, Binaya Raj Shivakoti, and Bijon Kumer Mitra*

**Chapter 9** Managing Climate Risk for the Water Sector with Tools and Decision Support .................................................................. 239

*Julian Doczi*

**Chapter 10** Transboundary River Systems in the Context of Climate Change ..... 291

*Soni M. Pradhanang and Nihar R. Samal*

**Chapter 11** International Negotiations on Climate Change and Water .............. 331

*Binaya Raj Shivakoti and Sangam Shrestha*

**Index** ............................................................................................................. 359

# Foreword by Kuniyoshi Takeuchi

Climate and water are intrinsically related, and any change in the former is bound to have repercussions on the latter and vice versa. Climate change is a phenomenon that is becoming increasingly difficult to deny, with compelling evidence worldwide pointing to a distinct change in temperatures, magnitude and timings of rainfall, sea level rise, and other related variables. Furthermore, in many parts of the world, unprecedented floods and droughts are being observed more frequently. Alarmingly, a very recent study published in *Nature* suggests that, for the business-as-usual scenario, by 2047, more than half of the planet will experience average temperatures hotter than anything seen between 1860 and 2005. Under the circumstances, depending upon the specific impact of climate change, the threat to water resources across the globe is very real.

The interest in climate change impacts on water resources is not new. A plethora of studies conducted by leading international organizations, NGOs, universities, and independent researchers have all, in some way or other, expanded the scope of our understanding of this complex process. In recent times, because of progress in technology and improved knowledge, massive strides have been made in quantifying the relationship between water and climate change at much finer resolutions. Despite this, uncertainties still remain, and there is a need for further research to improve on the existing deficiencies. To be able to do so, it is important to have a thorough understanding of the existing knowledge, for which this book is an excellent gateway.

*Climate Change and Water Resources* is a well-compiled book that covers the various dimensions of water and climate change. The strength of the book lies in the gamut of topics included, which range from detection and quantification of the impacts of climate change on water resources to managing water resources under these conditions. The case studies described at the end of some chapters provide admirable insight into the scientific, engineering, social, and governance aspects of water resources under changing climate. Importantly, the book also touches on adaptation options, which are crucial to ensure the sustainability of water resources. The expertise and experience of the vast selection of authors and contributors has culminated in this comprehensive piece of literature for which there is much need.

This book covers all the major issues related to water and climate change. Readers will get a very good background, both theoretical and practical, into developing an understanding of the potential effect of climate change on water resources and the various means to address them. Because of the broad range of areas discussed, I am quite confident that this book will serve as a useful tool for students, researchers, and decision makers to develop and implement the knowledge gained here. I expect this book to make a significant contribution to the growing body of literature on climate change and water resources.

**Kuniyoshi Takeuchi**
*Director, International Centre for Water Hazard
and Risk Management (ICHARM)
Professor Emeritus, University of Yamanashi
Chair, International Union of Geodesy and Geophysics
(IUGG) GeoRisk Commission
Vice-chair, Science Committee ICSU-ISSC-UNISDR
Recipient of IAHS-UNESCO-WMO International Hydrology Prize, 2012*

# Foreword by Worsak Kanok-Nukulchai

Our world is continuously changing and becoming increasingly complex. Today's dynamic environment requires timely and innovative solutions to overcome the challenges of tomorrow, and central to these challenges is the notion of sustainable development. As pointed out by UNESCO: "If sustainable development is to mean anything, it must be based on an appropriate understanding of the environment, i.e., an environment where knowledge of water resources is basic to virtually all human endeavors." In recent years there has been compelling evidence to suggest that climate change does, and will continue to, impact water resources in a way that poses serious questions on the sustainability of water resources. As a result, there is a keen interest among the concerned stakeholders to generate better understanding of (a) the linkage between climate change and water resources, (b) methodologies for detecting and attributing climate change, (c) assessing impacts and vulnerabilities, (d) evaluating adaptation strategies, and (e) quantifying uncertainties. Given that water is a basic human need, which ranks high on local and global political agendas, research focus on climate change and water is expected to intensify in the future.

The Asian Institute of Technology (AIT) is a leading regional institution for higher education in technology and management whose mandate promotes research on sustainable development. Over the years, water education and research at AIT has kept pace with the ever-changing needs in the region and beyond and has dynamically evolved from being "engineering oriented" to "sustainable development oriented." In the process, the institute has been instrumental in addressing the research and capacity needs of key water-related and cross-cutting issues in the region. This book is the latest in the series of contributions and attempts to compile and disseminate a consolidated knowledge on climate change and water resources to a broad range of audience.

The editors of this book are well-known international experts in the field of climate change and water resources and have assembled a remarkable team of scientists and water professionals as contributors. A unique feature of this book is the variety of topics that are covered, which include fundamentals of climate and its variability/change, detection and attribution of this change, vulnerability, impact, adaptation, climate change economics, tools for managing risks, uncertainties, and a thorough review and analysis of international negotiations on climate change and water. This book is a one-stop knowledge hub that can cater to the needs and demands of students, researchers, practitioners, and policy makers alike.

Climate change is now a global phenomenon, and its impacts on water resources are always subject to uncertainty. This uncertainty can only be reduced when our understanding of this complex process is improved. I hope that this book will provide readers with a thorough knowledge of the climate–water linkage and stimulate them into exploring new avenues of research to improve upon the drawbacks of existing approaches and enter new territories.

**Worsak Kanok-Nukulchai, PhD (University of California, Berkeley)**
*President, Asian Institute of Technology (AIT), Pathumthani, Thailand*

# Preface

Climate change is one of the most significant phenomena of the twenty-first century and has gained a lot of attention in recent times. It has affected all dimensions of life from agriculture, food security, and energy to water-induced disasters. Studies have identified climate change and the proper management of water in the context of climate change as important and as extremely challenging. As a result, increasing scientific research is being carried out globally to understand these areas better and to apply the results so as to benefit the human race.

Knowledge of climate variations and climate change can be valuable for water resources management in agriculture, urban and industrial water supply, hydroelectric power generation, navigation, and recreation and in maintaining the ecosystem. Forecasting near-term climate change or identification of the state of the global climate system and its consequences can help managers develop adaptive strategies, implement mitigating policies, and make strategic investments in infrastructure and information sources for integrated water resources management.

Several types of researches related to climate change and water resources have been carried out and presented and published in many reports and journal publications. However, the information is dispersed and makes a comprehensive overview difficult and often impractical for scholars and practicing water resource engineers. A source is needed that will provide a general overview of the many interrelated aspects of climate variations, climate change, and connections to water resources and provide references to the literature for details on individual subjects. Recognizing this need, we have synthesized theories and principles of climate change and its connection to water resources management in this book.

This book is a compilation of established theories and principles and research articles spanning the wide array of climate change science, impacts on the water sector, and mitigation and adaptation strategies. The major objective of this book is to contribute to the development of a knowledge base in the field of climate change and water. The chapters have been contributed, reviewed, and edited by well-known experts in the field of climate change and water resources. Each chapter provides an analysis of the issues raised and is supported by appropriate examples. Composed in a textbook format, the book discusses not only the theoretical background of the topics but also provides explicit case studies.

The book will be helpful to a wide range of readers who are directly or indirectly working with climate change and water as it deals with a broad scope of related topics. It is expected to cater to the requirements of a wide range of readers. Undergraduate and postgraduate students, scientists and those working at research institutes, design engineers and implementing agencies, and planners as well as different governments can make use of this book to better contribute to society.

**Sangam Shrestha**
**Mukand S. Babel**
**Vishnu Prasad Pandey**

# Acknowledgments

The material for this book is a product of the lecture notes prepared for the course on climate change and water resources conducted for the postgraduate program of Water Engineering and Management (WEM) at the Asian Institute of Technology. A number of individuals have contributed to the preparation of this book. We extend our deepest gratitude to all of them, but because of space constraints, it is not possible to mention all their names here. Our sincere thanks to all the contributing authors who prepared the chapters despite their busy schedules and who were always supportive despite constant and frequent reminders.

We are particularly thankful to Utsav Bhattarai, who helped in collecting the data and information needed to prepare the proposal for the book. We would also like to thank Smriti Malla for her tireless efforts in communicating with the contributing authors and compiling and formatting the chapters.

The inspiration to publish this book came from the students of the aforementioned course, who highlighted the absence of such a book encompassing the themes of the course. Hence, we would also like to thank all the students and staff of WEM who gave us their full support and encouragement in this endeavor.

# Editors

**Dr. Sangam Shrestha** is as an associate dean of the School of Engineering and Technology and an assistant professor of water engineering and management at the Asian Institute of Technology (AIT), Thailand. He is also a visiting faculty of the University of Yamanashi, Japan, and a research fellow of the Institute for Global Environmental Strategies (IGES), Japan. His research interests are in the field of hydrology and water resources, including climate change impact assessment and adaptation on the water sector, water footprint assessment, and groundwater assessment and management.

After earning his PhD, Dr. Shrestha continued his research in conjunction with the Global Centre of Excellence (GCOE) project at the University of Yamanashi in Japan until 2007, where he was involved in the development and application of a material circulation model and groundwater research in Kathmandu Valley. He then worked as a policy researcher at IGES, where he was actively involved in research and outreach activities related to water and climate change adaptation and groundwater management in Asian cities. Dr. Shrestha has published more than two dozen peer-reviewed international journal articles and presented more than three dozen conference papers, ranging from hydrological modeling to climate change adaptation in the water sector. He recently published *Kathmandu Valley Groundwater Outlook*, which consists of the findings of scientific research as well as opinions of concerned authorities and experts on groundwater at Kathmandu Valley, Nepal.

His present work responsibilities at AIT include lecturing at the postgraduate and undergraduate levels, supervising research done by postgraduate students, and providing consulting services on water-related issues to government and donor agencies as well as research institutions. He has conducted several projects related to water resources management, climate change impacts, and adaptation and has received project awards from several international organizations, such as Asian Pacific Network for Global Change Research (APN), Canadian International Development Agency (CIDA), European Union (EU), Food and Agriculture Organization (FAO), International Foundation for Science (IFS), Institute for Global Environmental Strategies (IGES), and United Nations Environment Programme (UNEP).

**Dr. Mukand S. Babel** is a professor and coordinator of water engineering and management at the Asian Institute of Technology (AIT), Thailand. He also has been a visiting professor at Kyoto University, Japan, and University Teknologi MARA, Malaysia. In addition, he holds the directorship of the Asian Water Research and Education (AWARE) Center at AIT, a regional center of excellence on capacity building and research and outreach for sustainable water management. He also is a coordinator of the regional center of the United Nations University's Water Learning Centre (UNU-INWEH, Canada), which offers an e-learning diploma course on integrated water resources management (IWRM).

Professor Babel's professional experience in teaching, research, and consultancy spans 30 years, mainly in Asia. He specializes in hydrologic and water resources modeling as applied to integrated water resources management and teaches graduate-level courses on watershed hydrology, hydrologic and water resources modeling, integrated water resources management, water supply and sanitation, and floods and droughts. He has over 250 publications in international refereed journals, book chapters, and conference proceedings, and he conducts interdisciplinary research relating hydrology and water resources with the economic, environmental, and socioeconomic aspects of water to address diverse water problems and issues in Asia. Currently, his research activities include climate change impact and adaptation in the water sector.

Professor Babel has carried out several capacity building, research, and sponsored projects in collaboration with several international agencies/organizations, such as DANIDA, UNEP, ADB, WB, UNESCO, UNU, IGES, APN, FAO, and GWP, as well as various universities and institutions in Japan, South Korea, Europe, and the United States. Notably, he was an IWRM expert for support to capacity building at the Water Resources University, Vietnam, which was funded by DANIDA for six and a half years, from 2001 to 2007. Dr. Babel has also worked as a consultant to various governments (India, Thailand, Nepal, Bhutan, Indonesia, etc.) and in the private sector in Asia.

**Dr. Vishnu Prasad Pandey** has been on the faculty at the Asian Institute of Technology and Management (AITM) in Nepal since July 2013. Before joining AITM, he worked as a postdoctoral researcher at the International Center for River Basin Environment–University of Yamanashi (ICRE-UY), Japan. He received an IAH fellowship (2012) to participate in the IAH Congress at Niagara Falls, a Takeda Foundation in Japan (2011–2013) fellowship for research activities in Nepal, an IWA fellowship (2010) to participate in the IWA World Water Congress held in Montreal, and a United Nations University (UNU) in Japan fellowship to carry out a "vulnerability assessment of freshwater resources in the Bagmati River Basin" in Nepal. Dr. Pandey received the Best Paper Award at the 5th IWA Young Water Professional Conference held July 5–7, 2010, in Sydney, Australia.

His research interests include hydrological modeling, water resources assessment and management, climate change vulnerability and adaptation, groundwater development and management, virtual water and water footprint analysis, and GIS application in water resources management. His recent publications include *Kathmandu Valley Groundwater Outlook* and journal articles related to climate change impacts and adaptation published in *Hydrological Processes, Regional Environment Change*, and *Climate Change*. He has published nearly two dozen peer-reviewed articles in refereed journals and regularly reviews articles for several international journals.

# Contributors

**Anshul Agarwal**
Water Engineering and Management
School of Engineering and Technology
Asian Institute of Technology
Klong Luang, Thailand

**H. Annamalai**
International Pacific Research Center
School of Ocean and Earth Science and Technology
University of Hawaii
Honolulu, Hawaii

**Mukand S. Babel**
Water Engineering and Management
School of Engineering and Technology
Asian Institute of Technology
Klong Luang, Thailand

**Satish Bastola**
Center for Ocean-Atmospheric Prediction Studies
Florida State University
Tallahassee, Florida

and

Center of Research for Environment Energy and Water
Kathmandu, Nepal

**Julian Doczi**
Water Policy Programme
Overseas Development Institute
London, United Kingdom

**Futaba Kazama**
International Research Center for River Basin Environment
University of Yamanashi
Kofu, Japan

**So Kazama**
Department of Civil and Environmental Engineering
Graduate School of Engineering
Tohoku University
Sendai, Japan

**Anthony S. Kiem**
Faculty of Science and Information Technology
School of Environmental and Life Sciences
University of Newcastle
Newcastle, New South Wales, Australia

**Sujata Manandhar**
Department of Civil and Environmental Engineering
Graduate School of Engineering
Tohoku University
Sendai, Japan

and

International Research Center for River Basin Environment
University of Yamanashi
Kofu, Japan

**Bijon Kumer Mitra**
Natural Resources and Ecosystem Services
and
Freshwater Team
Institute for Global Environmental Strategies
Hayama, Japan

**Vishnu Prasad Pandey**
Civil and Infrastructure Engineering
 Group
Asian Institute of Technology and
 Management
Lalitpur, Nepal

and

International Research Center for River
 Basin Environment
University of Yamanashi
Kofu, Japan

**S.V.R.K. Prabhakar**
Natural Resources and Ecosystem
 Services
Institute for Global Environmental
 Strategies
Hayama, Japan

**Soni M. Pradhanang**
Department of Geography
Institute for Sustainable Cities
Hunter College
City University of New York
New York, New York

**Nihar R. Samal**
Department of Geography
Institute for Sustainable Cities
Hunter College
City University of New York
New York, New York

**Victor R. Shinde**
Water Engineering and Management
School of Engineering and Technology
Asian Institute of Technology
Klong Luang, Thailand

**Binaya Raj Shivakoti**
Institute for Global Environmental
 Strategies
Hayama, Japan

**Sangam Shrestha**
Water Engineering and Management
School of Engineering and Technology
Asian Institute of Technology
Klong Luang, Thailand

**Prasamsa Singh**
Department of Earth and Atmospheric
 Science
University of Alberta
Edmonton, Alberta, Canada

# 1 Global Climate System, Energy Balance, and the Hydrological Cycle

*Sangam Shrestha and Prasamsa Singh*

## CONTENTS

1.1 Introduction ..................................................................................................1
1.2 Components of Global Climate System........................................................3
    1.2.1 Atmosphere........................................................................................3
        1.2.1.1 General Circulation of the Atmosphere..............................4
        1.2.1.2 Atmosphere and Its Role in Climate...................................7
    1.2.2 Hydrosphere.......................................................................................7
        1.2.2.1 Oceanic Circulation ........................................................... 11
    1.2.3 Cryosphere ....................................................................................... 14
    1.2.4 Lithosphere ...................................................................................... 16
    1.2.5 Biosphere ......................................................................................... 17
1.3 Global Energy Balance ............................................................................... 17
    1.3.1 Greenhouse Effect ........................................................................... 19
1.4 Hydrological Cycle ..................................................................................... 21
    1.4.1 Reservoirs ........................................................................................22
    1.4.2 Flows (or Fluxes).............................................................................23
    1.4.3 Residence Times .............................................................................24
    1.4.4 Water Vapor ....................................................................................24
1.5 Summary .....................................................................................................27
References...........................................................................................................27

## 1.1 INTRODUCTION

In general, climate is defined as the mean and variability of relevant atmospheric variables such as temperature, humidity, precipitation, and wind. Therefore climate can be viewed as average weather over a long period of time. The climate in a given region is determined by both natural and anthropogenic (human-made) factors. The Intergovernmental Panel on Climate Change (IPCC) glossary definition is as follows:

> Climate in a narrow sense is usually defined as the *average weather*, or more rigorously, as the statistical description in terms of the mean and variability of relevant quantities over a period of time ranging from months to thousands or millions of years. The classical period is 30 years, as defined by the World Meteorological Organization (WMO).

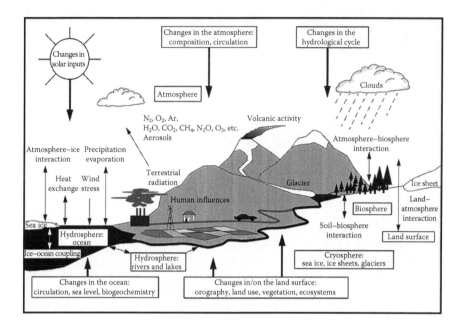

**FIGURE 1.1** Schematic view of the global climate system components (bold), their processes and interactions (thin arrows), and some aspects that may change (bold arrows). (From IPCC, *Climate Change 2001: The Scientific Basis*, Contribution of Working Group I to the Third Assessment Report of the Intergovernmental Panel on Climate Change, Cambridge University Press, Cambridge, U.K., 2001, 881pp.)

The physical climate system encompasses five fully interactive physical subsystems that contribute to creating a climate in a particular place or region. The subsystems (hereafter components) include the atmosphere, the hydrosphere (oceans, lakes, rivers), the cryosphere (sea ice, ice and snow on land, glaciers), the biosphere (surface vegetation and marine life), and the lithosphere (soil moisture, runoff) (Figure 1.1). These components interact with one another and with aspects of the Earth's biosphere to determine not only the day-to-day weather but also the long-term averages that we refer to as *climate*.

The climate system is driven by energy received from the sun (sunlight). Some of this energy is reflected back into space, but the rest is absorbed by the land and ocean and reemitted as radiant heat. Some of this radiant heat is absorbed and reemitted by the lower atmosphere in a process known as the *greenhouse effect*. The Earth's average temperature is determined by the overall balance between the amount of incoming energy from the sun and the amount of radiant heat that makes it through the atmosphere and is emitted to space.

A crucial feature of the climate system is that the sun's energy is not distributed uniformly, but rather is most intense at the equator and weakest at the poles. This nonuniform energy distribution leads to temperature differences, which the atmosphere and ocean act to reduce by transporting heat from the warm tropics to the cold polar regions. This nonuniform heating and the resulting heat transport give rise to ocean currents, atmospheric circulation, evaporation, and precipitation that we ultimately experience as weather. When the balance between incoming and outgoing

# Global Climate System, Energy Balance, and the Hydrological Cycle

energy is perturbed, this changes the amount of heat within the climate system and affects all those processes described earlier that transport heat around the globe.

Adequate understanding of interaction among the climate system components, distribution of solar radiation, global energy balance, and their influence or impact on the hydrological cycle is necessary for mainstreaming climate change adaptation in water resources management. Therefore, this chapter aims to describe the components of global climate system, energy balance, and hydrological cycle as they relate to climate change and water resource management.

## 1.2 COMPONENTS OF GLOBAL CLIMATE SYSTEM

### 1.2.1 ATMOSPHERE

The atmosphere is a layer of gases that surrounds the entire Earth. It consists mainly of nitrogen (78%), oxygen (21%), and a few other gaseous elements (1%). Water vapor accounts for roughly 0.25% of the atmosphere by mass. The concentration of water vapor (a greenhouse gas) varies significantly from around 10 ppmv in the coldest portions of the atmosphere to as much as 5% by volume in hot, humid air masses, and concentrations of other atmospheric gases are typically provided for dry air without any water vapor. The remaining gases are often referred to as trace gases, among which are the greenhouse gases such as carbon dioxide, methane, nitrous oxide, and ozone. Filtered air includes trace amounts of many other chemical compounds. Various industrial pollutants also may be present as gases or aerosol, such as chlorine (elemental or in compounds), fluorine compounds, and elemental mercury vapor. Sulfur compounds such as hydrogen sulfide and sulfur dioxide ($SO_2$) may be derived from natural sources or from industrial air pollution.

Earth's atmosphere can be divided into five main layers (Figure 1.2). These layers are mainly determined by whether temperature increases or decreases with altitude. These changes in physical properties result in a height-based division of the atmosphere into five parts: the exosphere, the thermosphere, the mesosphere, the stratosphere, and the troposphere. Two of them are relevant to climate change: the troposphere (from sea level to 10–15 km in altitude) and the stratosphere (from the top of the troposphere, to about 50 km altitude) (Salby 1992).

The troposphere contains the majority of the Earth's weather and is fundamentally driven by surface heating, which results in "the convective overturning of air (that) characterizes the region" (Salby 1992). The atmospheric circulation in the troposphere depends on imbalances between radiative heating at low and high latitudes. This uneven heating leads to a distribution of mass that "drives a meridional overturning of air, with rising motion at low latitudes and sinking motion at high latitudes." The atmosphere also circulates latitudinally, because of the Earth's rotation, so that at middle and high latitudes, the general thermal structure approximately parallels the latitudinal circles. The majority of the longitudinal (*meridional*) heat distribution is based on asymmetries in the instantaneous circulation, which leads to heat exchange between the tropics and the poles. At low latitudes, planetary rotation plays a smaller role; instead, the geographical distribution of atmospheric heating determines atmospheric circulation via latent heating and the resulting *Walker* (east–west overturning)

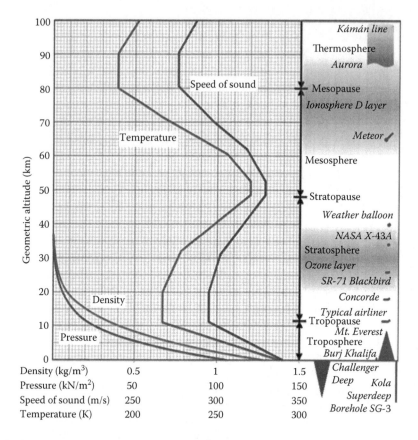

**FIGURE 1.2** Comparison of the 1962 US standard atmosphere graph of geometric altitude against density, pressure, the speed of sound, and temperature with approximate altitudes of various objects. (From Cmglee, http://commons.wikimedia.org/wiki/File:Comparison_US_standard_atmosphere_1962.svg, accessed September 16, 2013.)

and *Hadley* (north–south overturning) circulations. As a consequence of the atmospheric general circulation, temperatures at fixed altitudes in the troposphere decrease poleward from the equator, where they are at a maximum. Furthermore, pressure is only 10% of its surface value at an altitude of 10 km, while temperature falls almost linearly at a *lapse rate* of 6 K/km to roughly 220 K (−53°C) at the same altitude. The stratosphere differs significantly from the troposphere because of weak vertical motions and strong radiative processes. Although generally neglected in the past, the IPCC reports that stratospheric effects "can have a detectable and perhaps significant influence on tropospheric climate" (Stocker et al. 2001).

### 1.2.1.1 General Circulation of the Atmosphere

The high temperatures at the equator make the air there less dense. It thus tends to rise before being transported poleward at high altitudes in the troposphere. This motion is compensated for at the surface by an equatorward displacement of the air.

Global Climate System, Energy Balance, and the Hydrological Cycle 5

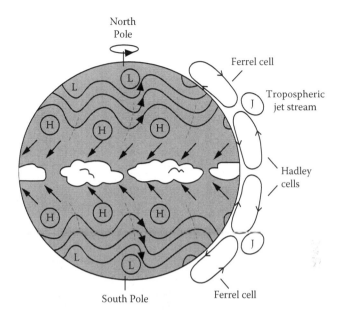

**FIGURE 1.3** Schematic representation of the annual mean general atmospheric circulation. H (L) represents high (low)-pressure systems. (From Wallace, J.M. and Hobbs, P.V., *Atmospheric Science: An Introductory Survey*, Vol. 92, International Geophysics Series, Academic Press, New York, 2006. Copyright 2006.)

On a motionless Earth, this big convection cell would reach the poles, inducing direct exchanges between the warmest and coldest places on Earth. However, because of the Earth's rotation, such an atmospheric structure would be unstable. Consequently, the two cells driven by the ascendance at the equator, called the Hadley cells, close with a downward branch at latitude of about 30° (Figure 1.3). The northern boundary of these cells is marked by strong westerly winds in the upper troposphere called the tropospheric jets. At the surface, the Earth's rotation is responsible for a deflection toward the right in the Northern Hemisphere and toward the left in the Southern Hemisphere (due to the Coriolis force) of the flow coming from the mid-latitudes to the equator. This gives rise to the easterly trade winds characteristics of the tropical regions (Figure 1.4).

The extratropical circulation is dominated at the surface by westerly winds whose zonal symmetry is perturbed by large wavelike patterns and the continuous succession of disturbances that governs the day-to-day variations in the weather in these regions. The dominant feature of the meridional circulation at those latitudes is the Ferrell cell (Figure 1.3), which is weaker than the Hadley cell. As it is characterized by rising motion in its poleward branch and downward motion in the equator branch, it is termed an indirect cell by contrast with the Hadley cell, which is termed a direct cell.

Outside a narrow equatorial band and above the surface boundary layer, the large-scale atmospheric circulation is close to geostrophic equilibrium. The surface pressure and winds are thus closely related. In the Northern Hemisphere, the winds rotate clockwise around a high pressure and counterclockwise around a

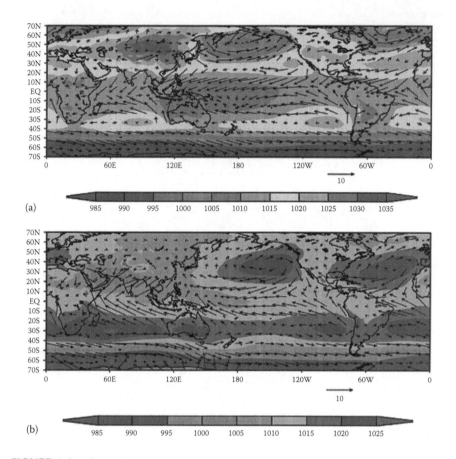

**FIGURE 1.4** 10 m winds (arrows, in m/s) and sea level pressure (in hPa) averaged over (a) December, January, and February and (b) June, July, and August. (From Goosse H. et al., *Introduction to Climate Dynamics and Climate Modeling*, September 2013, Online textbook available at http://www.climate.be/textbook.)

low pressure, while the reverse is true in the Southern Hemisphere. Consequently, the mid-latitude westerlies are associated with high pressure in the subtropics and low pressure at around 50°N–60°N. Rather than a continuous structure, this subtropical high-pressure belt is characterized by distinct high-pressure centers, often referred to as the name of a region close to their maximum (e.g., Azores high, St. Helena high). In the Northern Hemisphere, the low pressures at around 50°N–60°N manifest themselves on climatological maps as cyclonic centers called the Icelandic low and the Aleutian low. In the Southern Ocean, because of the absence of large land masses in the corresponding band of latitude, the pressure is more zonally homogenous, with a minimum in surface pressure around 60°S (Figure 1.4).

In the real atmosphere, the convergence of surface winds and the resulting ascendance does not occur exactly at the equator but in a band called the intertropical convergence zone (ITCZ). Because of the present geometry of the continents, it is located around 5°N, with some seasonal shifts. The presence of land surfaces also has a critical role in monsoon circulation. In summer, the continents warm faster than the oceans because of their lower thermal inertia. This induces a warming of the air close to the surface and a decrease in surface pressure there. This pressure difference between land and sea induces a transport of moist air from the sea to the land. In winter, the situation reverses, with high pressure over the cold continent and a flow generally from land to sea. Such a monsoonal circulation, with seasonal reversals of the wind direction, is present in many tropical areas of Africa, Asia, and Australia. Nevertheless, the most famous monsoon is probably the South Asian one that strongly affects the Indian subcontinent (Figure 1.5).

### 1.2.1.2 Atmosphere and Its Role in Climate

The atmosphere transfers heat energy and moisture across the Earth. Incoming solar radiation is redistributed from areas in which there is a surplus of heat (the equator) to areas where there is a heat deficit (the North and South Poles). This is achieved through a series of atmospheric cells: the Hadley cell, the Ferrel cell, and the polar cell (Figure 1.6). These operate in a similar way to, and indeed interact with, the ocean conveyor. For example, as the oceans at low latitudes are heated, water evaporates and is transported poleward as water vapor. This warm air eventually cools and subsides. Changes in temperature and $CO_2$ concentrations can lead to changes in the size of atmospheric cells (in particular, the Hadley cell is susceptible to these alterations), warming in the troposphere, and disproportionately strong warming in Arctic regions. The strong interactions between ocean and atmospheric dynamics, and the significant feedback mechanisms between them, mean that climate researchers must consider these Earth components as interlinked systems. The necessity to assess ocean–atmospheric changes at the global scale has implications for the way in which research is conducted. It is only by integrating paleo evidence of past changes, with present day monitoring, and projected models, that we can begin to understand such a complex system.

### 1.2.2 HYDROSPHERE

The hydrosphere describes the combined mass of water found on, under, and over the surface of a planet. Shiklomanov estimated that there are 1386 million cubic kilometers of water on Earth. This includes water in liquid and frozen forms in groundwaters, glaciers, oceans, lakes, and streams. Saline water accounts for 97.5% of this amount. Freshwater accounts for only 2.5%. Of this freshwater, 68.7% is in the "form of ice and permanent snow cover in the Antarctic, the Arctic, and in the mountainous regions." Next, 29.9% exists as fresh groundwaters. Only 0.26% of the total amount of freshwaters on the Earth is concentrated in lakes, reservoirs, and river systems where they are most easily accessible for our economic needs and absolutely vital for water ecosystems (UNESCO 1998). The total mass of the Earth's hydrosphere is about $1.4 \times 10^{18}$ tonnes, which is about 0.023% of the Earth's total mass. About $20 \times 10^{12}$ tonnes of this is in the Earth's atmosphere (the volume of 1 tonne of water is approximately 1 cm$^3$). Approximately 75% of the Earth's surface, an area of some 361 million square kilometers,

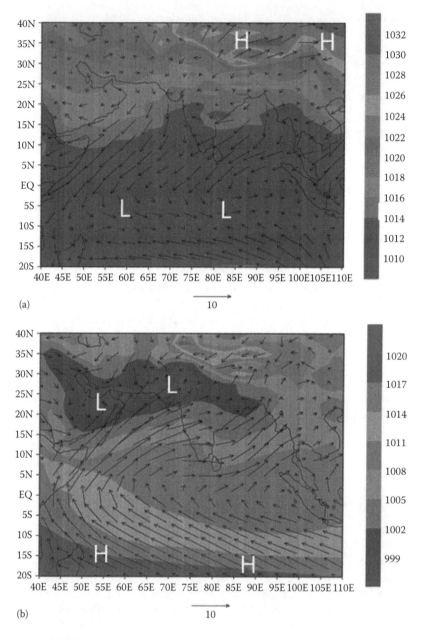

**FIGURE 1.5** 10 m winds (arrows, in m/s) and sea level pressure (in hPa) in (a) January and (b) July illustrating the wind reversal between the winter and the summer monsoon. In Northern latitude (H) represents higher wind pressure whereas (L) represents lower wind pressure in southern latitude in January. Likewise in July (L) represents lower wind pressure in northern latitude and (H) represents higher wind pressure in southern latitude. (From Goosse H. et al., *Introduction to Climate Dynamics and Climate Modeling*, September 2013, Online textbook available at http://www.climate.be/textbook.)

# Global Climate System, Energy Balance, and the Hydrological Cycle

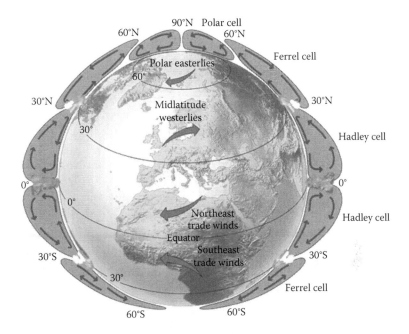

**FIGURE 1.6** Atmospheric cells that envelope the Earth in bands from the poles to the equator; the polar cell, Ferrel cell, and Hadley cell. (From Climatica, The Earth System, http://climatica.org.uk/climate-science-information/earth-system, accessed September 16, 2013.)

is covered by ocean. The average salinity of the Earth's oceans is about 35 g of salt per kilogram of seawater (3.5%) (Kennish 2001).

The ocean is an important part of the Earth system. It influences the transformation of energy and materials important to the climate system. On the most basic level, the ocean has shaped our atmosphere. Over millions of years, the concentration of gases in the atmosphere is determined by life. If life did not exist, especially life in the ocean, Earth would be very different. On a deeper level, oceanic microbes irreversibly altered the geochemistry of Earth and the biogeochemical cycles of H, C, N, O, and S. Together with the atmosphere, oceans regulate global temperatures, shape weather and climate patterns, and cycle elements through the biosphere. They also contain nearly all of the water on Earth's surface and are an important food source. Life on Earth originated in the oceans, and they are home to many unique ecosystems that are important sources of biodiversity, from coral reefs to polar sea ice communities.

Like the atmosphere, the oceans are not uniformly mixed but are structured in layers with distinct properties (Figure 1.7). Pressure increases with depth as the weight of the overlying air and water increases. Unlike the atmosphere, however, pressure changes at a linear rate rather than exponentially because water is considered as incompressible, so its mass is equally distributed throughout a vertical water column. Atmospheric pressure at sea level is 14.7 psi (also referred to as *one atmosphere*), which increases by an additional atmosphere for every 10 m of

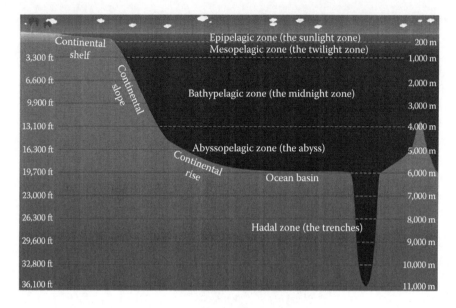

**FIGURE 1.7** Fiver layers of the ocean. (From National Weather Services, Profile of the Ocean, available at: http://www.srh.noaa.gov/, accessed September 16, 2013.)

descent underwater. This gradient is well known to scuba divers who have experienced painful *ear squeeze* from pressure differences between the air in their ears and the seawater around them.

The epipelagic, or sunlight, zone (so called because most visible light in the oceans is found here) comprises the first 200 m below the surface and is warm and mixed by winds and wave action. Surface waters account for about 2% of total worldwide ocean volume. At a depth of about 200 m, the continental shelf (the submerged border of the continents) begins to slope more sharply downward, marking the start of the mesopelagic, or twilight, zone. Here, water temperature falls rapidly with depth to less than 5°C at 1 km. This sharp transition, which is called the thermocline, inhibits vertical mixing between denser, colder water at depths and warmer water nearer the surface. About 18% of the total volume of the oceans is within this zone. Below 1 km, in the bathypelagic, or midnight, zone, water is almost uniformly cold, approximately 4°C. No sunlight penetrates to this level, and pressure at the bottom of the zone (around 4 km depth) is about 5880 psi. Little life exists at the abyssopelagic (abyssal) zone, which reaches to the ocean floor at a depth of about 6 km. Together, these cold, deep layers contain about 80% of the total volume of the ocean.

The deepest points in the ocean lie in long, narrow trenches that occur at convergence zones—points where two oceanic plates collide and one is driven beneath the other. This region is called the Hadal zone. The deepest oceanic trench measured to date is the Marianas Trench near the Philippines, which reaches more than 10 km below sea level. Highly specialized life forms, including fish, shrimps, sea cucumbers, and microbes, survive even at these depths.

# Global Climate System, Energy Balance, and the Hydrological Cycle

## 1.2.2.1 Oceanic Circulation

The surface ocean circulation is mainly driven by the winds. At mid-latitudes, the atmospheric westerlies induce eastward currents in the ocean, while the trade winds are responsible for westward currents in the tropics (Figure 1.8). Because of the presence of continental barriers, those currents form loops called the subtropical gyres. The surface currents in those gyres are intensified along the western boundaries of the oceans (the east coasts of continents) inducing well-known strong currents such as the Gulf Stream off the east coast of the United States and the Kuroshio off Japan. At higher latitudes in the Northern Hemisphere, the easterlies allow the formation of weaker subpolar gyres. In the Southern Ocean, because of the absence of continental barriers, a current that connects all the ocean basins can be maintained: the Antarctic circumpolar current (ACC). This is one of the strongest currents on Earth, which transports about 130 Sv (1 Sverdrup = $10^6$ m$^3$/s). All these currents run basically parallel to the surface winds. By contrast, the equatorial countercurrents, which are present at or just below the surface in all the ocean basins, run in the direction opposite to the trade winds.

Because of the Earth's rotation, the ocean transport induced by the wind is perpendicular to the wind stress (to the right in the Northern Hemisphere, to the left in the Southern Hemisphere). This transport, known as the Ekman transport, plays an important role in explaining the path of the wind-driven surface currents (Figure 1.8). Furthermore, along a coastline or if the transport has horizontal variations, this can lead to surface convergence/divergence that has to be compensated by vertical movements in the ocean. An important example is the equatorial

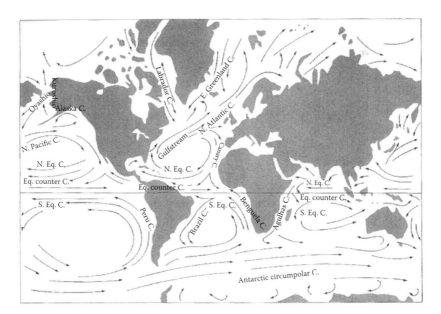

**FIGURE 1.8** A schematic representation of the major surface currents. Eq. is an abbreviation for equatorial, C. for current, N. for North, S. for South, and E. for East. (From Knauss, J.A., *Introduction to Physical Oceanography*, Waveland Press, Inc., Long Grove, IL, 2005, http://www.elic.ucl.ac.be/textbook/chapter1_node9.html, accessed September 16, 2013.)

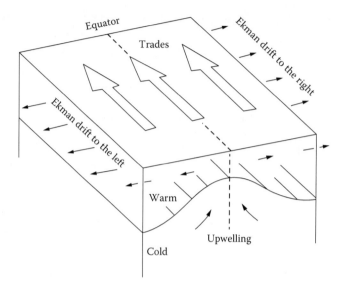

**FIGURE 1.9** A schematic representation of the equatorial upwelling. (From Cushman-Roisin, B., *Introduction to Geophysical Fluid Dynamics*, Prentice Hall, London, U.K., 1994, 319pp.)

upwelling (Figure 1.9). In the Northern Hemisphere, the Ekman transport is directed to the right of the easterly wind stress and is thus northward. By contrast, it is southward in the Southern Hemisphere. This results in a divergence at the surface at the equator that has to be compensated by an upwelling there. In coastal upwelling, the wind stress has to be parallel to coast, with the coast on the left when looking in the wind direction in the Northern Hemisphere (for instance, northerly winds along a coast oriented north–south). This causes an offshore transport and an upwelling to compensate for this transport.

At high latitudes, because of its low temperature and relatively high salinity, surface water can be dense enough to sink to great depths. This process, often referred to as deep oceanic convection, is only possible in a few places in the world, mainly in the North Atlantic and in the Southern Ocean. In the North Atlantic, the Labrador and Greenland–Norwegian Seas are the main sources of the North Atlantic deep water (NADW), which flows southward along the western boundary of the Atlantic toward the Southern Ocean. There, it is transported to the other oceanic basins after some mixing with ambient water masses. This deep water then slowly upwells toward the surface in the different oceanic basins. This is very schematically represented in Figure 1.10 by upward fluxes in the North Indian and North Pacific Oceans. However, while sinking occurs in very small regions, the upwelling is broadly distributed throughout the ocean. The return flow to the sinking regions is achieved through surface and intermediate depth circulation. In the Southern Ocean, the Antarctic bottom water (AABW) is mainly produced in the Weddell and Ross Seas. This water mass is colder and denser than the NADW and thus flows below it.

# Global Climate System, Energy Balance, and the Hydrological Cycle

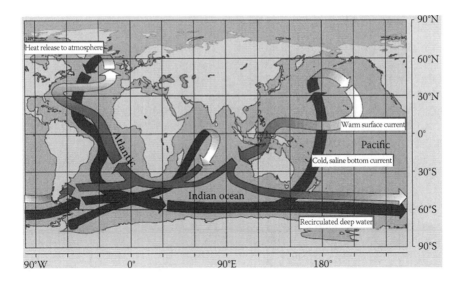

**FIGURE 1.10** A schematic representation of the oceanic thermohaline circulation. Arrows in black indicate cold, deep ocean currents; and dark gray arrows show shallow, warm water circulation patterns. (From Windows to the Universe, Thermohaline ocean circulation, http://www.windows2universe.org/earth/Water/thermohaline_ocean_circulation.html, accessed September 16, 2013.)

Note that, because of the mixing of water masses of different origins in the Southern Ocean, the water that enters the Pacific and Indian basins is generally called circumpolar deep water (CDW).

This large-scale circulation (Figure 1.10), which is associated with currents at all depths, is often called the oceanic thermohaline circulation as it is driven by temperature and salinity (and thus density) contrasts. However, winds also play a significant role in this circulation. First, they influence the surface circulation and thus the upper branch of the thermohaline circulation that feeds the regions where sinking occurs with dense enough surface waters. Secondly, because of the divergence of the Ekman transport, the winds influence the upwelling of deep water masses toward the surface in some regions. This plays a particularly important role in the Southern Ocean. Winds could also act as a local-/regional-preconditioning factor that favors deep convection.

The thermohaline circulation is quite slow. The time needed for water masses formed in the North Atlantic to reach the Southern Ocean is of the order of a century. If the whole cycle is taken into account, the time scale is estimated as several centuries to a few millennia, depending on the exact location and mechanism studied. On the other hand, this circulation transports huge amounts of water, salts, and energy. In particular, the rate of NADW formation is estimated to be around 15 Sv. Uncertainties are larger for the Southern Ocean, but the production rate of AABW is likely quite close to that of NADW. As a consequence, the thermohaline circulation has an important role in oceanographic as well as in climatology.

### 1.2.3 CRYOSPHERE

The cryosphere collectively describes the portions of the Earth's surface where water is in solid form, including sea ice, lake ice, river ice, snow cover, glaciers, ice caps and ice sheets, and frozen ground (which includes permafrost). Thus, there is a wide overlap with the hydrosphere. The cryosphere is an integral part of the global climate system with important linkages and feedbacks generated through its influence on surface energy and moisture fluxes, clouds, precipitation, hydrology, and atmospheric and oceanic circulation. Through these feedback processes, the cryosphere plays a significant role in the global climate and in climate model response to global changes.

Frozen water is found on the Earth's surface primarily as snow cover, freshwater ice in lakes and rivers, sea ice, glaciers, ice sheets, and frozen ground and permafrost (permanently frozen ground). The residence time of water in each of these cryospheric subsystems varies widely. Snow cover and freshwater ice are essentially seasonal, and most sea ice, except for ice in the central Arctic, lasts only a few years if it is not seasonal. A given water particle in glaciers, ice sheets, or ground ice, however, may remain frozen for 10–100,000 years or longer, and deep ice in parts of East Antarctica may have an age approaching 1 million years.

Most of the world's ice volume is in Antarctica, principally in the East Antarctic ice sheet. In terms of areal extent, however, Northern Hemisphere winter snow and ice extent comprise the largest area, amounting to an average 23% of hemispheric surface area in January (Table 1.1). The large areal extent and the important climatic roles of snow and ice, related to their unique physical properties, indicate that the ability to observe and model snow and ice cover extent, thickness, and physical properties (radiative and thermal properties) is of particular significance for climate research.

**TABLE 1.1**
**Areal Extent and Volume of Snow Cover and Sea Ice**

| Component | Maximum Area ($10^6$ km$^2$) | Minimum Area ($10^6$ km$^2$) | Maximum Ice Volume ($10^6$ km$^3$) | Minimum Ice Volume ($10^6$ km$^3$) |
|---|---|---|---|---|
| Northern Hemisphere snow cover | 46.5 (late January) | 3.9 (late August) | 0.002 | |
| Southern Hemisphere snow cover | 0.83 (late July) | 0.07 (early May) | | |
| Sea ice in the Northern Hemisphere | 14.0 (late March) | 6.0 (early September) | 0.05 | 0.02 |
| Sea ice in the Southern Hemisphere | 15.0 (late September) | 2.0 (late February) | 0.02 | 0.02 |

*Source:* Goosse H. et al., *Introduction to Climate Dynamics and Climate Modeling*, http://www.climate.be/textbook, accessed September 16, 2013.

There are several fundamental physical properties of snow and ice that modulate energy exchanges between the surface and the atmosphere. The most important properties are the surface reflectance (albedo), the ability to transfer heat (thermal diffusivity), and the ability to change state (latent heat). These physical properties, together with surface roughness, emissivity, and dielectric characteristics, have important implications for observing snow and ice from space. For example, surface roughness is often the dominant factor determining the strength of radar backscatter (Hall 1996). Physical properties such as crystal structure, density, length, and liquid water content are important factors affecting the transfers of heat and water and the scattering of microwave energy.

The surface reflectance of incoming solar radiation is important for the surface energy balance (SEB). It is the ratio of reflected to incident solar radiation, commonly referred to as albedo. Climatologists are primarily interested in albedo integrated over the shortwave portion of the electromagnetic spectrum (~300–3500 nm), which coincides with the main solar energy input. Typically, albedo values for nonmelting snow-covered surfaces are high (~80%–90%) except in the case of forests. The higher albedos for snow and ice cause rapid shifts in surface reflectivity in autumn and spring in high latitudes, but cloud cover spatially and temporally modulates the overall climatic significance of this increase. Planetary albedo is determined principally by cloud cover and by the small amount of total solar radiation received in high latitudes during winter months. Summer and autumn are times of high-average cloudiness over the Arctic Ocean so the albedo feedback associated with the large seasonal changes in sea ice extent is greatly reduced. Groisman et al. (1994) observed that snow cover exhibited the greatest influence on the Earth radiative balance in the spring (April–May) period when incoming solar radiation was greatest over snow-covered areas.

The thermal properties of cryospheric elements also have important climatic consequences. Snow and ice have much lower thermal diffusivities than air. Thermal diffusivity is a measure of the speed at which temperature waves can penetrate a substance. Snow and ice are many orders of magnitude less efficient at diffusing heat than air. Snow cover insulates the ground surface, and sea ice insulates the underlying ocean, decoupling the surface–atmosphere interface with respect to both heat and moisture fluxes. Even a thin skin of ice eliminates the flux of moisture from a water surface, whereas the flux of heat through thin ice continues to be substantial until it attains a thickness in excess of 30–40 cm. However, even a small amount of snow on top of the ice will dramatically reduce the heat flux and slow down the rate of ice growth. The insulating effect of snow also has major implications for the hydrological cycle. In nonpermafrost regions, the insulating effect of snow is such that only near-surface ground freezes and deepwater drainage are uninterrupted (Lynch-Stieglitz 1994).

While snow and ice act to insulate the surface from large energy losses in winter, they also act to retard warming in the spring and summer because of the large amount of energy required to melt ice (the latent heat of fusion, $3.34 \times 10^5$ J/kg at 0°C). However, the strong static stability of the atmosphere over areas of extensive snow or ice tends to confine the immediate cooling effect to a relatively shallow layer, so that associated atmospheric anomalies are usually short-lived and local to regional in scale (Cohen and Rind 1991). In some areas of the world such as Eurasia, however, the cooling associated with a heavy snowpack and moist spring soils is known to play a role in modulating the summer monsoon circulation (Vernekar et al. 1995). Gutzler and Preston (1997) have presented an evidence for a similar snow–summer circulation feedback over the southwestern United States.

The role of snow cover in modulating the monsoon is just one example of a short-term cryosphere–climate feedback involving the land surface and the atmosphere. From Figure 1.1, it can be seen that there are numerous cryosphere–climate feedbacks in the global climate system. These operate over a wide range of spatial and temporal scales from local seasonal cooling of air temperatures to hemispheric-scale variations in ice sheets over time scales of thousands of years. The feedback mechanisms involved are often complex and incompletely understood. For example, Curry and Webster (1999) showed that the so-called *simple* sea ice–albedo feedback involved complex interactions with lead fraction, melt ponds, ice thickness, snow cover, and sea ice extent.

### 1.2.4 Lithosphere

The lithosphere rests on a relatively ductile, partially molten layer known as the asthenosphere, which derives its name from the Greek word asthenes, meaning *without strength*. The asthenosphere extends to a depth of about 400 km in the mantle, over which the lithospheric plates slide along. Slow convection currents within the mantle, generated by radioactive decay of minerals, are the fundamental heat energy source that causes the lateral movements of the plates on top of the asthenosphere. According to the plate tectonic theory, there are approximately 20 lithospheric plates, each composed of a layer of continental crust or oceanic crust. Three types of plate boundaries—divergent, convergent, and transform fault—separate these plates. At divergent boundaries, tensional forces dominate the interaction between the lithospheric plates, and they move apart and new crust is created. At convergent boundaries, compression of lithospheric plate material dominates, and the plates move toward each other where crust is either destroyed by subduction or uplifted to form mountain chains. Lateral movements due to shearing forces between two lithospheric plates create transform fault boundaries. The lithosphere earthquakes and volcanic activities are mostly the result of lithosphere plate movement and are concentrated at the plate boundaries.

The lithosphere plates move at a rate of about 3 cm/year. The distribution and relative movement of the oceanic and continental plates across the latitude also have profoundly affected the global climate. The major contributing factors are differences in surface albedo, land area at high latitudes, the transfer of latent heat, restrictions on ocean currents, and the thermal inertia of continents and oceans. According to the present configuration of oceans and continents, the lithosphere low latitudes have a greater influence on surface albedo because the lower latitudes receive a greater amount of solar radiation than the higher latitudes.

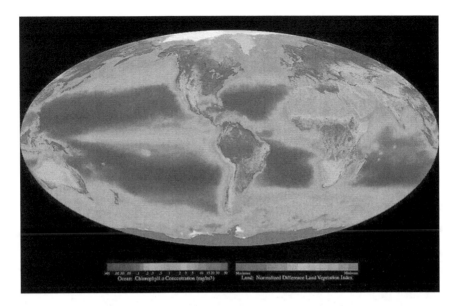

**FIGURE 1.11** This composite image gives an indication of the magnitude and distribution of global primary production, both oceanic (mg/m$^3$ chlorophyll $a$) and terrestrial (normalized difference land vegetation index). (From NASA, Goddard Space Flight Center and ORBIMAGE. The SeaWiFS Project. http://oceancolor.gsfc.nasa.gov/SeaWiFS/BACKGROUND/Gallery/index.html, accessed September 16, 2013.)

### 1.2.5 Biosphere

The biosphere is the global sum of all ecosystems. It can also be called the zone of life on Earth, a closed (apart from solar and cosmic radiation and heat from the interior of the Earth), and self-regulating system. From the broadest biophysiological point of view, the biosphere is the global ecological system integrating all living beings and their relationships, including their interaction with the elements of the lithosphere, hydrosphere, and atmosphere (Figure 1.11). The biosphere is postulated to have evolved, beginning through a process of biogenesis or biopoesis, at least some 3.5 billion years ago (Campbell et al. 2006).

## 1.3 GLOBAL ENERGY BALANCE

The global energy balance, the balance between incoming energy from the Sun and outgoing heat from the Earth, regulates the state of the Earth's climate. Any modifications to it, as a result of natural and man-made climate forcing, may cause the global climate to change.

Energy released from the Sun as electromagnetic radiation has a temperature of approximately 6000°C. At this temperature, electromagnetic radiation is emitted as shortwave light and ultraviolet energy. Electromagnetic radiation travels across space at the speed of light. When it reaches the Earth, some is reflected back to space by clouds, some is absorbed by the atmosphere, and some is absorbed at the Earth's surface (Figures 1.12 and 1.13).

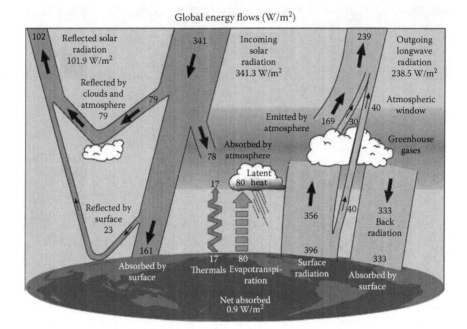

**FIGURE 1.12** Estimate of the Earth's annual and global mean energy balance for the March 2000–May 2004 period in W/m². (From Trenberth, K.E. et al., *Bull. Am. Meteorol. Soc.*, 90, 311, 2009.)

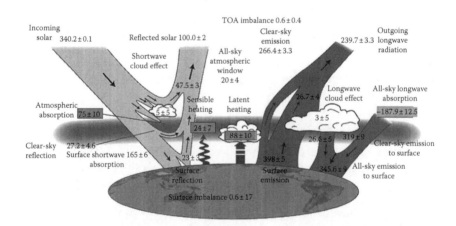

**FIGURE 1.13** Updated global annual mean energy budget of Earth for the approximate period 2000–2010. All fluxes are in W/m². Solar fluxes are in light gray shades and infrared fluxes dark gray. The four flux quantities in shaded rectangular boxes represent the principal components of the atmospheric energy balance. (From Stephens, G.L. et al., *Nat. Geosci.*, 5, 691, 2012.)

The Earth releases a lot of energy it has received from the Sun back to space. However, since the Earth is much cooler than the Sun, its radiating energy is of longer wavelength: infrared energy or heat. Sometimes, we can indirectly see heat radiation, for example, as heat shimmers rising from a tarmac road on a hot sunny day. The energy received by the Earth from the Sun balances the energy lost by the Earth back into space. In this way, the Earth maintains a stable average temperature and therefore a stable climate (although of course differences in climate exist at different locations around the world).

The Earth's atmosphere contains a number of greenhouse gases, which affect the Sun–Earth energy balance. The average global temperature is in fact 33°C higher than it should be. Greenhouse gases absorb electromagnetic radiation at some wavelengths but allow radiation at other wavelengths to pass through unimpeded. The atmosphere is mostly transparent in the visible light (which is why we can see the Sun), but significant blocking (through absorption) of ultraviolet radiation by the ozone layer, and infrared radiation by greenhouse gases, occurs.

The absorption of infrared radiation trying to escape from the Earth back to space is particularly important to the global energy balance. Such energy absorption by the greenhouse gases heats the atmosphere, and so the Earth stores more energy near its surface than it would if there was no atmosphere. The average surface temperature of the moon, about the same distance as the Earth from the Sun, is –18°C. The moon, of course, has no atmosphere. By contrast, the average surface temperature of the Earth is 15°C. This heating effect is called the natural greenhouse effect.

### 1.3.1 GREENHOUSE EFFECT

The greenhouse effect is a process by which thermal radiation from a planetary surface is absorbed by atmospheric greenhouse gases and is reradiated in all directions. Since part of this reradiation is back toward the surface and the lower atmosphere, it results in an elevation of the average surface temperature above what it would be in the absence of the gases (Figure 1.14).

Solar radiation at the frequencies of visible light largely passes through the atmosphere to warm the planetary surface, which then emits this energy at the lower frequencies of infrared thermal radiation. Infrared radiation is absorbed by greenhouse gases, which in turn reradiate much of the energy to the surface and lower atmosphere. The mechanism is named after the effect of solar radiation passing through glass and warming a greenhouse, but the way it retains heat is fundamentally different as a greenhouse works by reducing airflow, isolating the warm air inside the structure so that heat is not lost by convection (Schroeder 2000).

If an ideal thermally conductive blackbody was the same distance from the Sun as the Earth is, it would have a temperature of about 5.3°C. However, since the Earth reflects about 30% of the incoming sunlight, this idealized planet's effective temperature (the temperature of a blackbody that would emit the same amount of radiation) would be about –18°C. The surface temperature of this hypothetical planet is 33°C below Earth's actual surface temperature of approximately 14°C. The mechanism that produces this difference between the actual surface temperature and the effective temperature is due to the atmosphere and is known as the greenhouse effect (Smil 2003).

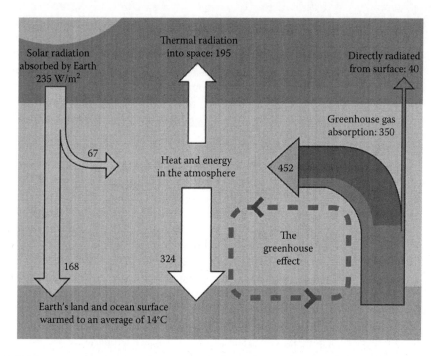

**FIGURE 1.14** A schematic representation of the energy flows between space, the atmosphere, and the Earth's surface, and creation of the greenhouse effect. Energy exchanges are expressed in watts per square meter (W/m²). (From Kiehl, J.T. and Trenberth, K.E., *Bull. Am. Meteorol. Soc.*, 78, 197, 1997.)

Natural greenhouse effect is making life possible on the Earth. However, human activities, primarily the burning of fossil fuels and clearing of forests, have intensified the natural greenhouse effect, causing global warming. Strengthening of the greenhouse effect through human activities is known as the enhanced (or anthropogenic) greenhouse effect. This increase in radiative forcing from human activity is attributable mainly to increased atmospheric $CO_2$ levels. According to the latest report from the IPCC, "most of the observed increase in globally averaged temperatures since the middle of the twentieth century is very likely due to the observed increase in anthropogenic greenhouse gas concentrations" (IPCC 2007).

$CO_2$ is produced by fossil fuel burning and other activities such as cement production and tropical deforestation. Measurements of $CO_2$ from the Mauna Loa observatory show that concentrations have increased from about 313 ppm in 1960 to about 389 ppm in 2010. It reached the 400 ppm milestone on May 9, 2013 (http://news.nationalgeographic.com/news/energy/2013/05/130510-earth-co2-milestone-400-ppm/). The current observed amount of $CO_2$ exceeds the geological record maxima (~300 ppm) from ice core data (Hansen 2005). The effect of combustion-produced carbon dioxide on the global climate, a special case of the greenhouse effect first described in 1896 by Svante Arrhenius, has also been called the Callendar effect.

Over the past 800,000 years, ice core data show that $CO_2$ has varied from values as low as 180 ppm to the preindustrial level of 270 ppm. Paleoclimatologists consider variations in $CO_2$ concentration to be a fundamental factor influencing climate variations over this time scale (Bowen 2005).

## 1.4 HYDROLOGICAL CYCLE

The global hydrological cycle (or water cycle) is a key component of Earth's climate system. Understanding the global hydrological cycle and how we use water is essential for planning a sustainable source of water for the future. The hydrological cycle describes the constant movement of water from ocean to atmosphere to the land surface and back to the ocean. On the global scale, the total amount of water does not change but where it's distributed does. A significant amount of the energy the Earth receives from the Sun is redistributed around the world by the hydrological cycle in the form of latent heat flux. Changes in the hydrological cycle have a direct impact on droughts, floods, water resources, and ecosystem services. Figure 1.15 summarizes, in a qualitative way, the water cycle. Reservoirs are places where water is "stored," or where it stays for some period of time. The oceans, glaciers and ice caps, lakes, and the atmosphere are some examples of reservoirs. Note that *water*, in this context, means the chemical substance $H_2O$, whether in liquid, solid, or gaseous form.

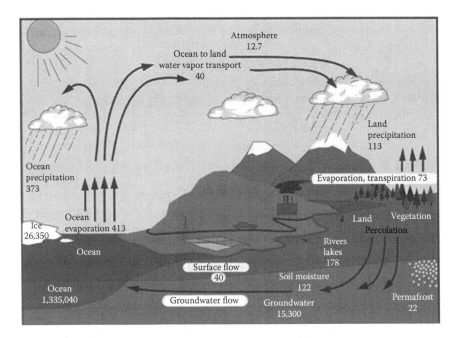

**FIGURE 1.15** The long-term mean global hydrological cycle. Estimates of the main water reservoirs in plain font (e.g., soil moisture) are given in $10^3$ km$^3$, and estimates of the flows between the reservoirs in italic (e.g., surface flow) are given in $10^3$ km$^3$/year. (From Trenberth, K.E. et al., *J. Hydrometeorol.*, 8(4), 758, 2007.)

Water in oceans and lakes is, of course liquid; but it is solid ice in glaciers, and gaseous water vapor in the atmosphere. Flows or pathways are the routes water takes between reservoirs. Evaporation moves water from the oceans to the atmosphere. Precipitation moves liquid (rain) or solid (snow) water from the sky back to the Earth's surface. Snowmelt runoff turns solid water into a liquid that flows down rivers to the sea.

Three basic concepts help us add quantitative aspects to our understanding of the water cycle. They are the following: (1) We can specify the quantity of water in a specific reservoir; (2) we can specify the rate of a given flow (or flux), commonly expressed in km$^3$/year; and (3) we can specify the average residence time that a water molecule spends in a given reservoir. Climate change may impact the reservoirs and flows in terms of reservoir sizes, flow rates, and residence times, and this ultimately affects the spatial and temporal availability of water resources.

### 1.4.1 Reservoirs

*Reservoirs* are places where water is *stored*, or where it stays for some period of time. The oceans, glaciers and ice caps, lakes, and the atmosphere are some examples of reservoirs. The oceans are by far the largest reservoir, containing between 1.3 and 1.4 × 10$^9$ km$^3$ of water, more than 95% of the total amount in the terrestrial water cycle. Some facts about water storage in various reservoirs of the global hydrological cycle are as follows:

- The overall water cycle contains between 1,386,000,000 and 1,460,000,000 km$^3$ of water in various states (liquid, solid, or gaseous).
- The vast majority of it, between 96.5% and 97.25%, is in the oceans.
- Only about 3% of Earth's water is freshwater. About two-thirds of that is frozen in the ice sheets near the poles and in glaciers. Mostly, the rest of the freshwater is underground; less than 1% of freshwater is on the surface in lakes, wetlands, and rivers.
- About 90% of the polar ice sheet and glacial ice is in Antarctica; most of the rest is in Greenland; a tiny fraction is locked up in mountain glaciers elsewhere.
- 434,000 km$^3$ of water evaporates from the oceans each year, while 71,000 km$^3$ (about one-sixth as much) rises into the air over land via evaporation and transpiration.
- About 80% of rainfall is over the oceans. However, more water evaporates from the oceans than falls upon them as rain. Conversely, precipitation onto land exceeds evapotranspiration from land. Runoff that flows in rivers to the seas makes up for most of this imbalance.

Table 1.2 depicts the size of the main water reservoirs. It is worth noting that only very little volume of water is contained in the atmosphere and in rivers.

## TABLE 1.2
### Reservoir Sizes and Fractions/Percentages

| Reservoir | Volume (km³) | Fraction or % of a Larger Reservoir |
| --- | --- | --- |
| All of the Earth's water | 1,386,000,000–1,460,000,000 | NA |
| Oceans | 1,338,000,000–1,400,000,000 | 97% of total water |
| Freshwater | 35,030,000 | 2.5%–3% of total water |
| Ice and snow | 43,400,000 | — |
| Ice caps, glaciers, and permanent snow | 24,064,000–29,000,000 | 68.7% of freshwater |
|  |  | ~2% of total water |
| Antarctic ice and snow | 29,000,000 | ~90% of all ice |
| Greenland | 3,000,000 | ~10% of all ice |
| Glaciers (not Greenland or Antarctica) | 100,000 | — |
| Groundwater (saline and fresh) | 23,400,000 | — |
| Groundwater (saline) | — | 54% of groundwater |
| Groundwater (fresh) | 10,530,000 | 30.1% of freshwater |
|  |  | 46% of groundwater |
| Surface water (fresh) | — | 0.3% of freshwater |
| Lakes | — | 87% of surface freshwater |
| Swamps | — | 11% of surface freshwater |
| Rivers | — | 2% of surface freshwater |
| Atmosphere | 12,000–15,000 | — |

### 1.4.2 FLOWS (OR FLUXES)

*Flows or pathways* are the routes water takes between reservoirs. Examples of the flows (or fluxes) include the following: Evaporation moves water from the oceans to the atmosphere; precipitation moves liquid (rain) or solid (snow) water from the sky back to the Earth's surface; and snowmelt runoff turns solid water into a liquid that flows down the rivers to the sea. Flows are commonly expressed in terms of km³/year.

Table 1.3 shows the flow rate and process from one reservoir to the other one. The following important points emerge from the table:

- More water is evaporated from the oceans than falls on to them as precipitation, and more water falls as precipitation on to the land masses than is evaporated. The balance is made up by river runoff. If the precipitation and evaporation budget did not work in this way, the land masses would progressively dry up, and oceans would progressively gain all of the world's water.
- The annual flux of water through the atmosphere is about 460,000 km³/year, about 35 times larger than the amount held in the atmosphere at any one time. This means that the average residence time of water in the atmosphere is very short. In contrast, the size of the ocean reservoir is over 3000 times larger than the annual flux to the atmosphere or from the atmosphere and land masses, so the average residence time of water in the oceans is very long.

### TABLE 1.3
### Flows of Water (between the Reservoirs)

| Process | From/to Reservoir | Flow Rate (km³/year) |
| --- | --- | --- |
| Precipitation | Atmosphere to ocean/land | 505,000 |
| Ocean precipitation | Atmosphere to ocean | 398,000 |
| Land precipitation | Atmosphere to land/surface | 96,000–107,000 |
| Evapotranspiration | Ocean and land/surface and plants to atmosphere | 505,000 |
| Ocean evaporation | Ocean to atmosphere | 434,000 |
| Land evaporation | Land/surface to atmosphere | 50,000 |
| Transpiration | Plants to atmosphere | 21,000 |
| Uptake by plants | Land/surface to biota | 21,000 |
| Runoff | Land/surface to ocean | 36,000 |
| Melting | Ice/snow to land/surface | 11,000 |
| Snowfall | Atmosphere to ice/snow | 11,000 |
| Percolation | Underground to and from land/surface | 100 |

### 1.4.3 Residence Times

The amount of time that a water molecule spends, on average, in a reservoir is called *residence time*. Water that evaporates into the atmosphere quickly falls back out as precipitation; the average atmospheric residence time is just 9 days. By contrast, once water reaches the ocean, it can stay there for a very long time; the average residence time for water in the oceans is more than 3000 years. It is important to realize that reported residence times are averages, and that the actual residence time for a given water molecule may be far from the average. Water vapor that reaches the stratosphere may remain there for a long time; water that flows into warm, shallow coastal waters from a river may evaporate and leave the ocean very quickly.

Table 1.4 shows the estimated mean residence times of a water molecule in different reservoirs. The estimated residence times range from 1 week to 10,000 years. These residence times are of great importance in analyzing the transport of pollutants, as well as nutrients, in the hydrologic cycle.

### 1.4.4 Water Vapor

Water vapor is of central importance to energy flows within the climate system, by modulating the transmission of radiative energy between the surface, atmosphere, and space and also through transferring latent heat from the surface (evaporation) to the atmosphere (precipitation) following transport of moisture within the atmosphere (Figure 1.16). There are many aspects of these processes that are well understood, based upon robust physics, detailed process modeling, and measurements ranging from laboratory experiments up to satellite observations of the entire globe (e.g., Held and Soden 2006; Sherwood et al. 2010a,b). One of the fundamental

### TABLE 1.4
### Estimated Mean Residence Times (in Reservoirs)

| Reservoir | Residence Time (Average) |
| --- | --- |
| Oceans | 3000–3230 years |
| Glaciers | 20–100 years |
| Seasonal snow cover | 2–6 months |
| Soil moisture | 1–2 months |
| Groundwater: shallow | 100–200 years |
| Groundwater: deep | 10,000 years |
| Lakes | 50–100 years |
| Rivers | 2–6 months |
| Atmosphere | 9 days |

controls on the climate system is the Clausius–Clapeyron equation (C–C equation), which provides a powerful constraint on how saturated moisture content varies with air temperature:

$$e_s(T) = 6.1094 \exp\left(\frac{17.625T}{T + 243.04}\right) \quad (1.1)$$

where
 $e_s(T)$ is the equilibrium or saturation vapor pressure in hPa
 $T$ is temperature in °C

This shows that when atmospheric temperature increases (e.g., due to greenhouse gases), the absolute humidity should also increase exponentially (assuming a constant relative humidity).

The amount of water held by the atmosphere is controlled by the saturation vapor pressure. The value of saturation vapor pressure depends only on the temperature of the air. According to the C–C equation, saturation vapor pressure increases rapidly with the temperature. The value at 32°C is about double the value at 21°C. An increase of 1°C at 20°C increases the water-holding capacity of the air by 6%; similarly, at 0°C and 80°C, the water-holding capacity increases by 8% and 20%, respectively.

Observations and modeling indicate that column-integrated water vapor, averaged over sufficiently large scales, increases approximately exponentially with atmospheric temperature (Raval and Ramanathan 1989) as predicted by Equation 1.1. The absorption of infrared radiation by water vapor increases approximately in proportion to the logarithm of its concentration. These two powerful constraints generate an amplifying effect on changes in the Earth's climate, enhancing the response of surface temperature to a radiative forcing or internal variability, and the resultant positive water vapor feedback is relatively *forgiving* in the sense that substantial excursions away from these basic physical constraints are required to alter its nature (Allan 2012).

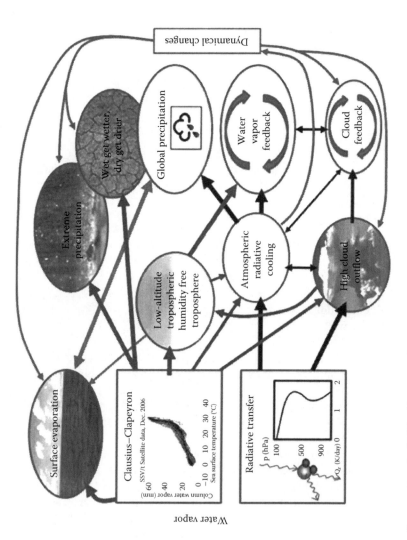

**FIGURE 1.16** Role of water vapor in the Earth's energy flows: A simplistic schematic representation of some of the links between water vapor and energetic processes in the climate system. Arrows denote the primary driving physical process. Thick arrows in black are radiative transfer, dark gray arrows are thermodynamics and thin arrows in gray are dynamics. (From Allan, R.P., *Surv. Geophys.*, 33, 557, 2012.)

## 1.5 SUMMARY

This chapter highlights the global climate system, the energy balance, and the hydrological cycle. The climate system is a complex, interactive system consisting of the atmosphere, lithosphere, snow and ice, oceans and other bodies of water, and living things. The atmospheric component of the climate system most obviously characterizes climate; climate is often defined as *average weather*. The climate system evolves in time under the influence of its own internal dynamics and changes in external factors (called as *forcing*) also affect climate. There are many feedback mechanisms in the climate system that can either amplify (*positive feedback*) or diminish (*negative feedback*) the effects of a change in climate forcing. For example, as rising concentrations of greenhouse gases warm the Earth's climate, snow and ice begin to melt. This melting reveals darker land and water surfaces that were beneath the snow and ice, and these darker surfaces absorb more of the Sun's heat, causing more warming, which causes more melting, and so on, in a self-reinforcing cycle.

The Earth atmosphere contains a number of greenhouse gases, which affect the Sun–Earth energy balance. Greenhouse gases absorb electromagnetic radiation at some wavelengths but allow radiation at other wavelengths to pass through unimpeded. The atmosphere is mostly transparent in the visible light (which is why we can see the Sun), but significant blocking (through absorption) of ultraviolet radiation by the ozone layer, and infrared radiation by greenhouse gases, occurs. The absorption of infrared radiation trying to escape from the Earth back to space is particularly important to the global energy balance. Such energy absorption by the greenhouse gases heats the atmosphere, and so the Earth stores more energy near its surface than it would if there was no atmosphere.

The global hydrological cycle is a key component of Earth's climate system. A significant amount of the energy the Earth receives from the Sun is redistributed around the world by the hydrological cycle in the form of latent heat flux. Changes in the hydrological cycle have a direct impact on droughts, floods, water resources, and ecosystem services. Water vapor is of central importance to energy flows within the climate system, by modulating the transmission of radiative energy between the surface, atmosphere, and space and also through transferring latent heat from the surface (evaporation) to the atmosphere (precipitation) following transport of moisture within the atmosphere.

## REFERENCES

Allan, R.P. 2012. The role of water vapor in earth's energy flows. *Surveys in Geophysics* 33: 557–564.

Bowen, M. 2005. *Thin Ice: Unlocking the Secrets of Climate in the World's Highest Mountains.* New York: Owl Books.

Campbell, N.A., Williamson, B., and Heyden, R.J. 2006. *Biology: Exploring Life.* Boston, MA: Pearson Prentice Hall.

Cmglee. http://commons.wikimedia.org/wiki/File:Comparison_US_standard_atmosphere_1962.svg, accessed September 16, 2013.

Cohen, J. and Rind, D. 1991. The effect of snow cover on the climate. *Journal of Climate* 4: 689–706.

Curry, J.A. and Webster, P.J. 1999. Thermodynamics of atmosphere and oceans. *International Geophysics* 65: 468.

Cushman-Roisin, B. 1994. *Introduction to Geophysical Fluid Dynamics*. London, U.K.: Prentice Hall, 319pp.

de Villiers, M. 2003. *Water: The Fate of Our Most Precious Resource*, 2nd edn. Toronto, Ontario, Canada: McClelland & Stewart, 453pp.

Groisman, P.Y., Karl, T.R., and Knight, R.W. 1994. Observed impact of snow cover on the heat balance and the rise of continental spring temperatures. *Science* 363: 198–200.

Goosse, H., Barriat, P.Y., Lefebvre, W., Loutre, M.F., and Zunz, V. September 2013. *Introduction to Climate Dynamics and Climate Modeling*. Online textbook available at http://www.climate.be/textbook.

Gutzler, D.S. and Preston, J.W. 1997. Evidence for a relationship between spring snow cover in North America and summer rainfall in New Mexico. *Geophysical Research Letters* 24: 2207–2210.

Hall, D.K. 1996. Remote sensing applications to hydrology: Imaging radar. *Hydrological Sciences* 41: 609–624.

Hansen, J. 2005. A slippery slope: How much global warming constitutes "dangerous anthropogenic interference"? *Climatic Change* 68: 269–279.

Held, I.M. and Soden, B.J. 2006. Robust responses of the hydrological cycle to global warming. *Journal of Climate* 19: 5686–5699.

IPCC. 2001. *Climate Change 2001: The Scientific Basis*. Contribution of Working Group I to the Third Assessment Report of the Intergovernmental Panel on Climate Change. Cambridge, U.K.: Cambridge University Press, 881pp.

IPCC. 2007. *Climate Change 2007: The Physical Science Basis*. Contribution of WG I to AR4 of the IPCC. Cambridge, U.K.: Cambridge University Press.

Kalnay, E., Kanamitsu, M., Kistler, R. et al. 1996. The NCEP/NCAR 40-year reanalysis project. *Bulletin of the American Meteorological Society* 77(3): 437–471.

Kennish, M.J. 2001. *Practical Handbook of Marine Science*. CRC Marine Science Series. Boca Raton, FL: CRC Press, p. 35.

Kiehl, J.T. and Trenberth, K.E. 1997. Earth's annual global mean energy budget. *Bulletin of the American Meteorological Society* 78: 197–208.

Knauss, J.A. 2005. *Introduction to Physical Oceanography*, Long Grove, IL: Waveland Press, Inc. Available at: http://www.elic.ucl.ac.be/textbook/chapter1_node9.html.

Lynch-Stieglitz, M. 1994. The development and validation of a simple snow model for the GISS GCM. *Journal of Climate* 7: 1842–1855.

NASA/Goddard Space Flight Center and ORBIMAGE. The SeaWiFS Project. Available at: http://oceancolor.gsfc.nasa.gov/SeaWiFS/BACKGROUND/Gallery/index.html.

National Weather Services. Profile of the Ocean. Available at: http://www.srh.noaa.gov/.

Raval, A. and Ramanathan, V. 1989. "Observational determination of the greenhouse effect." *Nature* 342(6251): 758–761.

Salby, M. 1992. The atmosphere. In: K. Trenberth ed., *Climate Systems Modeling*. Sponsored Jointly by UCAR and the Electric Power Research Institute (EPRI). Cambridge, U.K.: Cambridge University Press, pp. 53–115.

Schroeder, D.V. 2000. *An Introduction to Thermal Physics*. San Francisco, CA: Addison-Wesley, pp. 305–307.

Sherwood, S.C., Ingram, W., Tsushima, Y. et al. 2010a. Relative humidity changes in a warmer climate. *Journal of Geophysical Research* 115: D09104.

Sherwood, S.C., Roca, R., Weckwerth, T.M., and Andronova, N.G. 2010b. Tropospheric water vapor, convection, and climate. *Reviews of Geophysics* 48: RG2001.

Smil, V. 2003. *The Earth's Biosphere: Evolution, Dynamics, and Change*. Cambridge, U.K.: MIT Press, p. 107.

Stephens, G.L., Li, J., Wild, M. et al. 2012. An update on Earth's energy balance in light of the latest global observations. *Nature Geoscience* 5: 691–696.

Stocker, T.F. et al. 2001. Physical climate processes and feedbacks. In: J.T. Houghton et al., eds., *Climate Change 2001: The Scientific Basis*. Contributions of Working Group I to the Third Assessment Report of the Intergovernmental Panel on Climate Change. Cambridge, U.K.: Cambridge University Press.

Thermohaline Ocean Circulation. Available at: http://www.windows2universe.org/earth/Water/thermohaline_ocean_circulation.html.

Trenberth, K.E., Fasullo, J.T., and Kiehl, J. 2009. Earth's global energy budget. *Bulletin of the American Meteorological Society* 90: 311–323.

Trenberth, K.E., Smith, L., Qian, T.T., Dai, A.G., and Fasullo, J. 2007. Estimates of the global water budget and its annual cycle using observational and model data. *Journal of Hydrometeorology* 8(4): 758–769.

UNESCO. 1998. *World Water Resources: A New Appraisal and Assessment for the 21st Century: A Summary of the Monograph World Water Resources*. Paris, France: UNESCO.

Universitè catholique de Louvain. *Introduction to Climate Dynamics and Climate Modelling*. Available at: http://www.elic.ucl.ac.be/textbook/chapter1_node14.html#tab1x01.

Vernekar, A.D., Zhou, J., and Shukla, J. 1995. The effect of Eurasian snow cover on the Indian monsoon. *Journal of Climate* 8: 248–266.

Wallace J.M. and Hobbs, P.V. 2006. *Atmospheric Science: An Introductory Survey*, Vol. 92, International Geophysics Series. New York: Academic Press.

Wikimedia Commons. Available at: http://www.centennialofflight.gov/essay/Theories_of_Flight/atmosphere/TH1G1.htm.

Wikipedia. Greenhouse effect. Available at: http://en.wikipedia.org/wiki/File:Greenhouse_Effect.svg.

Windows to the Universe. Thermohaline ocean circulation. http://www.windows2universe.org/earth/Water/thermohaline_ocean_circulation.html, accessed September 16, 2013.

# 2 Climate Variability and Change

*Anthony S. Kiem*

## CONTENTS

2.1 Introduction ........................................................................................................31
    2.1.1 Definitions..............................................................................................32
2.2 Factors Responsible for Natural Climate Variability and Change ................33
    2.2.1 Ocean–Atmospheric Circulations and Interactions............................33
    2.2.2 Earth–Sun–Moon Interactions .............................................................46
    2.2.3 Atmospheric Chemistry, Aerosols, and Volcanic Eruptions .............49
    2.2.4 Geological Drivers of Climate Change ...............................................52
2.3 Factors Responsible for Anthropogenic Climate Change ...........................54
    2.3.1 Human-Induced Increases to Greenhouse Gases and the Enhanced Greenhouse Effect .................................................54
    2.3.2 Human-Induced Increases to Aerosols and Other Pollutants............54
    2.3.3 Land-Use Change (Including Urbanization) ......................................58
2.4 Summary ............................................................................................................59
References................................................................................................................60

## 2.1 INTRODUCTION

Up until relatively recently, water management and planning processes, including planning for extreme events such as droughts and floods, were largely based on the assumption of a *stationary* climate—that is, the assumption that the historical record (of climate and streamflow) is representative of current and future conditions. More specifically, as described by Milly et al. (2008), stationarity implies:

> that any variable (e.g. annual streamflow or annual flood peak) has a time-invariant (1-year-periodic) probability density function (pdf), whose properties can be estimated from the instrument record.... The pdfs, in turn, are used to evaluate and manage risks to water supplies, waterworks, and floodplains....

Traditionally, planning processes for ensuring reliable water supplies and an effective response to drought and floods were progressively developed based on this assumption. This approach does not account for potential nonstationarity in the physical climatological mechanisms that actually deliver rainfall and, in particular, climate extremes. As acknowledged by Milly et al. (2008), "low-frequency, internal variability (e.g., the Atlantic multi-decadal oscillation) enhanced by the slow dynamics of the oceans and ice sheets pose a challenge to the assumption of stationarity." Other examples include the Pacific decadal oscillation (PDO), which is also known as the interdecadal Pacific oscillation (IPO) and is associated with low-frequency hydroclimatic variability and has been shown to affect the likelihood of extreme events (e.g., Franks and Kuczera 2002; Kiem et al. 2003).

Further challenges to stationarity arise from externally forced changes in the climate system, arising from natural factors such as volcanic eruptions and from anthropogenic influences such as land-use change, increased aerosols, and increased greenhouse gases (GHGs).

While it is accepted that the assumption of stationarity is no longer appropriate (e.g., Milly et al. 2008), indeed if it ever was, an improved basis for planning must necessarily be founded on an improved understanding of the key influences or *drivers* of climate, their interactions, and how these might be expected to change into the future.

The hydroclimate is driven by a variety of physical processes that operate on a range of spatial and temporal timescales. In order to better manage water resources in the future, information on how the various hydroclimate drivers are likely to behave and interact into the future is required. Climate models (be they general circulation models [GCMs] or dynamical models aimed at seasonal forecasting) attempt to address this need, but their inability to replicate the characteristics of key climatic drivers, and their interactions, is well documented (e.g., Blöschl and Montanari 2010; Di Baldassarre et al. 2010; Kiem and Verdon-Kidd 2011; Koutsoyiannis et al. 2008, 2009; Montanari et al. 2010; Stainforth et al. 2007; Verdon-Kidd and Kiem 2010). This chapter reviews what is theorized and what is known about links between the known drivers of hydroclimatic variability and change and the underlying mechanisms that may determine how they behave.

### 2.1.1 Definitions

*Climate*: Intergovernmental Panel on Climate Change (IPCC) (2007) states that climate in a narrow sense is usually defined as the average weather or, more rigorously, as the statistical description in terms of the mean and variability of relevant quantities over a period of time ranging from months to thousands or millions of years. These quantities are most often surface variables such as temperature, precipitation, and wind. Climate in a wider sense is the state, including a statistical description, of the climate system.

*Natural climate variability or change*: Climate variability or change refers to a statistically significant variation either in the mean state of the climate or in its variability, persisting for an extended period (typically from seasons to decades or longer). Natural climate variability or change may be due to natural internal processes or external forcings (e.g., solar variability).

*Anthropogenic climate change*: The United Nations Framework Convention on Climate Change (UNFCCC), in its Article 1, defines climate change as "a change of climate which is attributed directly or indirectly to human activity that alters the composition of the global atmosphere and which is in addition to natural climate variability observed over comparable time periods." Therefore, there is a distinction between anthropogenic climate change (i.e., due to persistent anthropogenic changes in the composition of the atmosphere or in land use) and natural climate variability or change.

*Greenhouse gases (GHGs)*: Gases that absorb infrared radiation in the spectrum emitted by the Earth. These include water vapor ($H_2O$), carbon dioxide ($CO_2$), methane ($CH_4$), nitrous oxide ($N_2O$), ozone ($O_3$), and others. GHGs can be emitted by natural processes (e.g., volcanoes, cattle) or anthropogenic/human sources (e.g., coal fired power plants). GHGs cause the greenhouse effect. Increases in human-induced GHGs enhance the greenhouse effect and this leads to global warming and anthropogenic climate change.

## 2.2 FACTORS RESPONSIBLE FOR NATURAL CLIMATE VARIABILITY AND CHANGE

### 2.2.1 Ocean–Atmospheric Circulations and Interactions

*El Niño/Southern Oscillation (ENSO)*: On an interannual scale, *El Niño/Southern Oscillation* (ENSO), and the various different *flavors* of ENSO including ENSO Modoki, is the key climate driver in many countries. As a result of intensive research over the last 20 years, we have developed a good understanding of the basic physical features and processes involved in the ENSO cycle and how it evolves once it has begun. ENSO is an ocean–atmospheric climate pattern that occurs across the tropical Pacific Ocean. It is characterized by quasiperiodic (i.e., every 3–5 years) variations in the sea surface temperature (SST) of the tropical eastern Pacific Ocean. Under normal or ENSO neutral conditions, air rises in the west Pacific, flows eastward in the upper atmosphere, descends in the east Pacific, and flows westward along the surface of the tropical Pacific (i.e., the easterly trade winds). This is known as the Walker circulation (see Figures 2.1 and 2.2). Under neutral ENSO conditions (top of Figure 2.1 and middle of Figure 2.2), the typical easterly equatorial trade winds result in warm surface water pooling in the west Pacific and cold water upwelling along the South American coast. El Niño conditions (bottom of Figure 2.1 and left of Figure 2.2) are associated with a relaxing (or reversal) of the equatorial trade winds (weakening or reversal of the Walker circulation) that, in turn, results in warm surface water migrating toward the South American coast and reduced cold water upwelling in the east Pacific. La Niña conditions (right in Figure 2.2) are essentially the opposite of El Niño with a strengthened Walker circulation and stronger equatorial trade winds resulting in an enhancement of both the warm pool in the west Pacific and also the cold water upwelling in the east Pacific. Refer to Box 2.1 for further details. ENSO causes extreme weather (such as floods and droughts) in many regions of the world, particularly those bordering the Pacific Ocean (e.g., Gallant et al. 2012; Murphy and Timbal 2008), with the effects of ENSO including magnified fluctuations in streamflow volumes compared to rainfall (Chiew et al. 1998; Verdon et al. 2004b; Wooldridge et al. 2001), elevated flood risk during La Niña events (Kiem et al. 2003), and increased risk of drought (Kiem and Franks 2004) and bushfire (Verdon et al. 2004a) during El Niño events.

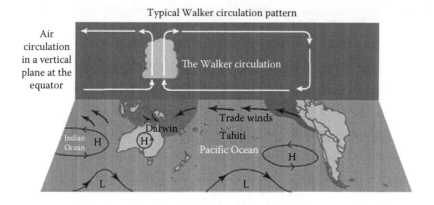

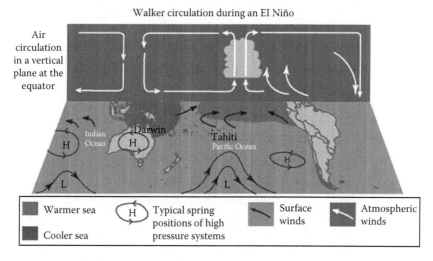

**FIGURE 2.1** The Walker circulation. (From cawcr.gov.au. Accessed June, 2013.)

*ENSO Modoki*: A pattern of central equatorial Pacific SST variations has also been identified by Ashok et al. (2007) and defined as an *ENSO Modoki* (*Modoki* is a Japanese word meaning *a similar but different thing*). An El Niño (La Niña) Modoki is characterized by warm (cool) central Pacific waters flanked by anomalously cool (warm) SSTs to the west and east, separating the Walker circulation into two distinct circulations. Since mid-2009, a succession of papers has revealed significant relationships between ENSO Modoki and Australian rainfall (e.g., Ashok et al. 2009; Cai and Cowan 2009; Taschetto and England 2009; Taschetto et al. 2009). These studies demonstrate that, compared with conventional ENSO episodes, ENSO Modoki is associated with markedly different Australian rainfall anomalies in terms of location, seasonality and magnitude of impact.

*Trade winds and the Hadley and Walker circulations*: The Hadley circulation (or cell) is the atmospheric circulation pattern that dominates the tropics and subtropics. The Hadley cell is a closed circulation loop, which begins at the equator (where average

# Climate Variability and Change

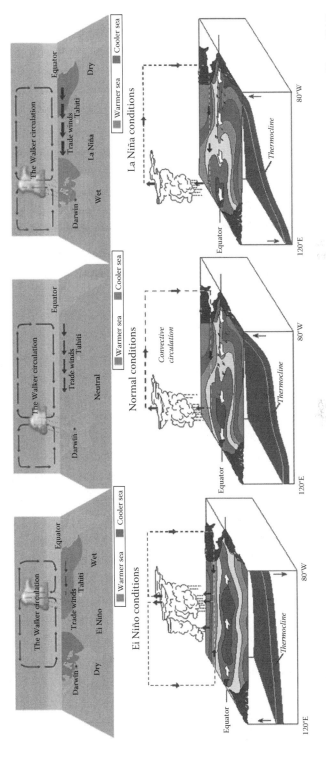

**FIGURE 2.2** Two schematic diagrams of the ENSO phenomenon. (Adapted from www.pmel.noaa.gov/tao/elnino/nino-home.html. Accessed June, 2013.)

## BOX 2.1 CHARACTERISTICS OF TROPICAL PACIFIC CLIMATE AND ENSO

*Bjerknes feedback*: A positive feedback loop that helps to control the state of the tropical Pacific and amplifies incipient El Niño events. The easterly trade winds in the tropical Pacific induce a surface zonal current and upwelling that maintain the east–west cold–warm SST gradient. The SST gradient in turn focuses atmospheric convection toward the west and drives the Walker circulation, which enhances the trades. Changes in any one of these elements tend to be amplified by this feedback loop.

*Thermocline*: A region of strong vertical temperature gradient in the upper few hundred meters of the tropical Pacific Ocean. Key aspects of the equatorial thermocline are its zonal, or east–west, slope (normally the thermocline is deeper in the west than in the east); its zonal-mean depth (indicative of the heat content of the upper ocean); and its intensity (which measures the strength of the temperature contrast between the surface and deep ocean). The thermocline is dynamically active during the ENSO cycle and is key to its evolution.

*Upwelling*: Upward movement of cold ocean water from depth, which in the equatorial Pacific is driven by the easterly *from the east* trade winds. Upwelling acts to cool the surface and supply nutrients to surface ecosystems.

*Surface zonal current*: The mean easterly trade winds drive a surface easterly current that pushes warm surface waters into the west Pacific. When the trades weaken during El Niño, the warm west Pacific water sloshes eastward, weakening the zonal SST contrast across the Pacific and thus further weakening the trades—a positive feedback for the ENSO cycle.

*Walker circulation*: The zonally oriented component of the tropical Pacific atmospheric circulation. Under normal conditions, air rises in convective towers in the west Pacific, flows eastward, descends in the east Pacific, and then flows westward at the surface as live trade winds.

*Delayed ocean adjustment*: During El Niño, a sharpening of the poleward gradient of the tropical Pacific trade winds enhances the poleward transport of upper-ocean waters from the equator toward the subtropics. This gradually discharges the reservoir of warm surface waters from the equatorial zone, eventually resulting in a shallower thermocline, cooler upwelling, and a restored westward flow of cold east Pacific water. This slow component of the ocean adjustment helps to transition the El Niño into normal or La Niña conditions.

*Atmospheric feedbacks*: Variations in SSTs and atmospheric circulation drive variations in clouds and surface winds, which impact the fluxes of heat and momentum between the atmosphere and ocean. Fluxes arc generally partitioned into SW and LW surface radiative fluxes and fluxes of sensible heat and latent heat. The properties of precipitation, convection, and stratiform clouds are all important in the heat balance of the mean climate and the ENSO cycle.

*Intraseasonal variability*: Atmospheric variability within seasons, often associated with the MJO and other organized modes of variability, can induce anomalies in surface winds in the west Pacific, which induce ocean Kelvin waves that propagate to the east and deepen the thermocline. These may either initiate or amplify the development of an El Niño event.

*Small-scale features*: Tropical instability waves are generated in the equatorial oceans because of a fluid dynamical instability. Although it is thought that they play a relatively minor role in ENSO dynamics, most CGCMs used for long-term climate projections do not simulate tropical instability waves well, so this is a current area of research.

*Source*: Collins, M. et al., *Nat. Geosci.*, 3, 391, 2010.

solar radiation is greatest) with warm, moist air lifted in equatorial low-pressure areas (the Intertropical Convergence Zone, ITCZ) to the tropopause and carried poleward. At about 30°N/S latitude, it descends in a high-pressure area. Some of the descending air travels toward the equator along the Earth's surface, closing the loop of the Hadley cell and creating the trade winds (which, due to the Coriolis force, predominantly blow from the east to southeast [northeast] in the Southern [Northern] Hemisphere). Under normal conditions, the trade winds cause a westward motion of warm equatorial Pacific surface waters toward the western Pacific resulting in an east to west temperature asymmetry (warm in the west and cold in the east). Deep atmospheric convection and heavy rainfall occur in the western Pacific over the warm water, whereas there is net atmospheric subsidence over the colder water in the eastern Pacific (i.e., the Walker circulation [Figure 2.1]). Subtropical/tropical circulation in the Pacific thus includes two orthogonal, but coupled, components: Hadley cell transport crossing latitude lines toward the equator at the surface and away from the equator in the upper troposphere and Walker circulation transport parallel to the equator in a westward (eastward) direction at the surface (in the upper troposphere). The combined effect of the Hadley cell, Walker circulation, and rotating Earth dynamics is to cause intense convergence and vertical circulation at the ITCZ (near 0° latitude), over both the land and the sea. The strength of the trade winds varies in phase with the strength of the Walker circulation and positioning and also the strength and size of the Hadley cell (e.g., Caballero and Anderson 2009; Frierson et al. 2007; Quan et al. 2004). Therefore, strengthening and weakening of the Hadley cell and Walker circulations play a crucial role in reinforcing El Niño/La Niña perturbations to the mean tropical Pacific ocean–atmosphere climatology. There has recently been some suggestion that the Walker circulation is weakening and that the Hadley cell may be intensifying and expanding due to anthropogenic global warming (there are numerous papers on this issue available at http://agwobserver.wordpress.com/2010/07/27/papers-on-hadley-cell-expansion/). Observed data indicate a Hadley cell expansion of about 5°–8° latitude between 1979 and 2005 (Reichler 2009; Seidel and Randel 2007). Climate model projections indicate that the subsiding limb of the Hadley cell could shift north/south by up to 2° latitude over the twenty-first century.

The direct effect of this would likely be a shifting of the dry/arid zone typically associated with the subsiding limb of the Hadley cell. However, the influence an expansion of the Hadley cell, and/or changes to the Walker circulation, will have on ENSO, or any of the other large-scale climate modes, is unknown (e.g., Power and Smith 2007)—partly because the climate models used to perform the required experiments are not yet capable of realistically representing the complexities or variability associated with the Hadley cell, Walker circulation, or trade wind interactions (i.e., the driving mechanisms behind ENSO) (e.g., Blöschl and Montanari 2010; Di Baldassarre et al. 2010; Kiem and Verdon-Kidd 2011; Koutsoyiannis et al. 2008, 2009; Montanari et al. 2010; Stainforth et al. 2007; Verdon-Kidd and Kiem 2010).

*Oceanic Rossby and Kelvin waves*: Also important in the development and/or evolution of ENSO, but not specifically mentioned in Box 2.1, are large internal waves (waves that have their peaks under the ocean surface) across the Pacific Ocean. These internal waves are referred to as oceanic Rossby and Kelvin waves (Rossby waves are also known as planetary waves or long waves and also occur in the atmosphere). During El Niño events, deep convection and heat transfer to the troposphere is enhanced over the anomalously warm eastern tropical Pacific. This, combined with a breakdown in the easterly (east to west) trade winds, generates oceanic Rossby waves and Kelvin waves. Oceanic Rossby waves are large-scale motions whose restoring force is the variation in the Coriolis effect with latitude, and therefore they always travel from east to west, following the parallels. Their horizontal scale is of the order of hundreds of kilometers, while the amplitude of the oscillation at the sea surface is just a few centimeters. Kelvin waves describe the buildup of warm water that moves across the equatorial Pacific from west to east during an El Niño. A Kelvin wave will cross the Pacific in about 70 days, while oceanic Rossby waves take several months (or even years) to cross the Pacific Ocean—and therefore are thought to be connected to the long-term evolution of ENSO or even a driver of interdecadal ENSO variability.

*Interdecadal ENSO*: ENSO is an irregular, interannual oscillation of equatorial Pacific SST and the overlying atmospheric circulation. However, the characteristics of ENSO also appear to be modulated on longer, interdecadal timescales, by a mode of variability that affects the wider Pacific Basin that is known as either the PDO (Mantua et al. 1997), if referring to northern Pacific Ocean variability, or the IPO (Power et al. 1999a), if referring to basin-wide Pacific Ocean variability. Links between the IPO phenomena and climate variability include decadal and annual-scale fluctuations in rainfall, maximum temperature, streamflows, drought, flood, bushfire, and wheat crop yield (Kiem et al. 2003; Kiem and Franks 2004; Power et al. 1999a; Verdon et al. 2004a,b). The IPO primarily influences the eastern Australian climate during the austral spring, summer, and autumn by inducing variations in the South Pacific Convergence Zone, which tends to be active during these months (Folland et al. 2002). The IPO regulates the climate indirectly by modulating both the magnitude and frequency of ENSO impacts (Kiem et al. 2003; Power et al. 1999b; Verdon et al. 2004b). This dual modulation manifests in the historical record as periods of two to three decades during which either El Niño (if IPO is positive) or La Niña (if IPO is negative) events tend to dominate. In addition, when the IPO is in a warm (i.e., positive) phase, the relationship between ENSO and rainfall is weakened, while it is strengthened during the cool phase (Power et al. 1999a; Verdon et al. 2004b). Verdon and Franks (2006) confirmed that the relationships between IPO/PDO phase and the frequency of ENSO events are consistent over the past 450 years by examining paleoclimate reconstructions of the two climate modes. However, the recent results of Gallant and Gergis (2011) suggest that the relationship between SEA streamflow and IPO may be more complicated, with an apparent decoupling between Pacific decadal-scale variability and River Murray flow in the early 1800s. The possibility that the relationship between IPO and SEA hydroclimate is more complex and nonstationary than has been observed over the instrumental period must be considered and made a priority for future investigation. Currently, the mechanisms responsible for the Pacific Ocean's interdecadal variability are unclear, and there is even still some debate as to whether the IPO/PDO operates independently of ENSO (e.g., McGregor et al. 2007, 2008, 2009). Many hypotheses have been proposed (e.g., Latif 1998; Miller and Schneider 2000; Power and Colman 2006) including that instabilities in the atmosphere can drive internal atmospheric variability and heat flux variability on timescales of up to and beyond a decade, which may alter the surface climate (e.g., Frankignoul and Hasselmann 1977; Frankignoul et al. 1997; James and James 1992; Power et al. 1995). Interdecadal variability can also arise in association with long timescale changes in ENSO activity, which can be generated in the tropical Pacific either through nonlinear mechanisms (e.g., Jin et al. 1994; Rodgers et al. 2004; Schopf and Burgman 2006; Timmermann et al. 2003) or by the ocean–atmosphere's response to the stochastic nature of ENSO events (e.g., Newman et al. 2003; Power and Colman 2006). Variability generated in the Pacific extratropics can also influence the tropics on interdecadal timescales via the atmosphere (e.g., Barnett et al. 1999; Pierce et al. 2000) and/or via the ocean by (1) changing the temperature of the water moving into the tropics via the meridional overturning circulation (i.e., the system of surface and deep-ocean currents encompassing all ocean basins responsible for transporting large amounts of water, heat, salt, carbon, nutrients, and other substances around the Earth's oceans); (2) changing the rate at which water in the meridional overturning circulation is moved into the tropics; or (3) exciting oceanic Rossby waves by variations in the extratropical wind stress (for details, see McGregor et al. 2007, 2008, 2009 and papers referred to within). McGregor (2009) points out that, in reality, it is probable that all the previously mentioned mechanisms play a role but there is currently no clear consensus as to their relative importance in initiating or explaining interdecadal ENSO variability—more modeling and analysis are required to improve our understanding here. However, what McGregor (2009) did demonstrate was that extratropical Pacific Ocean exchanges (via oceanic Rossby waves) with the tropics can explain the interdecadal variability of ENSO events and that, for certain types of ENSO, this can enable ENSO conditions to be predicted up to 24 months in advance.

*Indian Ocean variability*: Interannual climate variability has also been linked to eastern Indian Ocean SST anomalies, particularly in the austral winter (June–August) and spring (September–November) (e.g., Ashok et al. 2000; Drosdowsky 1993; Nicholls 1989; Verdon and Franks 2005) and the Indian Ocean Dipole (IOD)

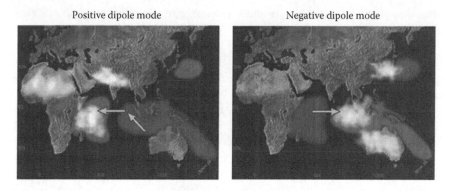

**FIGURE 2.3** Schematic representation of the two phases of the IOD. (From www.jamstec.go.jp/frsgc/research/d1/iod. Accessed June, 2013.)

(e.g., Ashok et al. 2003; Cai et al. 2009a,b; Meyers et al. 2007; Saji et al. 1999; Ummenhofer et al. 2009). The IOD is characterized by SST anomalies of opposite sign in the east and west of the Indian Ocean Basin, which are coincident with large-scale anomalous circulation patterns (see Figure 2.3). However, there is also evidence for the nonexistence of an equatorial IOD as an independent mode of climate variability, and many have argued that the IOD is the result of random variations in the east and west of the Indian Ocean Basin. The two nodes of the IOD are sometimes in phase (i.e., not always negatively correlated), indicating the lack of a consistent dipole structure (Dommenget and Latif 2001). Moreover, Dommenget (2007) and Dommenget and Jansen (2009) have shown that the dipole structure can be reconstructed by applying the same statistical technique used to decompose the mode to random noise, suggesting the IOD is simply an artifact of this technique and not a physical structure. The claims for the nonexistence of an equatorial IOD are also supported by numerous studies that show that the appearance of an IOD is simply a combination of a by-product of ENSO and random variations (e.g., Allan et al. 2001; Cadet 1985; Chambers et al. 1999; Nicholls 1984). The extreme phases of ENSO and IOD are also often *phase locked* meaning that there is a high probability that a La Niña (El Niño) will occur with warm (cool) eastern Indian Ocean SSTs (Meyers et al. 2007). As these coinciding phases are indicative of wet or dry conditions for both ENSO and the IOD, the result is often an amplification of the rainfall signature. The synchronicity between extreme ENSO and IOD variations suggests dependence between the two, and Allan et al. (2001) reported significant lag correlations in ENSO and IOD indices (with Pacific Ocean SST anomalies tending to lead Indian Ocean anomalies [Suppiah 1988]). Gallant et al. (2012) also found significant correlations of −0.57 between the austral spring Southern Oscillation Index (SOI, a well-known ENSO monitor) and the austral spring Dipole Mode Index (DMI, an indicator of the IOD) and −0.59 when the spring DMI was correlated against the austral winter SOI—indicating dependence between ENSO and Indian Ocean SSTs during the time when the Indian Ocean has its greatest influence. However, others have presented evidence that Indian Ocean SST anomalies can occur irrespective of the state of the tropical Pacific Ocean (Nicholls 1989; Saji et al. 1999;

Webster et al. 1999). Saji et al. (1999) reports that during the austral winter, there is a lack of statistically significant correlations between winter IOD and ENSO indices, potentially indicating independence. Webster et al. (1999) demonstrated that the Indian Ocean can exhibit strong ocean–atmosphere–land interactions that are self-maintaining and capable of producing large perturbations that are independent of ENSO. Fischer et al. (2005) used a climate model to study the triggers of the IOD and showed that two types may actually exist, one entirely independent of ENSO (i.e., an anomalous Hadley circulation over the eastern tropical Indian Ocean) and the other a consequence of tropical Pacific conditions (i.e., a zonal shift in the center of convection associated with developing El Niños). Importantly, this study (Fischer et al. 2005) indicates that the IOD may be partially linked to ENSO, but it also exhibits variability independent of the state of the tropical Pacific that is either random or driven by an as yet unknown cause. In light of these recent findings, it seems that although the Indian Ocean likely has an effect on hydroclimatic variability, especially during the austral winter and spring, the existence of a dipole mechanism as an atmosphere/ocean interaction is questionable.

*Southern and northern annular modes*: The southern annular mode (SAM) and the northern annular mode (NAM) are the leading modes of atmospheric variability over the extratropics. Also known as the Antarctic (or Arctic) oscillation and the high-latitude mode, the SAM and the NAM represent an exchange of mass (sea-level pressure seesaw) between the midlatitudes (~45°) and the polar region (>60°) (Thompson and Wallace 2000; Thompson et al. 2000). Various definitions of SAM and NAM have been proposed: a popular one is the normalized difference in the zonal mean sea-level pressure between 40° and 65°. As expected, the sea-level pressure pattern associated with SAM (or NAM) is a nearly annular (i.e., hemispheric) pattern with a large low-pressure anomaly centered on the South (or North) Pole and a ring of high-pressure anomalies at midlatitudes. This leads to an important zonal (i.e., parallel to the latitude bands) wind anomaly in a broadband around 55° with stronger westerlies indicative of a high SAM or NAM index (i.e., a positive SAM or NAM phase). Variations in the SAM or NAM effectively describe variations in the position of the midlatitude storm tracks (Thompson and Solomon 2002; Thompson and Wallace 2000). SAM and NAM typically fluctuate with a periodicity of about 10–14 days, though lower frequency cycles are also evident.

*Indonesian Throughflow (ITF)*: The Indonesian Throughflow (ITF) is an ocean current that transports water between the Pacific Ocean and the Indian Ocean through the Indonesian archipelago (Figure 2.4). In the north, the current enters the Indonesian seas through the Makassar Strait and Malacca Straits. The Indonesian seas function like a basin, and the ITF continues southerly and exits through the Lombok Strait, Ombai Strait, and Timor Passage. The ITF links two oceans and in so doing provides a pathway for modifying the stratification within each of these oceans as well as sea–air fluxes that impact on such climate phenomena as ENSO, IOD, and the Asian Monsoon. While a number of measurement programs have recently been undertaken in the Indonesian region, a serious shortcoming is that the data cover different time periods and depths in the different passages of the complex pathways linking the Pacific and Indian Oceans. This has led to ambiguity about the mean and variable

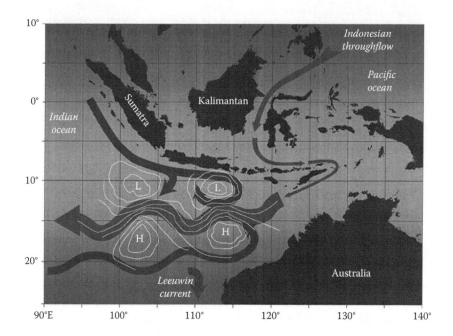

**FIGURE 2.4** Schematic of the ITF. (From www.incois.gov.in. Accessed June, 2013.)

nature of the ITF and about the transformation of the thermohaline and transport profiles within the interior seas. Therefore, while it is physically plausible (and even expected) that the ITF is the link between the Pacific and Indian Oceans (and the primary driving force behind Indian Ocean variability), some uncertainty still remains.

*Jet stream*: Jet streams are fast flowing, narrow air currents found near the tropopause, the transition between the troposphere (where temperature decreases with altitude) and the stratosphere (where temperature increases with altitude). Jet streams are caused by a combination of a planet's rotation on its axis and atmospheric heating (by solar radiation). The major jet streams on Earth are westerly winds (flowing west to east). Their paths typically have a meandering shape: jet streams may start, stop, split into two or more parts, combine into one stream, or flow in various directions including the opposite direction of most of the jet. The strongest jet streams are the polar jets, at around 7–12 km above sea level, and the higher and somewhat weaker subtropical jets at around 10–16 km above sea level. The Northern Hemisphere and the Southern Hemisphere each have both a polar jet and a subtropical jet. The Northern Hemisphere polar jet flows over the middle to northern latitudes of North America, Europe, and Asia and their intervening oceans, while the Southern Hemisphere polar jet circles Antarctica. As mentioned, jet stream paths typically have a meandering shape, and these meanders themselves propagate east, at lower speeds than that of the actual wind within the flow. Each large meander, or wave, within the jet stream is known as a Rossby wave (or planetary wave [see next paragraph]). Rossby waves are caused by changes in the Coriolis effect with latitude and propagate westward with respect to the flow in which they are embedded, which slows down the eastward migration of upper-level troughs and ridges across the globe when compared to their embedded shortwave troughs.

*Planetary wave (or atmospheric Rossby waves or wave train)*: Planetary waves are large (i.e., a full longitude circle), slow-moving waves generated in the troposphere by ocean–land temperature contrasts and topographic forcing (winds flowing over mountains). Not to be confused with oceanic Rossby waves, which move along the ocean thermocline, the atmospheric planetary or Rossby waves have wavelike form in the atmosphere in the longitudinal and vertical directions and often also in the latitudinal direction. Planetary (atmospheric Rossby) waves are basically disturbances in the jet stream and the number of disturbances around a latitude band determines the wave number. Planetary (atmospheric Rossby) waves are important because they have significant influence on wind speeds (e.g., trade winds, the jet stream), temperature, and the distribution of ozone, rainfall, and other characteristics of the lower and middle atmosphere. The terms *barotropic* and *baroclinic* Rossby waves are used to distinguish their vertical structure. Barotropic Rossby waves are air masses whose atmospheric pressure does not vary with height. Baroclinic waves are air masses whose pressure does vary with height. Baroclinic waves propagate much more slowly than barotropic waves, with speeds of only a few centimeters per second or less.

*Quasibiennial oscillation (QBO)*: The quasibiennial oscillation (QBO) is a quasiperiodic oscillation of the equatorial zonal wind between easterlies and westerlies in the tropical stratosphere with a mean period of 28–29 months. The alternating wind regimes develop at the top of the lower stratosphere and propagate downward at a rate of about 1 km per month until they are dissipated at the tropical tropopause. The QBO essentially describes the tendency for upper atmospheric winds over the equatorial Pacific to change direction approximately each year. It appears that this may act as a climatic *switch* during El Niño years (Angell 1992; Kane 1992; Sasi 1994). That is, if the QBO early in an El Niño year is in its positive (easterly) phase, then rainfall in Australia is less likely to be significantly below average than if the QBO is in its negative (westerly) phase. In general, the relationship between the ENSO and Australian rainfall tends to be stronger (during both El Niño and La Niña events) when the QBO is negative and the mechanism appears to be via the interference with the Walker circulation. What actually drives the QBO is an open research question but studies have shown that it may be regulated by solar activity (Labitzke 2005, 2007; Labitzke et al. 2006; Lu and Jarvis 2011).

*Northern Hemisphere climate modes*: In addition to the ocean–atmospheric patterns already mentioned, which are mostly Southern Hemisphere focused, there are numerous similar patterns or cycles that exist in the Northern Hemisphere (see Marshall et al. [2001] and Hurrell et al. [2003] for a summary). The most important is probably the North Atlantic oscillation (NAO) that refers to large-scale changes in the atmospheric Rossby wave and jet stream patterns. These in turn influence surface ocean conditions and regional climates on interannual to decadal timescales (e.g., Marshall et al. 2001; Seager et al. 2010). Significant coherence between NAO and, at least, ENSO (e.g., Huang et al. 1998; Seager et al. 2010) has also been shown.

Also important in the Northern Hemisphere is the Pacific North American (PNA) teleconnection pattern, which describes the primary patterns of atmospheric flow across the United States and Canada (e.g., Leathers et al. 1991). The positive phase of the PNA pattern features anomalously strong Aleutian lows and high pressure over western North America, coinciding with an enhancement and eastward shift of the East Asian jet stream. Changes in the PNA pattern are related to both ENSO and the PDO (e.g., Yarnal and Diaz 1986). However, positive and negative PNA patterns are not unique to either El Niño or La Niña events—roughly 50% of positive and negative PNA patterns occur during years when ENSO is not a significant factor, suggesting an unknown mechanism or interaction is contributing. While the Northern Hemisphere climate modes are summarized only briefly here, there are three important points that arise: (1) as with the Southern Hemisphere, understanding of the mechanisms, and their interactions, that drive Northern Hemisphere hydroclimatic variability is limited; (2) understanding into how variability in Northern Hemisphere climate modes is related to variability in Southern Hemisphere climate is in its infancy; and (3) insights into what actually drives variability in the Northern Hemisphere climate modes and what might occur in the future are speculative at best. These issues should be addressed, preferably in a collaborative manner, if advances in our holistic understanding of the Earth's climate system are to be made.

*Surface and deeper ocean circulations*: A critical factor in the Earth's ocean–atmospheric circulation processes is the thermohaline circulation (THC) or great ocean conveyor belt (Figure 2.5). The THC refers to the part of the large-scale ocean circulation that is driven by global density gradients created by surface heat and

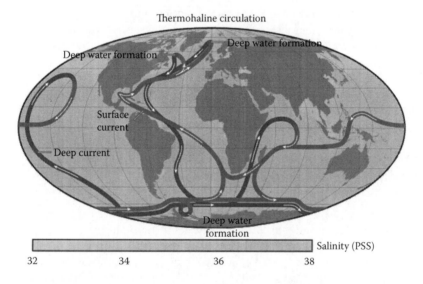

**FIGURE 2.5** A summary of the path of the THC or great ocean conveyor. Blue paths represent deep-water currents, while red paths represent surface currents. A summary of the path of the THC or Great Ocean Conveyor. Dark gray paths represent deep-water currents, while lighter gray paths represent surface currents (http://en.wikipedia.org/wiki/Thermohaline_circulation. Accessed June, 2013.)

freshwater fluxes. Wind-driven surface currents (such as the Gulf Stream) head poleward from the equatorial Atlantic Ocean, cooling all the while and eventually sinking at high latitudes (forming North Atlantic Deep Water). This dense water then flows into the ocean basins. While the bulk of it upwells in the Southern Ocean, the oldest waters (with a transit time of around 1600 years) upwell in the North Pacific (Primeau 2005). Extensive mixing therefore takes place between the ocean basins, reducing differences between them and making the Earth's ocean a global system. The moving water transports both energy (in the form of heat) and matter (solids, dissolved substances, and gases) around the globe. As such, the THC has a large impact on the climate of the Earth (e.g., Rashid et al. 2010). Due to the large time lags, and amounts of heat involved, the THC is likely related to interannual to multidecadal climate variability in some way, and several papers have recently been published that contain important insights along these lines (e.g., Boning et al. 2008; Froyland et al. 2007; Fukamachi et al. 2010; Holbrook et al. 2009, 2011; Rashid et al. 2010; Ridgway 2007; Ridgway and Dunn 2007). Froyland et al. (2007) showed how massive swirling structures in the ocean, the largest known as gyres, can be thousands of kilometers across and can extend down as deep as 500 m or more. The water in the gyres does not mix well with the rest of the ocean, so for long periods, these gyres can trap pollutants, nutrients, drifting plants, and animals and become physical barriers that divert even major ocean currents. The most well-known gyre is the Gulf Stream in the North Atlantic that pumps heat toward Europe, warming the atmosphere and giving the region a relatively mild climate (i.e., compare Portugal's climate to that of Nova Scotia, in Canada, which is roughly the same latitude). After releasing heat to the atmosphere, the waters recirculate toward the equator where they regain heat and rejoin the flow into the Gulf Stream. In this way, the ocean's gyres play a fundamental role in pumping heat poleward and cooler waters back to the tropics. This moderates Earth's climate extremes, reducing the equator-to-pole temperature gradients that would otherwise persist on an ocean-free planet. Froyland et al. (2007) showed how the East Australia Current has a similar, although more modest, impact on local climate on Australia's east coast (e.g., Holbrook et al. 2009, 2011). With respect to ocean gyres, Froyland et al. (2007) conclude "we're only just _beginning to get a grip on understanding their size, scale and functions, but we are sure that they have a major effect on marine biology and on the way that heat and carbon are distributed around the planet by the oceans." Ridgway (2007) and Ridgway and Dunn (2007) also investigated deep-ocean pathways (or super gyres) and discovered what is thought to be the previously undetected link between the ocean–atmospheric circulation patterns between the three Southern Hemisphere ocean basins. Prior to this work, the conventional understanding of Southern Hemisphere midlatitude circulation comprised basin-wide but quite distinct gyres contained within the Indian, Pacific, and Atlantic Oceans. However, the work by Ridgway (2007) and Ridgway and Dunn (2007), involving analysis of observations as well as model simulations, demonstrated that these gyres are connected. The implications are that these relationships must be properly understood and accounted for in order to understand how climate drivers initiate, evolve, and impact via teleconnections. More recently, Fukamachi et al. (2010) discovered another deep-ocean current (equivalent in volume to 40 Amazon Rivers) that flows more than 3 km below the surface of the Indian

Ocean sector of the Southern Ocean. The research showed that this is yet another important pathway in the global network of ocean currents that influences climate patterns (as well as ocean salinity, nutrient levels, etc.).

### 2.2.2 Earth–Sun–Moon Interactions

The previous section summarized some of the Earth's internal ocean–atmospheric interactions that may drive the key climate phenomena responsible for significant interannual to multidecadal hydroclimatic variability. However, scientists, astronomers, and philosophers have long speculated on the relationship between the Earth's climate and external forcing from the wider solar system in which it exists (e.g., Brougham 1803; Burroughs 1992; Eddy 1976; Herman and Goldberg 1978; Herschel 1801a,b; Hoyt and Schatten 1997). Mackey (2007) contains a comprehensive review of more than 200 years of research into this field and the summary of findings suggests that "the variable output of the Sun, the Sun's gravitational relationship between the Earth (and the Moon) and Earth's variable orbital relationship with the Sun, regulate the Earth's climate." It should be noted however that while there is a long history of finding cycles in climate data and claiming links to lunar and/or solar cycles, many of these investigations have been criticized on statistical grounds due to the fact that the physical mechanisms behind the solar–climate associations are often not clear (e.g., Pittock 1978, 1983, 2009). A review by Gray et al. (2010) addresses this by providing a summary of our current understanding of solar variability, solar–terrestrial interactions, and the mechanisms determining the response of the Earth's climate system. The review by Gray et al. (2010) represents the best current source of information on this issue with some of the main points expanded in the following.

Fairbridge (1961) emphasized, and later Mackey (2007) and Gray et al. (2010) reemphasized, that in order to understand and quantify Earth–Sun–Moon interactions, the totality of the various components of these interactions must be considered, which include

- Variations in the brightness of the Sun (i.e., the solar cycle or the solar magnetic activity cycle)
- The Sun's variable output of radiation across the electromagnetic spectrum
- Variations in the way the Sun disperses matter (i.e., objects that take up space and have mass)
- The variable gravitational force the Sun exerts on the Earth, the Sun exerts on the Moon, and the Moon exerts on the Earth as a system
- The strength of the Earth's geomagnetic field
- The heliosphere and cosmic radiation
- Variations in the shape of the Sun
- Variations in the quantity, intensity, and distribution of the Sun's output that is received by the Earth from the Sun (e.g., driven by variations in the distance the Earth is from the Sun as a result of the Earth's orbital cycles)
- The interactions between all of these processes

*Relationships between Earth–Sun–Moon interactions and key ocean–atmospheric phenomena*: During 1976, the vertical structure of the Pacific tropical thermocline changed significantly resulting in a significant reduction in the volume of deeper, colder water moving to the surface (e.g., Guilderson and Schrag 1998). Over just 1 year, Pacific Ocean surface temperatures changed from cooler than normal to warmer than normal, the phase of the IPO (and PDO) switched from negative to positive (e.g., Power et al. 1999a) and the frequency of El Niño events increased (e.g., Kiem et al. 2003). There has not yet been a satisfactory explanation for this large-scale change in the Pacific Ocean or similar changes occurring around the same time in the Indian and North Atlantic Oceans; however, it is consistent with the Keeling and Whorf (1997) hypothesis about the lunisolar tides churning of the Earth's oceans as recently demonstrated by Yasuda (2009). Similarly, Ramos da Silva and Avissar (2005) showed that lunisolar tides have been unambiguously correlated with the Arctic oscillation since the 1960s.

Further, despite the enormous amount of research into the dynamics and impacts of the ENSO, there is still no overall accepted theory that explains every aspect of the initiation and development of the ENSO phenomenon. Similarly, detailed theories do not exist that explain the physical causes of non-ENSO-related annual to interdecadal climate variability, such as that indicated by Indian Ocean SSTs, or longer-term decadal to multidecadal climate variability, such as that indicated by the IPO or PDO. However, studies do exist that suggest that solar activity (i.e., total solar irradiance [TSI]) plays a major role in controlling not only the ENSO phenomenon (e.g., Kirov et al. 2002; Landscheidt 2000) but also climate variability occurring on decadal/multidecadal timescales (Reid 2000; Waple 1999). Landscheidt (2001) and Kirov et al. (2002) also showed that the features of solar variability that are related to ENSO also have a significant influence on the NAO (see Marshall et al. [2001] and Hurrell et al. [2003] for a summary of the NAO and its impacts), while Zaitseva et al. (2003) found that the intensity of the NAO also depends on solar activity.

Coughlin and Kung (2004), Cordero and Nathan (2005), and Labitzke et al. (2006) also showed that the solar cycle is closely related to the strength of the QBO circulation and the length of the QBO waves, which in turn is related to ENSO. Nugroho and Yatini (2006) also reported that the sun strongly influences the IOD during the December to February monsoon wet season.

Therefore, although the physical mechanisms and causal relationships are still a matter of controversy, it appears that there is a link between solar (and lunar) variability and the dominant modes of seasonal to multidecadal climate variability occurring in both the Northern and Southern Hemispheres (e.g., Franks 2002; Gray et al. 2010; Reid 2000; White et al. 1997). Importantly, as summarized by Mackey (2007) and Gray et al. (2010), there have been numerous studies that have been suggesting for some time that the period from 2010 to 2040 will be associated with much less solar activity than recent decades (i.e., a tendency toward decreasing global average temperatures in the absence of other influences). The solar cycle (Sunspot Cycle #24), which began in approximately 2010, will be much weaker (i.e., cooler) than the previous solar cycle and similar to Sunspot Cycle #14 (1902–1913), which was the coldest in the last 100 years (see Figure 2.6). Sunspot Cycle #25 and #26 are expected to be weaker (i.e., cooler) again, with some suggesting the Earth's climate will

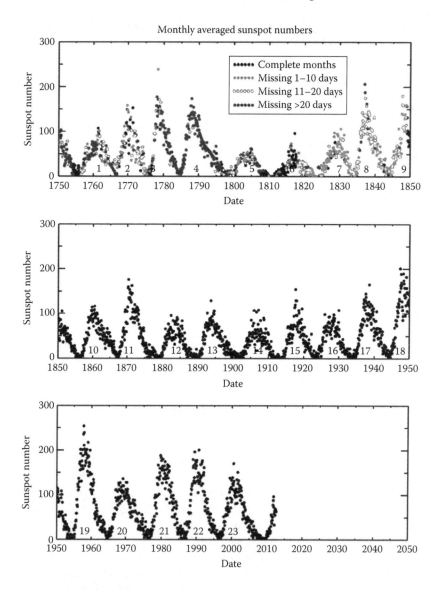

**FIGURE 2.6** Time series of the sun spot cycle (1750–2011). (From http://solarscience.msfc.nasa.gov/SunspotCycle.shtml. Accessed June, 2013.)

return to conditions similar to that experienced in the Dalton Minimum cold period (~1790s–1820s). The more active solar cycles, and associated warmer conditions, are expected to return around 2050 and persist for the remainder of the century. Note that these expectations appear to be in line with what is being observed in multidecadal climate cycles such as the IPO/PDO and NAO. As such, it is important to realize that any anthropogenically induced global warming (and the flow on effects that warming has on the climate) experienced over the period of weaker solar activity (~2010–2040) will likely be magnified when the more active solar cycles return around 2050.

Ignorance of Earth–Sun–Moon interactions could result in a false sense of security over the coming decades if recent warming trends stabilize or reverse, and as such, it is recommended that urgent research is required into how this solar-induced warming and cooling (1) influences the large-scale climate drivers and (2) enhances or suppresses the projected impacts associated with an anthropogenically enhanced greenhouse effect. Importantly, it is also noted that the solar variability field and the climate and hydrology fields very rarely collaborate or communicate and, as such, important insights are not shared across the different research fields. The final section of this chapter includes some discussion as to who could fund and coordinate the multidisciplinary collaborations that are required.

### 2.2.3 Atmospheric Chemistry, Aerosols, and Volcanic Eruptions

The atmosphere is composed of a mixture of invisible gases and a large number of suspended microscopic solid particles and water droplets (including cloud droplets and precipitation). Small solid particles and water droplets (excluding cloud droplets and precipitation) are collectively known as aerosols. Molecules of the gases and aerosols can be exchanged between the atmosphere and the Earth's surface by physical processes (e.g., evaporation, volcanic eruptions, continental weathering, wind blowing over dusty regions) or biological processes (e.g., evapotranspiration, photosynthesis, bushfires) or anthropogenic processes (e.g., burning fossil fuels, industrial activities).

This exchange of gases and aerosols between the atmosphere and Earth's surface (i.e., variability in atmospheric composition) influences climate by modulating the radiation budget. The main radiative effect of the gases is through the greenhouse effect, while aerosols may either heat or cool the surface, depending on their optical properties, which affect the solar and thermal radiation (e.g., Isaksen et al. 2009). The emission of chlorofluorocarbons (CFCs) and other chlorine and bromine compounds not only has an impact on the radiative forcing but has also led to the depletion of the stratospheric ozone layer. As discussed previously, it follows that this variability in atmospheric composition influences circulation patterns in the ocean and atmosphere.

The coupling between climate and atmospheric composition results from the basic structure of the Earth–atmosphere climate system and the fundamental processes within it. The composition of the atmosphere is determined by natural and human-related emissions and the energy that flows into, out of, and within the atmosphere. The principal source of this energy is sunlight at ultraviolet, visible, and near-infrared wavelengths. This incoming energy is balanced at the top of the atmosphere by the outgoing emission of infrared radiation from the Earth's surface and from the atmosphere. The structure of the troposphere (with temperature generally decreasing with altitude) is largely determined by the energy absorbed at or near the Earth's surface, which leads to the evaporation of water and the presence of reflecting clouds. Through many interactions, the composition and chemistry of the atmosphere are inherently connected to the climate system and the importance of climate–chemistry interactions has been recognized for more than 20 years (e.g., Ramanathan et al. 1987).

Interactions between climate and atmospheric chemistry provide important coupling mechanisms in the Earth system (see Chapter 7.1.3 in IPCC [2007] for further details). For example, the concentration of tropospheric ozone has increased substantially since the pre-industrial era, especially in polluted areas of the world, and has contributed to radiative warming. Emissions of chemical ozone precursors (carbon monoxide, methane, nonmethane hydrocarbons, nitrogen oxides) have increased as a result of larger use of fossil fuel, more frequent biomass burning, and more intense agricultural practices.

*Aerosols*: Aerosols influence climate in two main ways, referred to as direct forcing and indirect forcing. In the direct forcing mechanism, aerosols reflect sunlight back to space, thus cooling the planet. The indirect effect involves aerosol particles acting as (additional) cloud condensation nuclei, spreading the cloud's liquid water over more, smaller, droplets. This makes clouds more reflective and longer lasting. Calculating the direct forcing effect is a relatively straightforward exercise. However, computing the details of cloud microphysics requires a detailed understanding of the dynamical processes, moving water vapor through the atmosphere, and the physical mechanisms involved in the formation and growth of cloud particles, including heating and cooling by solar and infrared radiation. As such, the response of precipitation to aerosols attracts much attention and remains uncertain (Cui et al. 2011; IPCC 2007; Rosenfeld et al. 2008; Rotstayn et al. 2009).

Sources of aerosols include sea salt (due to evaporation of sea water), volcanic eruptions, bushfires, wind blowing over dusty areas, continental weathering, and human and industrial activities. Therefore, the concentration and influence of aerosols vary significantly from region to region. However, in some cases, it has been suggested that aerosols may be influential enough to alter large-scale climate patterns—and, via teleconnections, even global-scale hydroclimatic variability (see Rotstayn et al. [2009] for a comprehensive review). The cases relating to aerosols emanating from volcanic eruptions and the Asian brown cloud are discussed briefly here.

*Volcanic eruptions*: Volcanic eruptions can alter the climate of the Earth, via both the gases and the aerosols that are emitted, for both short and long periods of time. Usually volcanic eruptions have a cooling effect, for example, average global temperatures dropped about 0.5°C for about 2 years after the eruption of Mount Pinatubo in 1991, and it is also known that very cold temperatures caused crop failures and famine in North America and Europe for 2 years following the eruption of Tambora in 1815. The amount of cooling depends on the amount of dust put into the air, and the duration of the cooling depends on the size of the dust particles. Particles the size of sand grains fall out of the air in a matter of a few minutes and stay close to the volcano and have minimal effect on the climate. Tiny dust-size particles thrown into the lower atmosphere will float around for hours or days, causing darkness and cooling directly beneath the ash cloud, but these particles are quickly washed out of the air by the abundant water and rain present in the lower atmosphere. However, dust particles ejected higher into the dry upper atmosphere, the stratosphere, can remain for weeks to months before they finally settle. These particles block sunlight and cause some cooling over large areas of the Earth. Volcanoes that release large amounts of sulfur compounds (e.g., sulfur dioxide [$SO_2$]) affect the climate more strongly than those that eject just dust since the sulfur compounds rise easily into the stratosphere. Once there, they combine with the (limited) water available to form a haze of tiny droplets of sulfuric acid that are very light in color and reflect a great deal of sunlight for their size. Although the droplets eventually grow large enough to fall to the Earth, the stratosphere is so dry that it takes time, months, or even years to happen. Consequently, reflective hazes of sulfur droplets can cause significant cooling of the Earth for as long as 2 years after a major sulfur-bearing eruption. Volcanoes also release large amounts of water and carbon dioxide. When in the form of gases in the atmosphere, water and carbon dioxides absorb heat radiation (infrared) emitted by the ground and hold it in the atmosphere. This causes the air below to get warmer, which is counter to the cooling of the sulfur compounds in the atmosphere. However, there are very large amounts of water and carbon dioxide in the atmosphere already, and even a large eruption does not change the global amounts enough to affect the eruption-induced cooling. In addition, the water generally condenses out of the atmosphere as rain in a few hours to a few days, and the carbon dioxide quickly dissolves in the ocean or is absorbed by plants. Consequently, the sulfur compounds have a greater short-term effect, and cooling dominates. However, over long periods of time (thousands or millions of years), multiple eruptions of giant volcanoes can raise the carbon dioxide levels enough to cause significant global warming (see Section 2.2.4). There is compelling evidence of the role that volcanic eruptions can play and that they are influential enough to impact the behavior of ocean–atmospheric climate modes (e.g., Stenchikov et al. 2007); however, a complete understanding or definite conclusions have not yet been obtained.

*The Asian brown cloud*: The Asian brown cloud is a layer of air pollution (i.e., aerosols) that covers parts of South Asia, namely, the northern Indian Ocean, India, and Pakistan, especially between January and March. The Asian brown cloud changes rainfall patterns associated with the Asian Monsoon, resulting in a weakened Indian monsoon and reduced (increased) rainfall in northern (southern) China (Ramanathan et al. 2008). Rotstayn et al. (2007) also demonstrated that Australian rainfall and cloudiness is affected by the Asian brown cloud. Rotstayn et al. (2007) suggest that the Asian haze (1) alters the meridional temperature and pressure gradients over the tropical Indian Ocean, thereby increasing the tendency of monsoonal winds to flow toward Australia, and (2) based on modeled results, makes the simulated pattern of surface-temperature change in the tropical Pacific more like La Niña. Rotstayn et al. (2007) also suggest that one of the reasons that global climate model simulations fail to realistically simulate precipitation over much of the world, especially Australia, is that the climate models do not include forcing by Asian aerosols. Therefore, in summary, observations and modeling studies indicate that aerosols associated with the Asian brown cloud are important and alter ocean–atmospheric circulation patterns—and, as a result, potentially influence global-scale climate. Rotstayn et al. (2007) conclude "further research is essential to more accurately quantify the role of Asian aerosols in forcing Australian climate change." The more recent review by Rotstayn et al. (2009) contains a useful summary of the influence of Northern Hemisphere aerosols on climate drivers that affect Australia (e.g., the SAM).

## 2.2.4 GEOLOGICAL DRIVERS OF CLIMATE CHANGE

Climatic changes occurring over millions to billions of years were at least partly due to changes in sizes and locations of the Earth's continents. At shorter timescales, the modification of the Earth's surface (e.g., deforestation, changes to vegetation) can greatly influence the distribution of solar radiation. Instrumental records are typically too short to document these changes. However, geological records include evidence of climate changes occurring on timescales from hundreds to billions of years. The last 65 million years (the Cenozoic Era) is a period of dramatic global climate changes (Figure 2.7). The period from 65 to about 35 million years ago (Ma) (broadly equivalent to the Eocene and Paleocene Epochs) saw a significantly warmer world, with no Antarctic ice sheet, sea levels perhaps some 60 m higher than present and atmospheric $CO_2$ concentrations more than three times modern levels. Indeed, until about 15 Ma, Nothofagus trees were still growing in Antarctica at latitude 78°S (Lewis et al. 2007; Pillans 2011).

From about 35 Ma (latest Eocene), the Antarctic ice sheet progressively increased in size, global temperatures decreased, sea levels fell, and central Australia became progressively more arid as the Australian continent moved northward to straddle the zone of subtropical high pressure that is characterized by midlatitude deserts on both sides of the equator. The so-called Quaternary period (the last 2.6 million years) includes ice ages that represent the time when major Northern Hemisphere ice sheets waxed and waned in response to variations in the Earth's orbital parameters (see Section 2.2.2). The major glacial–interglacial fluctuations of the Quaternary

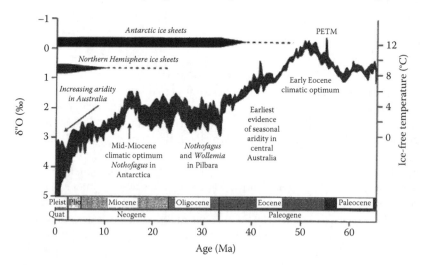

**FIGURE 2.7** Global climate during the Cenozoic Era (the last 65 million years) based on oxygen-isotope analyses of deep-sea foraminifera. The deep-ocean temperature scale on the right applies only to the time before major Antarctic ice sheets formed (i.e., prior to 35 million years ago), a period sometimes referred to as a *Greenhouse World* when atmospheric $CO_2$ levels exceeded 1000 parts per million (ppm) (cf ~390 ppm today) and global temperatures were significantly warmer than today. The unit on the x-axis is Ma, which refers to mega annum or *million years ago*. (From Pillans, B., *Aust. Geol.*, 150, 29, 2011.)

have been reconstructed via detailed analyses of deep-sea sediment cores and, more recently, of ice cores more than 3 km deep, retrieved from Antarctica. Air bubbles trapped in the ice allow past concentrations of various trace gases in the atmosphere, including $CO_2$ to be determined. The isotopic composition of the ice can also reveal past temperature changes extending back to 800,000 years ago (e.g., Lüthi et al. 2008). Highly detailed records of Quaternary climate are also available from speleothems in caves located in southeast China (Cheng et al. 2009; Wang et al. 2001). The oxygen-isotope ratio in the Chinese speleothems reflects the summer–winter precipitation ratio, and Wang et al. (2001) used this to demonstrate that the strength of the summer monsoon correlates with summer insolation values at 30°N (the latitude of the caves), punctuated by millennial-scale events that are also recorded in Greenland ice cores and North Atlantic sediments. These results confirm that there were massive and rapid changes, linked across the Northern and Southern Hemispheres, in ocean and atmospheric circulation patterns.

Pillans (2011) note the following important points that can be derived from looking at geological records:

- Interglacials (periods when global climate is similar to present) occur only about 15% of the time. In other words, during much of the last 800,000 years, global climate was significantly colder than present. This is in line with the discussion at the end of Section 2.2.2 and raises the question of whether it should be global warming or global cooling we should be worried about—perhaps anthropogenically increased $CO_2$ is the only thing that can prevent or lessen the effects of the anticipated orbital cycle-induced cooling.
- Not all interglacials are identical in structure and duration, but they are all characterized by rapid warming at the end of the previous glacial interval—these times of rapid change are referred to as glacial terminations.
- $CO_2$ concentrations are currently at 390 ppm and rising. This is the highest $CO_2$ concentrations in the last 800,000 years (prior to the current maximum, the previous maximum over the last 800,000 years was about 300 ppm). However, much of the last 550 million years has had $CO_2$ concentrations well in excess of 400 ppm.

In summary, on timescales of millions of years, major controls on global and regional climates include such things as atmospheric composition and plate tectonics. For example, high-latitude land masses in both hemispheres were a prerequisite for glaciation in both hemispheres in the Quaternary. On timescales of hundreds to thousands of years, other factors also become important, including variations in the Earth's orbit, solar variability, volcanic eruptions and earthquakes, sea-level changes, and changes in ocean and atmosphere circulation systems.

The Quaternary and earlier periods offer immense opportunities for understanding global climate change. For example, the likelihood of a return to $CO_2$ concentrations in excess of 400 ppm, whether anthropogenically or otherwise induced, and all the climatic changes associated with that is a real possibility (though this topic is also the subject of immense debate). Regardless, given that 400 ppm or greater $CO_2$ concentrations is a possibility, geological eras with high $CO_2$ concentrations may prove to be useful analogues for determining the possible implications of increased $CO_2$ concentrations on the drivers and impacts of climatic variability. Similarly, the analysis of the time lags associated with past tectonic shifts, earthquakes, volcanic eruptions (including undersea volcanoes), and large-ice sheet melts/growth episodes and corresponding alterations to the climate may provide some useful insights. Of concern though is the fact that the rate of change of greenhouse concentration (including $CO_2$) since 1950 is unprecedented in at least the last 100,000 years and possibly the last million years or more.

## 2.3 FACTORS RESPONSIBLE FOR ANTHROPOGENIC CLIMATE CHANGE

### 2.3.1 Human-Induced Increases to Greenhouse Gases and the Enhanced Greenhouse Effect

Human activities, in particular those involving the combustion of fossil fuels for industrial or domestic usage and biomass burning, produce GHGs and aerosols that affect the composition of the atmosphere and enhance the greenhouse effect. For about a 1000 years before the Industrial Revolution (~1760–1820), the amount of GHGs in the atmosphere remained relatively constant. Since then, the concentration of various GHGs has increased. For example, the amount of $CO_2$ has increased by more than 30% since pre-industrial times and is still increasing at an unprecedented rate of on average 0.4% per year, mainly due to the combustion of fossil fuels and deforestation. The concentration of other natural radiatively active atmospheric components, such as methane and nitrous oxide, is increasing as well due to agricultural, industrial, and other activities (see Table 2.1). The concentration of the nitrogen oxides (NO and $NO_2$) and of carbon monoxide (CO) is also increasing. Although these gases are not GHGs, they play a role in the atmospheric chemistry and have led to an increase in tropospheric ozone, a GHG, by 40% since pre-industrial times. It is also important to note that $NO_2$ is an absorber of visible solar radiation.

CFCs and some other halogen compounds do not occur naturally in the atmosphere but have been introduced by human activities. Besides their depleting effect on the stratospheric ozone layer, they are strong GHGs. Their greenhouse effect is only partly compensated for by the depletion of the ozone layer that causes a negative forcing of the surface–troposphere system. All these gases, except tropospheric ozone and its precursors, have long to very long atmospheric lifetimes and therefore become well-mixed throughout the atmosphere.

The increased concentration of GHGs in the atmosphere enhances the absorption and emission of infrared radiation. This effect is called the enhanced greenhouse effect, which is discussed in detail in Chapter 6 of IPCC (2007).

### 2.3.2 Human-Induced Increases to Aerosols and Other Pollutants

Human industrial, energy-related, and land-use activities increase the amount of aerosol in the atmosphere, in the form of mineral dust, soot, sulfates, and nitrates. The effect of the increasing amount of aerosols on the radiative forcing is complex and not yet well known (see Figure 2.8). The direct effect is the scattering of part of

## TABLE 2.1
### GHGs Influenced by Human Activities

| Greenhouse Gases | Principal Sources | Sinks | Lifetime in Atmosphere | Atmospheric Concentration (1998) | Annual Rate of Growth (1998) | Proportional Contribution to Greenhouse Warming (%) |
|---|---|---|---|---|---|---|
| Carbon dioxide ($CO_2$) | Fossil fuel burning, deforestation, biomass burning, gas flaring, cement production | Photosynthesis, ocean surface | 5–200 years | 365 ppmv | 0.4% | 60 |
| Methane ($CH_4$) | Natural wetlands, rice paddies, ruminant animals, natural gas drilling, venting and transmission, biomass burning, coal mining | Reaction with tropospheric hydroxyl (OH), removal by soils | 12 years | 1745 ppbv | 0.4% | 20 |
| Halocarbons (includes CFCs, HFCs, HCFCs, perfluorocarbons) | Industrial production and consumer goods (e.g., aerosol propellants, refrigerants, foam-blowing agents, solvents, fire retardants) | Varies (e.g., CFCs, HCFCs: removal by stratospheric photolysis, HCHC, HFC: reaction with tropospheric hydroxyl [OH]) | 2–50,000 years (e.g., CFC-11: 45 years, HFC-23: 260 years, $CF_4$: >50,000 years) | Varies (e.g., CFC-11: 268 pptv, HFC-23: 14 pptv, $CF_4$: 80 pptv) | Varies, most CFCs now decreasing or stable but HFCs and perfluorocarbons growing (e.g., CFC-11: −0.5%, HFC-23: +4%, $CF_4$: +1.3%) | 14 |
| Nitrous oxide ($N_2O$) | Biological sources in oceans and soils, combustion, biomass burning, fertilizer | Removal by soils stratospheric photolysis | 114 years | 314 ppbv | 0.25% | 6 |

*Source:* http://www.bom.gov.au/info/GreenhouseEffectAndClimateChange.pdf.
*Note:* ppmv, parts per million by volume; ppbv, parts per billion by volume; pptv, parts per trillion by volume.

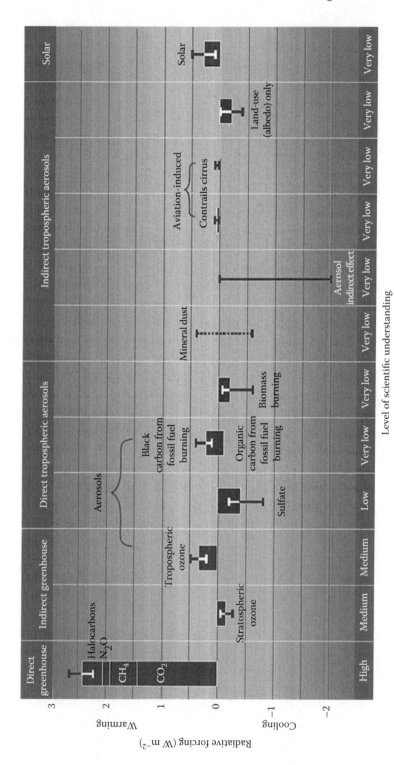

**FIGURE 2.8** Contribution of various factors to global, annual mean radiative forcing (W m$^{-2}$) since the mid-1700s. Range of uncertainty is indicated by vertical lines about the bars. Words across the bottom axis indicate the level of scientific understanding associated with each of the estimates. (From http://www.bom.gov.au/info/GreenhouseEffectAndClimateChange.pdf. Accessed June, 2013.)

the incoming solar radiation back into space. This causes a negative radiative forcing that may partly, and locally even completely, offset the enhanced greenhouse effect. However, due to their short atmospheric lifetime, the radiative forcing is very inhomogeneous in space and in time. This complicates their effect on the highly nonlinear climate system. Some aerosols, such as soot, absorb solar radiation directly, leading to local heating of the atmosphere, or absorb and emit infrared radiation, adding to the enhanced greenhouse effect.

Aerosols may also affect the number, density, and size of cloud droplets. This may change the amount and optical properties of clouds and hence their reflection and absorption. It may also have an impact on the formation of precipitation. As discussed in Chapter 5 of IPCC (2007), these are potentially important indirect effects of aerosols, resulting probably in a negative radiative forcing of as yet very uncertain magnitude (see Figure 2.8).

In addition to human-induced increases to aerosols, increases in atmospheric pollutants are also important. For example, the decrease in concentration of stratospheric ozone in the 1980s and 1990s due to manufactured halocarbons (which produced a slight cooling) has slowed down since the late 1990s; recent changes in the growth rate of atmospheric $CH_4$ are not well understood, but indications are that there have been changes in source strengths; nitrous oxide continues to increase in the atmosphere, primarily as a result of agricultural activities; and photochemical production of the hydroxyl radical (OH), which efficiently destroys many atmospheric compounds, occurs in the presence of ozone and water vapor and should be enhanced in an atmosphere with increased water vapor, as projected under future global warming. The importance of these relationships and effects is not yet well quantified (see Figure 2.8).

Changes to atmospheric gas concentration directly affect radiative forcing but also initiate, enhance, suppresse, or reverse *feedbacks* in the climate system. The strong nonlinearity of these feedbacks and the many processes involved make it difficult to quantify the total implications. Absolute concentrations are important and so are rates of change. What we do know is that an increase of atmospheric GHG concentrations leads to an average increase of the temperature of the lower atmosphere (i.e., surface atmosphere or troposphere). The response of the upper atmosphere (i.e., stratosphere) is entirely different. The stratosphere is characterized by a radiative balance between absorption of solar radiation, mainly by ozone, and emission of infrared radiation mainly by carbon dioxide. An increase in the carbon dioxide concentration in the stratosphere therefore leads to an increase of the emission of infrared radiation and, thus, to a cooling of the stratosphere.

Ozone is also particularly important. Interactions between ozone and climate have been subjects of discussion ever since the early 1970s when scientists first suggested that human-produced chemicals could destroy the ozone shield in the stratosphere. The discussion intensified in 1985 when atmospheric scientists discovered an ozone *hole* in the stratosphere over Antarctica. This lead to the phasing out of ozone-depleting chemicals such as CFCs and halons and predictions that the stratospheric ozone layer would recover to 1980 ozone levels by 2050. However, it may not be as simple as this. Ozone's impact on climate consists primarily of changes in temperature. The more ozone in a given parcel of air, the more heat it retains.

However, it is also mostly agreed that warmer (cooler) temperature will be associated with increased (decreased) water vapor in the lower atmosphere (troposphere), which would tend to produce more (less) ozone. Ozone also generates heat in the stratosphere, both by absorbing the sun's ultraviolet radiation and by absorbing upwelling infrared radiation from the lower troposphere. Consequently, decreased ozone in the stratosphere results in lower stratospheric temperatures. Observations show that over recent decades, the mid to upper stratosphere (from 30 to 50 km above the Earth's surface) has cooled by 1°C–6°C. This stratospheric cooling has taken place at the same time that GHG amounts, and temperatures, in the lower atmosphere (troposphere) have risen. Stratospheric cooling may have been taking place over recent decades for a number of reasons. One reason may be that the presence of ozone itself generates heat (via absorption of solar radiation) and ozone depletion cools the stratosphere (which leads to further ozone depletion) and warms the troposphere (as less incoming solar radiation is absorbed in the stratosphere, meaning more solar radiation reaches the troposphere). Another contributing factor to the stratospheric cooling may be that rising amounts of GHGs in the troposphere are retaining heat that would normally warm the stratosphere (and a colder stratosphere enhances ozone depletion, which results in a positive feedback loop that enables more solar energy to pass through to the troposphere that enhances the surface warming and stratospheric cooling). Therefore, there is some debate that suggests that the observed stratospheric cooling due to ozone loss (and/or increases in GHGs in the troposphere) may delay the recovery of the ozone layer (even if/when the ozone-depleting chemicals (e.g., CFCs and halons) are totally phased out). In any case, the role of ozone is (1) obviously important and (2) not well understood (e.g., Son et al. 2012).

### 2.3.3 Land-Use Change (Including Urbanization)

Land-use change refers to a change in the use or management of land. Such change may result from various human activities such as changes in agriculture and irrigation, deforestation, reforestation, and afforestation but also from urbanization or traffic. Land-use change results in changing the physical and biological properties of the land surface. Such effects change the radiative forcing and have a potential impact on regional and global climate.

Physical processes and feedbacks caused by land-use change, which may have an impact on the climate, include changes in albedo and surface roughness and the exchange between land and atmosphere of water vapor and GHGs. Land-use change may also affect the climate system through biological processes and feedbacks involving the terrestrial vegetation, which may lead to changes in the sources and sinks of carbon in its various forms. These climatic consequences of land-use change are discussed and evaluated in Section 4 of Chapter 7 in IPCC (2007).

Urbanization is another kind of land-use change. This may affect the local wind climate through its influence on the surface roughness. It may also create a local climate substantially warmer than the surrounding countryside—an urban heat island (UHI). The UHI effect refers to the observation that towns and cities tend to be warmer than their rural surroundings due to physical differences between the urban and natural landscapes. The concrete and asphalt of the urban environment tend to reduce a city's reflectivity compared with the natural environment. This increases the amount of solar radiation absorbed at the surface. Cities also tend to have fewer trees than the rural surroundings and hence the cooling effects of shade and evapotranspiration are reduced. The cooling effects of winds can also be reduced by city buildings.

## 2.4 SUMMARY

In terms of utilizing climate science insights to improve water resource security and better manage hydroclimate-related extremes, there is a need for improved understandings of the physics (particularly interactions between the various natural and anthropogenic climate drivers) and rigorous quantification (or at least acknowledgement) of the uncertainties. This will require continued investigation into the hydroclimatic impacts of the major climate modes, particularly the role of the Southern Ocean that has been largely ignored in comparison to the research efforts focused on Pacific and Indian Ocean variability and the associated teleconnections across Northern and Southern Hemispheres.

However, a fundamental knowledge gap exists in that we do not currently know enough about what drives the climate drivers? There are numerous theories and suspicions, supported in some cases by analyses of historical data (including preinstrumental information) or climate modeling experiments, as to what the causal mechanisms are behind some of the key climate modes. But, as indicated by their contradictory nature and the number of questions that still remain, these theories are speculative at best and there is a lack of clear and agreed insight into what drives the drivers. There are some things we just do not know and there are also likely to be unknown unknowns, especially relating to climate feedbacks, the role of the deep ocean, and the role of anthropogenically increased $CO_2$.

Innovative thinking is obviously needed. However, the type of research required to address some of the questions raised in this chapter is inherently *risky* and unlikely to return *practically useful* results—and within the constraints of a typical 2- to 3-year research grant is unlikely to produce much at all. Major scientific investigation, including a reframing of the questions we ask and the physical assumptions that are made, is required. This type of investigation should consider all the possibilities for what drives the drivers and involve scientists from all the relevant fields (e.g., atmospheric scientists, climate, meteorology, hydrology, oceanography, space physics, solar physics, geologists, chemists).

Undoubtedly climate models play an important role. However, the models are never going to be perfect and are only useful if their outputs are physically plausible and realistic. For water resources management under climate change, it is crucial to understanding that insights into climate variability and change, and in particular modeling the climate, will always be associated with assumptions and uncertainties: uncertainty around explaining past variability (i.e., instrumental and preinstrumental); uncertainty around attributing recent variability, in particular over the last few decades where anthropogenic climate change impacts may be playing a role; and uncertainty over what is in store for the future. It must be realized that in terms of climate-related risk, uncertainty is the reality of existence. Therefore, in order to adapt to climate variability, and in particular when attempting to adapt to an unknown future where previous and existing variability and uncertainty will likely be compounded by anthropogenic global warming, decision making under uncertainty is required. Various ways to do this have been proposed (e.g., Kiem and Verdon-Kidd 2011; McMahon et al. 2008; Verdon-Kidd and Kiem 2010) with the common theme being that the uncertainty must be quantifiable to enable win–win or no-regrets or optimized strategies and solutions—this is essential in a non-ergodic (i.e., continually changing) world (North 1999). In terms of water resources management under a variable and changing climate, we at least need to put ourselves in a position where we do not keep making the same mistakes (Cullen et al. 2002). Mistakes will continue to be made but insight should be gained from previous mistakes and the mistakes of the future should at least be new mistakes. Referring to and learning from the literature surrounding the management of *wicked problems* (e.g., Churchman 1967; Rittel and Webber 1973) may be a good place to start (refer to Australian Public Service Commission [2007] for information on how to do this).

## REFERENCES

Allan, R.J., Chambers, D.P., Drosdowsky, W. et al. 2001. Is there an Indian Ocean Dipole, and is it independent of the El Niño–Southern Oscillation? *CLIVAR Exchanges* 6(3): 18–22.

Angell, J.K. 1992. Evidence of a relation between El Nino and QBO, and for an El Nino in 1991–92. *Geophysical Research Letters* 19(3): 285–288, doi:10.1029/91GL02731.

Ashok, K., Behera, S.K., Rao, S.A., Weng, H., and Yamagata, T. 2007. El Niño Modoki and its possible teleconnection. *Journal of Geophysical Research* 112: C11007, doi:10.1029/2006JC003798.

Ashok, K., Guan, Z., and Yamagata, T. 2003. Influence of the Indian Ocean Dipole on the Australian winter rainfall. *Geophysical Research Letters* 30: 1821, doi:10.1029/2003GL017926.

Ashok, K., Reason, C.J.C., and Meyers, G.A. 2000. Variability in the tropical southeast Indian Ocean and links with southeast Australian winter rainfall. *Geophysical Research Letters* 27: 3977–3980.

Ashok, K., Tam, C.Y., and Lee, W.J. 2009. ENSO Modoki impact on the Southern Hemisphere storm track activity during extended austral winter. *Geophysical Research Letters* 36: L12705, doi:10.1029/2009GL038847.

Australian Public Service Commission. 2007. *Tackling Wicked Problems—A Public Policy Perspective*. Commonwealth of Australia, Canberra, Australian Capital Territory, Australia.

Barnett, T.P., Pierce, D.W., Saravanan, R., Schneider, N., Dommenget, D., and Latif, M. 1999. Origins of midlatitude Pacific decadal variability. *Geophysical Research Letters* 26: 1453–1456.

Blöschl, G. and Montanari, A. 2010. Climate change impacts—Throwing the dice? *Hydrological Processes* 24: 374–381, doi:10.1002/hyp.7574.

Boning, C.W., Dispert, A., Visbeck, M., Rintoul, S.R., and Schwarzkopf, F.U. 2008. The response of the Antarctic circumpolar current to recent climate change. *Nature Geoscience* 1: 864–869.

Brougham, H. 1803. Art. XV. Observations on the two lately discovered Celestial Bodies By William Herschel, L.L.D. F.R.S. From Phil. Trans. RS 1802," *Edinburgh Review*, Vol 1. pp. 426–431, January 1803.

Burroughs, W.J. 1992. *Weather Cycles Real or Imaginary?* Cambridge University Press, Cambridge, U.K.

Caballero, R. and Anderson, B.T. 2009. Impact of midlatitude stationary waves on regional Hadley cells and ENSO. *Geophysical Research Letters* 36: L17704, doi:10.1029/2009GL039668.

Cadet, D. 1985. The Southern Oscillation over the Indian Ocean. *International Journal of Climatology* 5: 189–212.

Cai, W. and Cowan, T. 2009. La Niña Modoki impacts Australia autumn rainfall variability. *Geophysical Research Letters* 36: L12805, doi:10.1029/2009GL037885.

Cai, W., Cowan, T., and Sullivan, A. 2009a. Recent unprecedented skewness towards positive Indian Ocean Dipole occurrences and its impact on Australian rainfall. *Geophysical Research Letters* 36: L11705, doi:10.1029/2009GL037604.

Cai, W., Sullivan, A., and Cowan, T. 2009b. How rare are the 2006–2008 positive Indian Ocean Dipole events? An IPCC AR4 climate model perspective. *Geophysical Research Letters* 36: L08702, doi:10.1029/2009GL037982.

Chambers, D.P., Tapley, B.D., and Stewart, R.H. 1999. Anomalous warming in the Indian Ocean coincident with El Niño. *Journal of Geophysical Research* 104: 3035–3047.

Cheng, H., Edwards, R.L., Broecker, W.S. et al. 2009. Ice age terminations. *Science* 326(5950): 248–252, doi:10.1126/science.1177840.

Chiew, F.H.S., Piechota, T.C., Dracup, J.A., and McMahon, T.A. 1998. El Niño Southern Oscillation and Australian rainfall, streamflow and drought—Links and potential for forecasting. *Journal of Hydrology* 204(1–4): 138–149.

Churchman, C.W. 1967. Wicked problems. *Management Science* 14: 141–142.

Collins, M., An, S.I., Cai, W. et al. 2010. The impact of global warming on the tropical Pacific and El Niño. *Nature Geoscience* 3: 391–397.

Cordero, E.C. and Nathan, T.R. 2005. A new pathway for communicating the 11-year solar cycle signal to the QBO. *Geophysical Research Letters* 32: L18805, doi:10.1029/2005GL023696.

Coughlin, K. and Kung, K.K. 2004. Eleven-year solar cycle signal throughout the lower atmosphere. *Journal of Geophysical Research* 109: D21105, doi:10.1029/2004JD004873.

Cui, Z., Davies, S., Carslaw, K.S., and Blyth, A.M. 2011. The response of precipitation to aerosol through riming and melting in deep convective clouds. *Atmospheric Chemistry and Physics* 11: 3495–3510.

Cullen, P., Flannery, T., Harding, R. et al. 2002. *Blueprint for a Living Continent: A Way Forward from the Wentworth Group of Concerned Scientists.* World Wide Fund for Nature, Sydney, New South Wales, Australia.

Di Baldassarre, G., Montanari, A., Lins, H., Koutsoyiannis, D., Brandimarte, L., and Blöschl, G. 2010. Flood fatalities in Africa: From diagnosis to mitigation. *Geophysical Research Letters* 37: L22402, doi:10.1029/2010GL045467.

Dommenget, D. 2007. Evaluating EOF modes against a stochastic null hypothesis. *Climate Dynamics* 28: 517–531.

Dommenget, D. and Jansen, M. 2009. Predictions of Indian Ocean SST indices with a simple statistical model: A null hypothesis. *Journal of Climate* 22: 4930–4938.

Dommenget, D. and Latif, M. 2001. A cautionary note on the interpretation of EOFs. *Journal of Climate* 15(2): 216–225.

Drosdowsky, W. 1993. Potential predictability of winter rainfall over southern and eastern Australian using Indian Ocean sea-surface temperature anomalies. *Australian Meteorological Magazine* 42: 1–6.

Eddy, J.A. 1976. The Maunder minimum. *Science* 192: 1189–1202.

Fairbridge, R.W. 1961. Solar variations, climatic change, and related geophysical problems. *Annals of the New York Academy of Science* 95: 1–740.

Fischer, A.S., Terray, P., Guilyardi, E., Gualdi, S., and Delecluse, P. 2005. Two independent triggers for the Indian Ocean Dipole/zonal mode in a coupled GCM. *Journal of Climate* 18: 3428–3449.

Folland, C.K., Renwick, J.A., Salinger, M.J., and Mullan, A.B. 2002. Relative influences of the Interdecadal Pacific Oscillation and ENSO on the South Pacific Convergence Zone. *Geophysical Research Letters* 29(13): 1643.

Frankignoul, C. and Hasselmann, K. 1977. Stochastic climate models. Part II: Application to sea surface temperature variability and thermocline variability. *Tellus* 29: 284–305.

Frankignoul, C., Muller, P., and Zorita, E. 1997. A simple model of the decadal response of the ocean to stochastic wind forcing. *Journal of Physical Oceanography* 27: 1533–1546.

Franks, S.W. 2002. Assessing hydrological change: Deterministic general circulation models or spurious solar correlation? *Hydrological Processes* 16: 559–564.

Franks, S.W. and Kuczera, G. 2002. Flood frequency analysis: Evidence and implications of secular climate variability, New South Wales. *Water Resources Research* 38(5): 20-1–20-7.

Frierson, D.M.W., Jian, L., and Chen, G. 2007. Width of the Hadley cell in simple and comprehensive general circulation models. *Geophysical Research Letters* 34(18): L18804, doi:10.1029/2007GL031115.

Froyland, G., Padberg, K., England, M.H., and Tréguier, A.M. 2007. Detection of coherent oceanic structures via transfer operators. *Physical Review Letters* 98: 224503, doi:10.1103/PhysRevLett.98.224503.

Fukamachi, Y., Rintoul, S.R., Church, J.A., Aoki, S., Sokolov, S., Rosenberg, M.A., and Wakatsuchi, M. 2010. Strong export of Antarctic Bottom Water east of the Kerguelen plateau. *Nature Geoscience* 3(5): 327–331.

Gallant, A.J.E. and Gergis, J. 2011. An experimental streamflow reconstruction for the River Murray, Australia, 1783–1988. *Water Resources Research* 47: W00G04, doi:10.1029/2010WR009832.

Gallant, A.J.E., Kiem, A.S., Verdon-Kidd, D.C., Stone, R.C., and Karoly, D.J. 2012. Understanding hydroclimate processes in the Murray–Darling Basin for natural resources management. *Hydrology and Earth System Sciences* 16: 2049–2068.

Gray, L.J., Beer, J., Geller, M. et al. 2010. Solar influences on climate. *Reviews of Geophysics* 48: RG4001, doi:10.1029/2009RG000282.

Guilderson, T.P. and Schrag, D.P. 1998. Abrupt shift in subsurface temperatures in the tropical pacific associated with changes in El Nino. *Science* 281: 240–243.

Herman, J.R. and Goldberg, R.A. 1978. *Sun, Weather and Climate*. NASA Special Publication, Vol. SP-426. National Aeronautics and Space Administration, Washington, DC.

Herschel, W. 1801a. Observations tending to investigate the nature of the Sun, in order to find the causes or symptoms of its variable emission of light and heat; with remarks on the use that may possibly be drawn from solar observations. *Philosophical Transactions of the Royal Society of London* 91: 265–318, http://www.jstor.org/stable/107097.

Herschel, W. 1801b. Additional observations tending to investigate the symptoms of the variable emission of the light and heat of the Sun; with trials to set aside darkening glasses, by transmitting the solar rays through liquids; and a few remarks to remove objections that might be made against some of the arguments contained in the former paper. *Philosophical Transactions of the Royal Society of London* 91: 354–362, http://www.jstor.org/stable/107100.

Holbrook, N.J., Goodwin, I.D., McGregor, S., Molina, E., and Power, S.B. 2011. ENSO to multi-decadal time scale changes in East Australian Current transports and Fort Denison sea level: Oceanic Rossby waves as the connecting mechanism. *Deep Sea Research II* 58: 547–558.

Holbrook, N.J., Goodwin, I.D., McGregor, S., and Power, S.B. 2009. Baroclinic modeling of the South Pacific Gyre. *Proceedings of the Extended Abstracts of the Ninth International Conference on Southern Ocean Meteorology and Oceanography*, Melbourne, Victoria, Australia.

Hoyt, D.V. and Schatten, K.H. 1997. *The Role of the Sun in Climate Change*. Oxford University Press, New York.

Huang, J., Higuchi, K., and Shabbar, A. 1998. The relationship between the North Atlantic Oscillation and El Nino–Southern Oscillation. *Geophysical Research Letters* 25: 2707–2710.

Hurrell, J.W., Kushnir, Y., Ottersen, G., and Visbeck, M. 2003. *The North Atlantic Oscillation: Climate Significance and Environmental Impact*. American Geophysical Union, Washington, DC.

IPCC. 2007. *Climate Change 2007: The Physical Science Basis*. Contribution of Working Group I to the Fourth Assessment Report of the Intergovernmental Panel on Climate Change. Cambridge University Press, Cambridge, U.K.

Isaksen, I.S.A., Granier, C., Myhre, G. et al. 2009. Atmospheric composition change: Climate–chemistry interactions. *Atmospheric Environment* 43(33): 5138–5192.

James, I.N. and James, P.M. 1992. Ultra low frequency variability of flow in a simple atmospheric circulation model. *Quarterly Journal of the Royal Meteorological Society* 118: 1211–1233.

Jin, F.F., Neelin, J.D., and Ghil, M. 1994. El Niño on the Devil's Staircase: Annual subharmonic steps to chaos. *Science* 264: 70–72.

Kane, R.P. 1992. Relationship between QBOs of stratospheric winds, ENSO variability and other atmospheric parameters. *International Journal of Climatology* 12: 435–447.

Keeling, C.D. and Whorf, T.P. 1997. Possible forcing of global temperature by the oceanic tides. *Proceedings of the National Academy of Sciences of the United States of America* 94: 8321–8328.

Kiem, A.S. and Franks, S.W. 2004. Multi-decadal variability of drought risk—Eastern Australia. *Hydrological Processes* 18(11): 2039–2050.

Kiem, A.S., Franks, S.W., and Kuczera, G. 2003. Multi-decadal variability of flood risk. *Geophysical Research Letters* 30(2): 1035, doi:10.1029/2002GL015992.

Kiem, A.S. and Verdon-Kidd, D.C. 2011. Steps towards 'useful' hydroclimatic scenarios for water resource management in the Murray–Darling Basin. *Water Resources Research* 47: W00G06, doi:10.1029/2010WR009803.

Kirov, B., Georgieva, K., and Javaraiah, J. 2002. 22-Year periodicity in solar rotation, solar wind parameters and Earth rotation. *Proceedings of the 10th European Solar Physics Meeting "Solar Variability: From Core to Outer Frontiers,"* Prague, Czech Republic.

Koutsoyiannis, D., Efstratiadis, A., Mamassis, N., and Christofides, A. 2008. On the credibility of climate predictions. *Hydrological Sciences Journal* 53(4): 671–684.

Koutsoyiannis, D., Montanari, A., Lins, H.F., and Cohn, T.A. 2009. Climate, hydrology and freshwater: Towards an interactive incorporation of hydrological experience into climate research—DISCUSSION of The implications of projected climate change for freshwater resources and their management. *Hydrological Sciences Journal* 54(2): 394–405.

Labitzke, K. 2005. On the solar cycle–QBO-relationship: A summary. *Journal of Atmospheric, Solar and Terrestrial Physics* 67(1–2): 45–54.

Labitzke, K. 2007. Effects of the solar cycle on the Earth's atmosphere. In: Kamide, Y. and Chian, A. (eds.), *Handbook of the Solar-Terrestrial Environment*, Chapter 18, pp. 445–466. Springer, Berlin, Germany.

Labitzke, K., Kunze, M., and Brinnimann, S. 2006. Sunspots, the QBO, and the stratosphere in the North Polar Region 20 years later. *Meteorologische Zeitschrift* 15(3): 355–363.

Landscheidt, T. 2000. Solar forcing of El Nino and La Nina. *Proceedings of the First Solar and Space Weather Euroconference—The Solar Cycle and Terrestrial Climate*, Santa Cruz de Tenerife, Tenerife, Spain.

Landscheidt, T. 2001. Solar eruptions linked to North Atlantic Oscillation, http://www.john-daly.com/theodor/solarnao.htm.

Latif, M. 1998. Dynamics of interdecadal variability in coupled ocean–atmosphere models. *Journal of Climate* 11: 602–624.

Leathers, D.J., Yarnal, B., and Palecki, M.A. 1991. The Pacific/North American teleconnection pattern the United States climate. Part I: Regional temperature and precipitation associations. *Journal of Climate* 4: 517–528.

Lewis, A.R., Marchant, D.R., Ashworth, A.C., Hemming, S.R., and Machlus, M.L. 2007. Major middle Miocene global climate change: Evidence from East Antarctica and the Transantarctic Mountains. *Geological Society of America Bulletin* 119: 1449–1461.

Lu, H. and Jarvis, M.J. 2011. Is the stratospheric quasi-biennial oscillation affected by solar wind dynamic pressure via an annual cycle modulation? *Journal of Geophysical Research—Atmospheres* 116: D06117, doi:10.1029/2010JD014781.

Lüthi, D., Le Floch, M., Bereiter, B. et al. 2008. High-resolution carbon dioxide concentration record 650,000–800,000 years before present. *Nature* 453: 379–382, doi:10.1038/nature06949.

Mackey, R. 2007. Rhodes Fairbridge and the idea that the solar system regulates the Earth's climate. *Journal of Coastal Research* Special Issue 50; *Proceedings of the International Coastal Symposium (ICS2007)*, Gold Coast, Queensland, Australia, pp. 955–968.

Mantua, N.J., Hare, S.R., Zhang, Y., Wallace, J.M., and Francis, R.C. 1997. A Pacific interdecadal climate oscillation with impacts on salmon production. *Bulletin of the American Meteorological Society* 78(6): 1069–1079.

Marshall, J., Kushner, Y., Battisti, D., Chang, P., Czaja, A., Dickson, R., Hurrell, J., McCartney, M., Saravanan, R., and Visbeck, M. 2001. North Atlantic climate variability: Phenomena, impacts and mechanisms. *International Journal of Climatology* 21(15): 1863–1898.

McGregor, S. 2009. Mechanisms forcing interdecadal variability of the El Niño–Southern Oscillation, PhD thesis, Department of Environment and Geography, Faculty of Science, Macquarie University, Sydney, New South Wales, Australia.

McGregor, S., Holbrook, N.J., and Power, S. 2007. Interdecadal SST variability in the equatorial Pacific Ocean. Part I: The role of off-equatorial wind stresses and oceanic Rossby waves. *Journal of Climate* 20: 2643–2658.

McGregor, S., Holbrook, N.J., and Power, S. 2008. Interdecadal SST variability in the equatorial Pacific Ocean. Part II: The role of equatorial/off-equatorial wind stresses in a hybrid coupled model. *Journal of Climate* 21: 4242–4256.

McGregor, S., Holbrook, N.J., and Power, S. 2009. The response of a stochastically forced ENSO model to observed off-equatorial wind-stress forcing. *Journal of Climate* 22: 2512–2525.

McMahon, T.A., Kiem, A.S., Peel, M.C., Jordan, P.W., and Pegram, G.G.S. 2008. A new approach to stochastically generating six-monthly rainfall sequences based on Empirical Model Decomposition. *Journal of Hydrometeorology* 9: 1377–1389.

Meyers, G., McIntosh, P., Pigot, L., and Pook, M. 2007. The years of El Niño, La Niña, and interactions with the tropical Indian Ocean. *Journal of Climate* 20: 2872–2880.

Miller, A.J. and Schneider, N. 2000. Interdecadal climate regime dynamics in the North Pacific Ocean: Theories, observations and ecosystem impacts. *Progress in Oceanography* 47: 355–379.

Milly, P.C.D., Betancourt, J., Falkenmark, M. et al. 2008. Stationarity is dead: Whither water management? *Science* 319: 573–574, doi:10.1126/science.1151915.

Montanari, A., Blöschl, G., Sivapalan, M., and Savenije, H. 2010. Getting on target. *Public Service Review: Science and Technology* 7: 167–169.

Murphy, B.F. and Timbal, B. 2008. A review of recent climate variability and climate change in southeastern Australia. *International Journal of Climatology* 28(7): 859–879.

Newman, M., Compo, G.P., and Alexander, M.A. 2003. ENSO forced variability of the Pacific decadal oscillation. *Journal of Climate* 16: 3853–3857.

Nicholls, N. 1984. The Southern Oscillation and the Indonesian sea surface temperature. *Monthly Weather Review* 112: 424–432.

Nicholls, N. 1989. Sea surface temperatures and Australian winter rainfall. *Journal of Climate* 2: 965–973.

North, D.C. 1999. Dealing with a non-ergodic world: Institutional economics, property rights, and the global environment. *Duke Environmental Law and Policy Forum* 10(1): 1–12, http://www.law.duke.edu/journals/10DELPFNorth.

Nugroho, J.T. and Yatini, C.Y. 2006. Indication of solar signal in Indian Ocean Dipole (IOD) phenomena over Indonesia. *Proceedings of the Second UN/NASA Workshop on International Heliophysical Year and Basic Space Science*, Indian Institute of Astrophysics, Bangalore, India.

Pierce, D.W., Barnett, T.P., and Latif, M. 2000. Connections between the Pacific Ocean tropics and midlatitudes on decadal time scales. *Journal of Climate* 13: 1173–1194.

Pillans, B. 2011. Climate change—A view from the Quaternary. *The Australian Geologist* 150: 29–32.

Pittock, A.B. 1978. A critical look at long-term Sun–weather relationships. *Reviews of Geophysics* 16: 400–420.

Pittock, A.B. 1983. Solar variability, weather and climate: An update. *Quarterly Journal of the Royal Meteorological Society* 109: 23–55.

Pittock, A.B. 2009. Can solar variations explain variations in the Earth's climate? An editorial comment. *Climatic Change* 96(4): 483–487.

Power, S., Casey, T., Folland, C., Colman, A., and Mehta, V. 1999a. Inter-decadal modulation of the impact of ENSO on Australia. *Climate Dynamics* 15(5): 319–324.

Power, S. and Colman, R. 2006. Multi-year predictability in a coupled general circulation model. *Climate Dynamics* 26: 247–272.

Power, S., Tseitkin, F., Mehta, V., Lavery, B., Torok, S., and Holbrook, N. 1999b. Decadal climate variability in Australia during the twentieth century. *International Journal of Climatology* 19: 169–184.

Power, S.B. and Smith, I.N. 2007. Weakening of the Walker Circulation and apparent dominance of El Niño both reach record levels, but has ENSO really changed? *Geophysical Research Letters* 34: L18702, doi:10.1029/2007GL030854.

Power, S.B., Tseitkin, F., Dix, M., Kleeman, R., Colman, R., and Holland, D. 1995. Stochastic variability at the air sea interface on decadal timescales. *Geophysical Research Letters* 22: 2593–2596.

Primeau, F. 2005. Characterizing transport between the surface mixed layer and the ocean interior with a forward and adjoint global ocean transport model. *Journal of Physical Oceanography* 35(4): 545–564.

Quan, X.W., Diaz, H.F., and Hoerling, M.P. 2004. Changes in the Tropical Hadley cell since 1950—The Hadley circulation: Present, past, and future. *Advances in Global Change Research* 21: 85–120.

Ramanathan, V., Agrawal, M., Akimoto, H. et al. 2008. *Atmospheric Brown Clouds: Regional Assessment Report with Focus on Asia.* United Nations Environment Programme, Nairobi, Kenya.

Ramanathan, V., Callis, L., Cess, R. et al. 1987. Climate–chemical interactions and effects of changing atmospheric trace gases. *Review of Geophysics and Space Physics* 25: 1441–1482.

Ramos da Silva, R. and Avissar, R. 2005. The impacts of the Luni-Solar oscillation on the Arctic oscillation. *Geophysical Research Letters* 32: L22703, doi:10.1029/2005GL023418.

Rashid, H., Power, S.B., and Knight, J.R. 2010. Impact of multidecadal fluctuations in the Atlantic thermohaline circulation on Indo-Pacific climate variability in a coupled GCM. *Journal of Climate* 23: 4038–4044.

Reichler, T. 2009. Changes in the atmospheric circulation as indicator of climate change. In: Letcher, T.M. (ed.), *Changes in the Atmospheric Circulation as Indicator of Climate Change*, pp. 145–164. Elsevier BV, Amsterdam, the Netherlands.

Reid, G.C. 2000. Solar variability and the earth's climate: Introduction and review. In: Friis-Christensen, E., Fröhlich, C., Haigh, J.D., Schűssler, M., and Von Steiger, R. (eds.), *Space Science Review* 94(1–2): 1–11.

Ridgway, K.R. 2007. Long-term trend and decadal variability of the southward penetration of the East Australian Current. *Geophysical Research Letters* 34: L13613, doi:10.1029/2007GL030393.

Ridgway, K.R. and Dunn, J.R. 2007. Observational evidence for a Southern Hemisphere oceanic supergyre. *Geophysical Research Letters* 34: L13612, doi:10.1029/2007GL030392.

Rittel, H. and Webber, M.M. 1973. Dilemmas in a general theory of planning. *Policy Sciences* 4: 155–169.

Rodgers, K.B., Friederichs, P., and Latif, M. 2004. Tropical Pacific decadal variability and its relation to decadal modulations of ENSO. *Journal of Climate* 17: 3761–3774.

Rosenfeld, D., Lohmann, U., Raga, G.B. et al. 2008. Flood or drought: How do aerosols affect precipitation? *Science* 321: 1309–1313.

Rotstayn, L.D., Cai, W., Dix, M.R. et al. 2007. Have Australian rainfall and cloudiness increased due to the remote effects of Asian anthropogenic aerosols? *Journal of Geophysical Research* 112: D09202, doi:10.1029/2006JD007712.

Rotstayn, L.D., Keywood, M.D., and Forgan, B.W. 2009. Possible impacts of anthropogenic and natural aerosols on Australian climate: A review. *International Journal of Climatology* 29(4): 461–479.

Saji, N.H., Goswami, B.N., Vinayachandran, P.N., and Yamagata, T. 1999. A dipole mode in the tropical Indian Ocean. *Nature* 401: 360–363.

Sasi, M.N. 1994. A relationship between equatorial lower stratospheric QBO and El Niño. *Journal of Atmospheric and Terrestrial Physics* 56(12): 1563–1570.

Schopf, P.S. and Burgman, R.J. 2006. A simple mechanism for ENSO residuals and asymmetry. *Journal of Climate* 19: 3167–3179.

Seager, R., Kushnir, Y., Nakamura, J., Ting, M., and Naik, N. 2010. Northern Hemisphere winter snow anomalies: ENSO, NAO and the winter of 2009/10. *Geophysical Research Letters* 37: L14703, doi:10.1029/2010GL043830.

Seidel, D.J. and Randel, W.J. 2007. Recent widening of the tropical belt: Evidence from tropopause observations. *Journal of Geophysical Research* 112: D20113, doi:10.1029/2007JD008861.

Son, S.-W., Purich, A., Hendon, H.H., Kim, B.-M., and Polvani, L.M. 2013. Improved seasonal forecast using ozone hole variability?, Geophys. Res. Lett., 40, 6231–6235, doi: 10.1002/2013GL057731.

Stainforth, D.A., Allen, M.R., Tredger, E.R., and Smith, L.A. 2007. Confidence, uncertainty and decision-support relevance in climate predictions. *Philosophical Transactions of the Royal Society A* 365: 2145–2161.

Stenchikov, G., Delworth, T., and Wittenberg, A. 2007. Volcanic climate impacts and ENSO interactions. *AGU Spring Meeting Abstracts*, 88(23), Abstract A43D-09.

Suppiah, R. 1988. Relationships between Indian Ocean Sea Surface Temperature and the Rainfall of Sri Lanka. *Journal of the Meteorological Society of Japan*, 66(1), 103–112.

Taschetto, A.S. and England, M.H. 2009. El Niño Modoki impacts on Australian rainfall. *Journal of Climate* 22: 3167–3174.

Taschetto, A.S., Ummenhofer, C.C., Sen Gupta, A., and England, M.H. 2009. Effect of anomalous warming in the central Pacific on the Australian monsoon. *Geophysical Research Letters* 36: L12704, doi:10.1029/2009GL038416.

Thompson, D.W.J. and Solomon, S. 2002. Interpretation of recent Southern Hemisphere climate change. *Science* 296: 895–899.

Thompson, D.W.J. and Wallace, J.M. 2000. Annular modes in the extratropical circulation, part I: Month-to-month variability. *Journal of Climate* 13: 1000–1016.

Thompson, D.W.J., Wallace, J.M., and Hegerl, G.C. 2000. Annular modes in the extratropical circulation, part II: Trends. *Journal of Climate* 13: 1018–1036.

Timmermann, A., Jin, F.F., and Abshagen, J. 2003. A nonlinear theory for El Niño bursting. *Journal of the Atmospheric Sciences* 60: 152–165.

Ummenhofer, C.C., England, M.H., McIntosh, P.C. et al. 2009. What causes southeast Australia's worst droughts? *Geophysical Research Letters* 36: L04706, doi:10.1029/2008GL036801.

Verdon, D.C. and Franks, S.W. 2005. Indian Ocean sea surface temperature variability and winter rainfall: Eastern Australia. *Water Resources Research* 41: W09413, doi:10.1029/2004WR003845.

Verdon, D.C. and Franks, S.W. 2006. Long-term behaviour of ENSO—Interactions with the PDO over the past 400 years inferred from paleoclimate records. *Geophysical Research Letters* 33(6): L07612, doi:10.1029/2005GL025052.

Verdon, D.C., Kiem, A.S., and Franks, S.W. 2004a. Multi-decadal variability of forest fire risk—Eastern Australia. *International Journal of Wildland Fire* 13(2): 165–171.

Verdon, D.C., Wyatt, A.M., Kiem, A.S., and Franks, S.W. 2004b. Multi-decadal variability of rainfall and streamflow—Eastern Australia. *Water Resources Research* 40(10): W10201, doi:10.1029/2004WR003234.

Verdon-Kidd, D.C. and Kiem, A.S. 2010. Quantifying drought risk in a non-stationary climate. *Journal of Hydrometeorology* 11(4): 1020–1032.

Wang, Y.J., Cheng, H., Edwards, R.L. et al. 2001. A high-resolution absolute-dated late Pleistocene monsoon record from Hulu Cave, China. *Science* 294(5550): 2345–2348.

Waple, A.M. 1999. The sun–climate relationship in recent centuries: A review. *Progress in Physical Geography* 23(3): 309–328.

Webster, P.J., Moore, A.M., Loschnigg, J.P., and Leben, R.R. 1999. Coupled ocean–atmosphere dynamics in the Indian Ocean during 1997–98. *Nature* 401: 356–360.

White, W.B., Lean, J., Cayan, D.R., and Dettinger, M.D. 1997. Response of global upper ocean temperature to changing solar irradiance. *Journal of Geophysical Research* 102: 3255–3266.

Wooldridge, S.A., Franks, S.W., and Kalma, J.D. 2001. Hydrological implications of the Southern Oscillation: Variability of the rainfall–runoff relationship. *Hydrological Sciences Journal* 46(1): 73–88.

Yarnal, B. and Diaz, H.F. 1986. Relationships between extremes of the Southern Oscillation and the winter climate of the Anglo-American Pacific coast. *Journal of Climatology* 6(2): 197–219.

Yasuda, I. 2009. The 18.6-year period moon-tidal cycle in Pacific Decadal Oscillation reconstructed from tree-rings in western North America. *Geophysical Research Letters* 36: L05605, doi:10.1029/2008GL036880.

Zaitseva, S.A., Akhremtchik, N., Pudovkin, M.I., and Galtsova, Y.V. 2003. Long-term variations of the solar activity–lower atmosphere relationship. *International Journal of Geomagnetism and Aeronomy* 4(2): 167–174.

# 3 Detection and Attribution of Climate Change

*H. Annamalai*

## CONTENTS

3.1 Introduction ........................................................................................................69
3.2 Detection of Climate Change ............................................................................71
3.3 Attribution of Climate Change ..........................................................................77
3.4 Summary ............................................................................................................78
Acknowledgments........................................................................................................79
References....................................................................................................................79

## 3.1 INTRODUCTION

This chapter is intended for nonspecialists in climate science and provides basic and elementary background on our current understanding of climate change, in particular the phrase *global warming* that has attracted the public attention in recent years. We begin by outlining the basic physics involved in conjunction with the competing influences of greenhouse gases (GHGs) and aerosols on the climate system. Undoubtedly, future state of the climate system can only be obtained by running climate models in supercomputers. A challenge for climate modeling community is to demonstrate credibility for the model-simulated future climate scenario. This can only be accomplished if the model simulations also detect a climate change signal as observed for the current climate. Further, confidence in climate model-projected future changes is enhanced only if we understand the interactive processes that make up the future climate system. Such a systematic approach boosts the reliability of the projected changes and any impact assessment studies then provides pathways for making meaningful policy changes.

Direct observations confirm that during the recent six to seven decades, increase in the atmospheric concentration of carbon dioxide ($CO_2$) is unprecedented (Figure 3.1). Natural sinks of carbon such as oceans (through biological and chemical processes) and plants (through photosynthesis) help offset the atmospheric concentrations of $CO_2$—yet the increase in the burning of fossil fuels (e.g., wood, coal, oil and natural gas) is the primary reason for the observed upward trend. Here trend is defined as *changes in time mean*. There is consensus among the scientific community that the recent observed monotonic increase in $CO_2$ is man-made

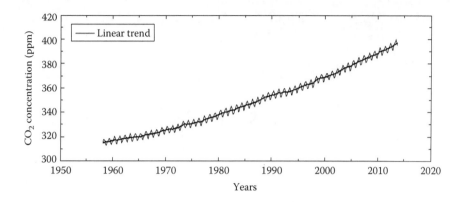

**FIGURE 3.1** Observed atmospheric concentration of carbon dioxide ($CO_2$).

(IPCC 2007), and the biggest concern and uncertainty is the trajectory the $CO_2$ trend will adopt into the future. Therefore, considering different pathways and different concentrations of $CO_2$, climate modelers perform a suite of century-long simulations for projecting future climate termed *future scenarios*. Due to uncertainty in the future concentration of $CO_2$, future climate model simulations are termed *projections* and not predictions. Even if the future emission of $CO_2$ is controlled, the present-day injected carbon in the atmosphere has a lifetime of about 100–200 years. Therefore, it is fair to say that humans have disturbed the climate system and the consequences are inevitable.

In simple words, GHGs are opaque to outgoing infrared radiation and are efficient in trapping the *heat* in the lower atmosphere. Effectively altering the radiative balance of the atmosphere and through simple physical process, $CO_2$ concentration leads to an increase in surface as well as atmospheric temperature. The subsequent consequences are manifold. The facts that evaporation from ocean surface increases nonlinearly with respect to surface temperature and that warmer air can hold more moisture lead to an increase in surface and atmospheric moisture or water vapor content. It should be noted that water vapor is a more efficient GHG than $CO_2$ and thus provides a *positive feedback* to global warming. Due to lack of sufficient observations, our current knowledge on the magnitude of this water vapor feedback is uncertain. Nevertheless, increase in moisture content promotes more rainfall implying *wet gets wetter* or, simply, regions receiving high rainfall today are expected to receive even higher rainfall in a warmer planet (Held and Soden 2006). The changes in rainfall patterns exert an impact on near-surface atmospheric circulation that in turn influences the ocean processes. Consequently, local ocean–atmosphere interaction processes are altered (Annamalai et al. 2013). Another direct consequence of surface warming is rise in sea level through *thermal expansion of water* and melting of ice sheets (Timmerman et al. 2010).

The radiative balance of the atmosphere is also influenced by the concentration and vertical distribution of aerosols (e.g., sulfate and black carbon aerosols), another important anthropogenic forcing agent. Since industrial revolution concentrations of sulfate aerosols (Charlson et al. 1992) as well as black carbon aerosols, particularly over Asia, have increased (Ramanathan et al. 2005; Bollasina et al. 2011). In contrast to the effect of GHGs, aerosols absorb and scatter incoming visible radiation. Thereby, they reduce the amount of radiation reaching the surface and hence *cool* the surface. The absorbing aerosols, depending on their vertical distributions, tend to heat the atmosphere and could *burn off* the clouds. Despite their short life time in the atmosphere (~few weeks) due to their impact on radiative balance, aerosols can influence climate at longer timescales (Charlson et al. 1992). In summary, changes in earth's energy budget (i.e., *radiative forcing*), either through changes in outgoing longwave radiation by GHGs or incoming shortwave radiation by aerosols, exerts changes in the climate system.

Detecting any climate change and attributing it to particular physical process is basically a statistical problem. Detection and attribution approaches attempt to separate out the variability associated with various components of anthropogenic forcing from natural climate variability. Detecting measurable change induced by anthropogenic forcing (GHG and aerosols) that is above background natural climate variability requires longer observational record. Attributing the detected signal to anthropogenic forcing requires modeling the climate system response to relative contributions of aerosol versus $CO_2$ and natural forcing. Detection and attribution studies face large errors due to lack of long and sustained observations of climate variables and due to our lack of complete understanding of the various interactive processes that make up climate system. In detection and attribution studies, finding optimal methods to enhance confidence in the detected signal and identifying the cause for that change is challenging.

The remainder of the chapter is organized as follows: From a suite of observations and century-long simulations of global climate models, long-term changes are detected (Section 3.2) and discussions involving constraints in observations and climate model integrations are provided. The discussions on attributions are provided next (Section 3.3). In Section 3.4, we close the chapter by addressing technological solutions to curb global warming and the need for educating people on the consequences of global warming.

## 3.2 DETECTION OF CLIMATE CHANGE

In this section, we begin with the definition of *detection*. Then, detected signals in observed and modeled global mean surface temperature and regional monsoon rainfall are presented. We close the section with uncertainties and caveats inherent with observations and climate models and provide cautionary notes in interpreting trends from limited data sets as the phase of the decadal variability may interfere with the signal.

*Detection* is the process of demonstrating that an observed change is significantly different (in a statistical sense) than can be explained by natural internal variability (Barnett et al. 1999; IPCC 2007). In other words, one needs to demonstrate that the observed climate change is unlikely to have occurred due to natural variability in the past. Here, natural variability refers to internal interactions within the climate system, such as El Niño–Southern Oscillation (ENSO), Pacific decadal variability, and forcing through changes in solar luminosity or volcanic activity. To consolidate detection of climate change in a particular climate variable then requires the knowledge about the amplitude and geographical patterns of natural variability and long observational record (~100–500 years) to confirm that such changes have not occurred in the past. The observational record needs to be homogenous in time, that is, free from artifacts due to changes in instruments or location, temporal sampling, observing procedures, etc. (Heger et al. 2006).

Estimates of natural variability require long and sustained observational records to account for decadal and multidecadal (10–30 years) variations. The danger is in seeking trends in short records as the phase of decadal–multidecadal variability can inflate the signal. Further, variance associated with natural variability has geographical preferences. To obtain reliable estimates of natural variability, one requires well-distributed spatial coverage. Climate models offer a surrogate for observations here. *Without* prescribing the estimated historical anthropogenic forcing agents such as GHGs and aerosols, climate models are run in *control mode*—where the model-generated variability is completely due to internal natural variability in the model world. The great advantage is that the models can be run for any number of years (depending on computer resources). The model generates many atmospheric and oceanic data sets and the data cover the entire globe. The present-day models are able to capture the structure of the observed decadal–multidecadal natural variability (e.g., Stouffer et al. 2000), but the models underestimate the magnitude of the largest spatial scales of variability (Barnett et al. 1999). In summary, due to lack of long and consistent observational records, it is clear that any observations, regional or global, employed to detect measurable climate change are bound to have uncertainties. Similarly, inherent problems exist in models. Within these known limitations, detecting a climate change signal and attributing that change to a specific mechanism is very demanding.

An ongoing question that is intensely debated in climate community is: Has the climate system started to respond to the observed unprecedented increase in $CO_2$ concentration (Figure 3.1)? In other words, based on the simple physics provided earlier, is there a steady increase in observed surface temperature? If yes, is there an imprint on the hydrological cycle? For example, has the monsoon rainfall over South Asia responded to anthropogenic forcing? Are these detected signals above natural climate variability? How reliable are the observations, and what is the confidence level in the detected signal? We provide plausible answers for these and other related questions.

Figure 3.2 shows observed global mean surface temperature (land + ocean) record for the last 140 years. The observed surface temperature record is long enough and has near-global coverage. In Figure 3.2, anomalies shown are relative to 1951–1980 mean. While year-to-year fluctuations associated with natural variability are apparent, in this global-averaged picture, the temperature depicts a steady increasing trend since about the beginning of the twentieth century. A closer inspection reveals that the temperature rose sharply during 1920–1945 and again from about 1970 to the present.

Figure 3.3 shows the spatial pattern of estimated linear trend (slope of the best fit line) in sea surface temperature (SST) over the globe from observations (Figure 3.3a)

# Detection and Attribution of Climate Change

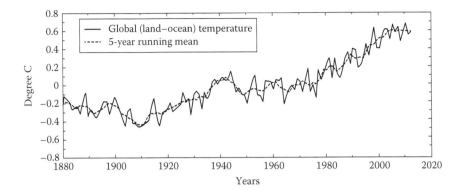

**FIGURE 3.2** Observed annual mean global (land + ocean) temperature anomalies with respect to the 1951–1980 climatology. Five-year running mean is shown in thin line.

and from climate model simulations (Figure 3.3b). Results are based on estimations during the period 1949–2000. Robust results from both observations and climate model simulations include the following: (i) in both the tropical Indian and western Pacific oceans (30°S–30°N; 40°E–170°E), the warming is pronounced; (ii) similar rise in SST tendency is also noticeable over Southern Atlantic and parts of tropical Pacific Oceans; and (iii) the overall warming pattern is accompanied with SST cooling in the North Pacific Ocean (30°N–50°N, 160°E–160°W) indicating that changes in local ocean–atmosphere interaction processes could potentially offset the direct impact of any external forcing. Another plausible interpretation is that decadal–multidecadal variability dominates the North Pacific Ocean and the trend signal may be influenced by such variability. Our results are consistent with those presented in Knutson et al. (2006), Compo and Sardeshmukh (2010), and Deser et al. (2010). One can argue that the model is realistically simulating the observed features and therefore its future projections are credible.

Are the identified warming trends in Figures 3.2 and 3.3 lie above natural variability? One quantitative measure of natural variability is the estimation of interannual standard deviation (year-to-year variability) of the variable considered. For the SST variable, this quantity (noise) is estimated first, and only if the trend (signal) exceeds it, they are plotted in Figure 3.3. One could note that despite uniform GHG forcing, rise in SST has geographical preferences. Despite observational uncertainties, surface temperature rise is not *entirely* due to natural variability but *likely* due to human intervention.

Sustained research by various authors has attempted to detect climate change signal in other climate variables such as rainfall, upper air temperature and moisture content in free atmosphere, sea level, and ocean heat content. Rainfall observations over land only show a trend toward less rainfall from 1951 to 2000 in the subtropics and tropics but more rainfall in the midlatitudes of the Northern Hemisphere, a signal that has been attributed to the increase in GHG concentrations (Zhang et al. 2007). Models also capture the large-scale tropical drying patterns (Neelin et al. 2006). This drying trend has also been noted in the seasonal mean (June–September) regional rainfall associated with the South Asian summer monsoon.

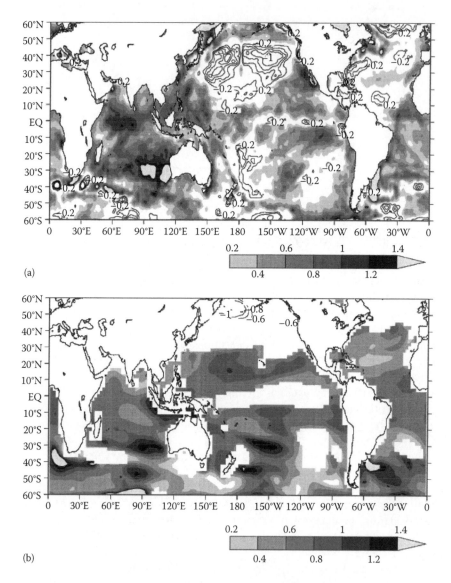

**FIGURE 3.3** (a) Linear trend in observed SST during boreal summer (June–September) from Hadley Center gridded data set for the period 1949–2000. Unit is (°C/52 years); (b) same as (a) but from a coupled model simulation. Positive values are progressively shaded while negative values are shown as contours.

Figure 3.4 shows estimated linear trend in boreal summer season rainfall over South Asia, from observations (Figure 3.4a) and climate model simulations (Figure 3.4b). In observations, the declining pattern over plains of central India and Indochina is consistent with other studies (Ramanathan et al. 2005; Meehl et al. 2008; Gautam et al. 2009; Bollasina et al. 2011). From the results presented here and elsewhere, one robust and significant result is that there is a declining tendency in monsoon rainfall over

Detection and Attribution of Climate Change 75

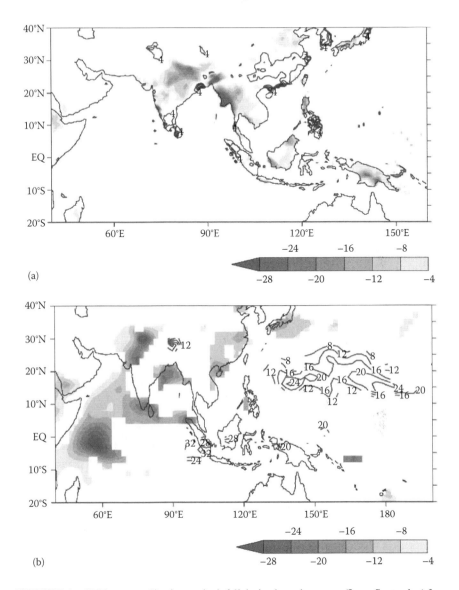

**FIGURE 3.4** (a) Linear trend in observed rainfall during boreal summer (June–September) from Climate Research Unit gridded data set for the period 1949–2000. Unit is (mm/month/52 years); (b) same as (a) but from a coupled model simulation. Negative values are shaded progressively while positive values are shown as contours.

central India. Due to observational uncertainties and different algorithms employed in re-gridding station rainfall observations, the magnitude of this declining varies among the products (Annamalai et al. 2013). This led Turner and Annamalai (2012) to suggest to reprocess all available observational regional rainfall products and to validate them against independent observations such as crop yields.

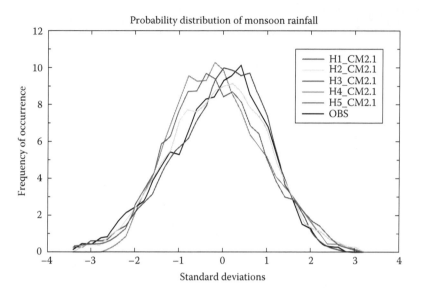

**FIGURE 3.5** Histogram of observed (black line) and coupled simulated anomalous monsoon rainfall events in standard deviations.

It needs to be borne in mind that century-long climate model simulations are not constrained by observations. They are free runs in which time evolving historical forcings of estimated (not observed) GHG and aerosol concentration along with land-use changes are prescribed. In these runs, the model is not expected to simulate specific observed events such as the 1997/1998 El Niño but is expected to realistically simulate the *distribution*. For example, Figure 3.5 shows the probability distribution of monsoon rainfall over India from observations (thick black line) and simulations from a coupled model (thin lines). The results presented in Figure 3.5 are based on the period 1861–2000. Compared to observations, the rainfall distributions of all the model simulations are realistic, aside from a slight tendency toward less normal rainfall. This result provides credence to examine long-term changes in monsoon rainfall in this model simulation.

Estimated linear trend (Figure 3.4b) from the model suggests drying tendency over South Asia, and the results are consistent with observations (Figure 3.4a). Concurrently, model simulations imply increased monsoon rainfall over the tropical western Pacific region. Annamalai et al. (2013) argue that this east–west shift (increased monsoon rainfall over tropical western Pacific and decreased rainfall over South Asia) is perhaps due to rise in tropical western Pacific SST (Figure 3.3).

In summary, various authors have reported detecting a climate change signal (monotonic increase or decrease) in a suite of observed ocean and atmosphere variables. Notable among them are a steady increasing trend in ocean heat content (e.g., Balmaseda et al. 2013), sea level rise (Timmermann et al. 2010), and tropospheric moisture content from satellite observations (Santer et al. 2007). While these results are encouraging, when the assessment is subject to different sources of the same variable, one encounter large uncertainties in the amplitude of the detected signal that obscure the findings. For example, SST trend estimated from five different sources do not agree in the amplitude of the trend in most places (Deser et al. 2010). This disagreement is higher for regional rainfall trends estimated from different products (Annamalai et al. 2013).

Estimation of trends from regularly spaced gridded data sets has certain caveats. It should be noted that equally spaced gridded data are generated by statistical interpolation methods from unequally spaced station data sets. Different authors use different interpolation techniques to generate gridded data. While proper weighting to number of observed stations falling within a grid is generally taken (e.g., Rajeevan et al. 2006), generation of very high-resolution (~25 km) gridded rainfall data, for example, is expected to introduce large errors. In other words, due to lack of sufficient direct observed station rainfall data at such high resolution, gridded data set is bound to be inflated by interpolation techniques. Caution must be exercised in using such data for detecting *regional* climate change-induced signal.

## 3.3 ATTRIBUTION OF CLIMATE CHANGE

Detection of a change in climate, however, does not necessarily imply it is due to anthropogenic factors or its causes are understood. The unequivocal *attribution* of climate change to anthropogenic causes would require controlled experimentation with the climate system in which the hypothesized agents of change are systematically varied in order to determine the climate's sensitivity to these agents (IPCC 2007). Simply, if we have detected a climate change signal, can we then attribute that change to specific anthropogenic forcing factors? Given the level of uncertainties in observations and models, elucidating cause and effect is very challenging but steady progress has been made. The climate community has been using controlled numerical model experiments to understand cause and effect at interannual and decadal time scales. For example, role of ENSO on global circulation anomalies at seasonal to interannual time scales was confirmed by performing idealized numerical experiments (e.g., Hoskins and Karoly 1981; Shukla and Wallace 1983). Such a framework is being followed to identify the effect of additional radiative forcing due to anthropogenic forcing. Model results are indeed encouraging.

Controlled experiments with the present-day climate models suggest that rise in surface temperature during the twentieth century is unlikely due to natural forcing alone (IPCC 2007). Models are capable of simulating the long-term changes in surface temperature only when the time evolving historical forcing associated with GHGs and aerosols are prescribed along with natural forcing (e.g., Barnett et al. 1999; Knutson et al. 2006).

As mentioned earlier, surface warming leads to rise in sea level either through thermal expansion of water or due to melting of glaciers/ice sheets. Researchers have examined climate model simulations to isolate the role of anthropogenic forcing on sea level rise. The consensus is that the models including anthropogenic and natural forcing simulate the observed thermal expansion since 1961 reasonably well (IPCC 2007). On the other hand, anthropogenic forcing dominates the surface temperature change simulated by models and has likely contributed to the observed warming of the upper ocean and widespread glacier retreat (IPCC 2007). While qualitative and not quantitative agreement prevails on the role of anthropogenic forcing on sea level rise, Woodworth et al. (2004) suggest that anthropogenic forcing has likely contributed at least one-quarter to one-half of the sea level rise during the second half of the twentieth century.

Annamalai et al. (2013) attempted to explain the dryness or declining tendency in monsoon rainfall over India. Both observational and climate model simulations (Figure 3.4) indicate that the center of monsoon rainfall has shifted eastward. Is the SST warming in the Indo-Pacific warm pool the cause for the monsoon declining over South Asia? By performing controlled climate model experiments, the authors demonstrated that SST rise over tropical western Pacific, perhaps induced by GHGs, contributed to the rainfall increase over tropical western Pacific and rainfall decrease over South Asia. The authors concluded that within the observational and model limitations, the regional monsoon rainfall may have started to respond to anthropogenic forcing.

## 3.4 SUMMARY

To detect climate change signal, the climate community is actively involved in generating or reprocessing sufficiently long time series of quality observational records. While doing so, measures of uncertainty or errors due to instrumental bias and location changes are being taken into account before a significant signal is detected. Such long observational records are available only for few variables precluding a clear assessment for consistency. In some cases, observational errors can be as large as the signal being sought. While climate model solutions offer an alternative platform, inadequate representations in model physics and uncertainty in the historical natural and anthropogenic forcing result in model systematic errors. Despite improvements to model formulations in the last two to three decades, the uncertainty range in both the current and future projected climate has not narrowed down. For example, Sperber et al. (2013) compared and contrasted the ability of the latest two generations of global climate models and noted that the systematic errors in simulating the boreal summer monsoon rainfall over Asia remain identical in the two generations. This warrants for large investments by federal and private institutions in climate science research to improve the climate models—the only reliable tools for systematic assessment of water and food security issues and proper planning of the future.

It is true that technology is the source of GHG emissions and it is also true that technological innovations are solutions to reduce GHG emissions. While in a developing economy reducing GHG emissions will not be an easy task, sustained technological innovations are needed to develop cleaner and climate-resilient tools. In a developing economy, income growth leads to more demands on energy and water. Without a revolution in energy technology, for example, societies will continue to pump a large amount of GHG into the atmosphere. To advance technology and to limit global warming in the future, timely action on research and development is required. Specifically, research on technological solutions needs to be taken up in developing countries.

Adaptation to climate variability and change, however, requires direct involvement of people from various walks of life. First and foremost, basic science of climate variability and change and their impacts need to be introduced to people. This can be accomplished, for instance, through focused working groups; seminars at high school, colleges, and government service sectors; and media at large. At all professional and nonprofessional levels, without a sustained educational and training program, it is impossible to appreciate the consequences of increased emissions of GHGs and to find new methods to reduce the emissions.

Finally, there exists the need for large investment by federal and corporate sectors in addressing and assessing the impact of global warming on all walks of societies.

## ACKNOWLEDGMENTS

This work was partly supported by a Department of Energy grant DE-FG02-07ER6445 and partly by the three institutional grants (JAMSTEC, NOAA, and NASA) of the IPRC.

## REFERENCES

Annamalai, H., Hafner, J., Sooraj, K.P., and Pillai, P. 2013. Global warming shifts the monsoon circulation, drying South Asia. *J. Climate* 26(9): 2701–2718.

Balmaseda, M.A., Trenberth, K.E., and Källén, E. 2013. Distinctive climate signals in reanalysis of global ocean heat content. *Geophys. Res. Lett.* 40: 1754–1759, doi: 10.1002/grl.50382.

Barnett, T.P. et al. 1999. Detection and attribution of recent climate change: A status report. *Bull. Am. Meteor. Soc.* 80: 2631–2659.

Bollasina, M.A., Ming, Y., and Ramaswamy, V. 2011. Anthropogenic aerosols and the weakening of the South Asian summer monsoon. *Science* 334: 502–505.

Charlson, R.J. et al. 1992. Climate forcing by anthropogenic aerosols. *Science* 255: 423–430.

Compo, G.P. and Sardeshmukh, P. 2010. Removing ENSO-related variations from the climate record. *J. Climate* 23: 1957–1978.

Deser, C. et al. 2010. Twentieth century tropical sea surface temperature trends revisited. *Geophys. Res. Lett.* 37: L10701, doi: 10.1029/2010GL043321.

Gautam, R., Hsu, N.C., Lau, K.M., and Kafatos, M. 2009. Aerosol and rainfall variability over the Indian monsoon region: Distributions, trends and coupling. *Ann. Geophys.* 27: 3691–3703.

Heger, G.C. et al. 2006. Climate change detection and attribution beyond mean temperature signals. *J. Climate* 19: 5058–5077.

Held, I.M. and Soden, B.J. 2006. Robust responses of the hydrological cycle to global warming. *J. Climate* 19: 5686–5699.

Hoskins, B.J. and Karoly, D.J. 1981. The steady linear response of a spherical atmosphere to thermal and orographic forcing. *J. Atmos. Sci.* 38: 1179–1196.

IPCC. 2007. *Climate Change 2007: The Physical Science Basis.* Contribution of WG I to AR4 of the IPCC. Cambridge, U.K.: Cambridge University Press.

Knutson, T. et al. 2006. Assessment of twentieth-century regional surface temperature trends using the GFDL CM2 coupled models. *J. Climate* 19: 1624–1651.

Meehl, G.A., Arblaster, J.M., and Collins, W. 2008. Effects of black carbon aerosols on the Indian monsoon. *J. Climate* 21: 2869–2882.

Neelin, J.D. et al. 2006. Tropical drying trends in global warming models and observations. *Proc. Natl. Acad. Sci. USA* 103: 6110–6115.

Rajeevan, M., Bhate, J., Kale, J.D., and Lal, B. 2006. High-resolution daily gridded rainfall data for the Indian region: Analysis of break and active monsoon spell. *Curr. Sci.* 91: 296–306.

Ramanathan, V. et al. 2005. Atmospheric brown clouds: Impacts on South Asian climate and hydrological cycle. *Proc. Natl. Acad. Sci. USA* 102: 5326–5333.

Santer, B.D. et al. 2007. Identification of human-induced changes in atmospheric moisture content. *Proc. Natl. Acad. Sci. USA* 104: 15248–15253.

Shukla, J. and Wallace, J.M. 1983. Numerical simulation of the atmospheric response to equatorial Pacific sea surface temperature anomalies. *J. Atmos. Sci.* 40: 1613–1640.

Sperber, K.R., Annamalai, H., Kang, I.S. et al. 2013. The Asian summer monsoon: An intercomparison of CMIP5 vs. CMIP3 simulations of the late 20th century. *Clim. Dyn.* 41: 2711–2744, doi: 10.1007/s00382-012-1607-6.

Stouffer, Ronald, J., Gabriele Hegerl, Simon Tett, 2000. A Comparison of Surface Air Temperature Variability in Three 1000-Yr Coupled Ocean–Atmosphere Model Integrations. *J. Climate* 13: 513–537.

Timmermann, A., McGregor, S., and Jin, F.F. 2010. Wind effects on past and future regional sea level trends in the Southern Indo-Pacific. *J. Climate* 23: 4429–4437, doi: 10.1175/2010JCLI3519.1.

Turner, A. and Annamalai, H. 2012. Climate change and the South Asian summer monsoon. *Nat. Climate Change* 2: 587–595.

Woodworth, P., Gregory, J., and Nicholls, R. 2004. *Long term sea level changes and their impacts, in The Sea*, vol. 12/13, edited by A. Robinson and K. Brink, chap. 19, Harvard Univ. Press, Cambridge, Mass.

Zhang, X. et al. 2007. Detection of human influence on twentieth-century precipitation trends. *Nature* 448: 461–465.

# 4 Uncertainty in Climate Change Studies

*Satish Bastola*

## CONTENTS

4.1 Introduction ..................................................................................................81
4.2 Sources of Uncertainties................................................................................83
4.3 Approaches to Uncertainty Analysis.............................................................85
     4.3.1 Quantifying Uncertainties in Climate Models.................................86
4.4 Uncertainty in Hydrological Impacts ............................................................89
     4.4.1 Downscaling Climate Model Output................................................89
     4.4.2 Bias Correction of Climate Model Output .......................................90
     4.4.3 Uncertainty in Hydrological Models ................................................91
4.5 Case Studies...................................................................................................92
     4.5.1 Quantifying Uncertainty in Future Simulations: Moy River Basin, Republic of Ireland ..............................................92
     4.5.2 Probabilistic Approach for Impact Assessment................................93
     4.5.3 Hydrological Impacts from CMIP3 and CMIP5 Experiments: Paint Rock River, Alabama, United States.......................................97
     4.5.4 Climate Change Sensitivity of the Moy River Basin, Republic of Ireland ........................................................................ 100
4.6 Conclusions................................................................................................. 103
Acknowledgments.................................................................................................. 104
References.............................................................................................................. 104

## 4.1 INTRODUCTION

The Intergovernmental Panel on Climate Change (IPCC 2007) concluded with high confidence that the atmosphere is warming because of anthropogenic activities. For the water sector, which is strongly impacted by climate variability, climate change and the associated uncertainty is a major concern. Decision makers require a quantitative estimate of future projection because it has significant consequences for the hydrological cycle and for water resource management (Buytaert et al. 2009; Christensen et al. 2004; Kay et al. 2009; Wood et al. 2004).

The hydrological impact of future changes is commonly evaluated by forcing locally calibrated hydrological models with output from global climate models (GCMs) (e.g., Bastola et al. 2011a,b; Chen et al. 2011; Kay et al. 2009). However, assessing climate change and its implications for river discharges has been a major challenge because of the range of uncertainties associated with such an approach. Uncertainties in future projection come predominately from three sources: scenario considered, climate models used, and natural variability within the climate system. If these associated uncertainties are robustly quantified, then they can help in formulating and designing robust adaptation strategies aimed at improving the resilience of water supply management and infrastructure to future risk.

Recent studies on impact and adaptation assessments of climate change have predominately taken two routes, the top-down and bottom-up approaches (Dessai and Hulme 2004; Wilby and Dessai 2010).

In a top-down approach, climate projections generated from climate models are propagated through a calibrated hydrological model. This approach, which gives a potential impact of climate change based on projection, has been a mainstay in impact studies literature. Though climate models are the dominant tools in simulating future scenarios, they have considerable uncertainties (Bastola et al. 2011a,b; Kay et al. 2009; Rowell 2006). High uncertainty implies that quantification is essential in impact studies and is usually accomplished using output from ensembles of GCMs. Despite the recent advances in climate models, they still have biases, which necessities an additional processing of the output, for example, downscaling and bias correction, before they are used in climate change impact studies.

Contrary to the top-down approach, the bottom-up approach uses historical data to assess the vulnerability of society. In the top-down approach, projections from the climate model and assumptions based on the evolution of future development are the major driving components. Therefore, the top-down approach is affected by different sources of uncertainty inherent to climate models and future development pathways. Consequently, the uncertainties impede the formulation of a robust and efficient adaptation and mitigation strategy. The large uncertainty has been the strongest argument against the top-down approach and has fueled exploration of the bottom-up approach in climate change impact studies (e.g., Brown et al. 2011; Lempert et al. 2004; Prudhomme et al. 2010; Wilby and Dessai 2010). Moreover, the assessments of potential impacts are difficult for decision makers to use (e.g., Stainforth et al. 2007; Wilby and Dessai 2010). A number of studies (e.g., Brown et al. 2011; Prudhomme et al. 2010; Wilby and Dessai 2010) implemented the bottom-up approach to facilitate decision making with climate projection. Brown et al. (2011) divided their approach into three stages: (a) identification of climate conditions and thresholds for performance that cause risk and, therefore, require adaptive actions (identified from discussion with stakeholders and historical data); (b) sensitivity analysis (determining the sensitivity of a system to future changes) and development of a climate response function that aids in visualizing how a system responds to future changes; and (c) informing decision makers on the residual risk and opportunities. Typically, for the hydrological system, the responses are modeled using hydrological models. The sensitivity domain is then designed to represent future climate space.

The top-down approach focuses on a natural system's vulnerability whereas a bottom-up approach emphasizes society's vulnerability to climate change. Dessai and

Hulme (2004) pointed out that the selection of a particular route, among many other factors, depends upon the type and scale of the domain, time scale, planning horizons, and economic status of the region or country. The bottom-up approach focuses more on a shorter planning horizon whereas the top-down approaches focus on units that have a longer planning horizon, such as bridges and dams. Developed countries focus on a top-down approach whereas developing countries focus more on societal vulnerability.

Wilby and Dessai (2010) proposed a conceptual framework for a scenario-neutral approach, which placed adaptation option appraisal at the core of the impact studies. The framework requires identification of the most significant risk and a range of possible low-regret adaptation measures, for example, safety margin, changing decision time horizon, and reversible and flexible options, and then appraises each option and ranks the preferred adaption options on the basis of their performance under present or future climate conditions.

Different studies have approached the problem of quantifying uncertainty and addressing the issues related to uncertainty in a number of different ways, ranging from very simple, for example, best guess framework, to a more complex approach based on a robust mathematical framework.

In this chapter, deterministic scenario-based and probabilistic approaches are discussed through case studies. These case studies are based on propagation of output from a range of climate models through a suite of calibrated hydrological models. The case study for the deterministic scenario-based approach is based on locally relevant climate change scenarios, derived from a few climate models. The probabilistic approach is demonstrated using a wide range of outputs from the climate models participating in the Coupled Model Intercomparison Project (CMIP3) and CMIP5 experiments and using stochastic weather generators (WGENS). In addition to a top-down approach, a sensitivity-based approach is also discussed through a case study.

## 4.2 SOURCES OF UNCERTAINTIES

Mathematical models emulate complex natural processes through numerous assumptions and approximations. Therefore, uncertainty is unavoidable and has many sources, for example, lack of knowledge, poor observation, and natural variability. As output from a mathematical model is often used in decision making in various sectors, the magnitude of uncertainty has direct implications for the decision-making processes. Though the impact of uncertainty in decision-making processes is well acknowledged in the literature, there is no consensus on the definition of uncertainty (Klir and Wierman 1999; Stewart 2000; Young 2001). Uncertainty can refer to imprecision, vagueness, or disagreement. Not all sources of uncertainty are reducible. The uncertainty that arises from lack of knowledge and from data collection procedures is widely referred to as epistemic uncertainty. This type of uncertainty is usually quantified by using plausible alternate models. Advances in understanding the system can reduce epistemic uncertainty. The uncertainties that arise from variability in the natural processes are called aleatory uncertainty. Outcomes vary each time an experiment is run under identical conditions and are usually quantified using the probability density function (PDF). Aleatory uncertainties are irreducible and an exploitation-based strategy is used to manage such uncertainty.

Climate model represents a complex natural process. However, it is affected by three different sources of uncertainties, namely, uncertainty in future emission scenarios, uncertainty in model response, and natural variability. Future emissions cannot be predicted in a deterministic way, and therefore, they are prescribed on the basis of projected global economy, population, and technological change. The interactions between the different aspects of development result in distinct uses of energy and, subsequently, in varying levels of future emission of greenhouse gases. The IPCC (2007) used four storylines to characterize diverse pathways of economic development, which are differentiated according to the use of technology and heterogeneity in the regional development pattern. High-end scenarios, for example, A2 and A1F1, assume rapid economic development and a heterogeneous world. Low-end scenarios, for example, B2 and B1, assume that development is environmental friendly and uniform across the world. The different Special Report on Emissions Scenarios (SRES), which are no-climate-policy scenarios, are described in Nakicenovic et al. (2000). As future climate could be modulated by today's climate policy, the need for a new type of scenario that also accounts for different mitigation measures resulted in the development of Representative Concentrated Pathways (RCP) (Moss et al. 2010). The climate change experiment with a suite of climate models run with SRES was called the CMIP3 experiment (Meehl et al. 2007), and the experiment run with four Representative Concentration Pathways (RCP) scenarios was called the CMIP5 experiment (Taylor et al., 2012). An overview of the CMIP3 and CMIP5 experiments can be found in Taylor et al. (2012). Future simulations, which depend upon unpredictable technological development and change in socioeconomic behavior, are uncertain.

The uncertainty in projection also comes from limited understanding of the different interacting components of a climate system. Because of simplification and parameterization of different subgrid scale processes, there is always uncertainty about how climate models respond to future changes. Most of the small-scale processes and their interactions are parameterized in climate models. Different parameterization schemes simulate the subgrid scale processes in different ways, resulting in parameter uncertainties. The perturbed physics ensembles (PPEs) experiment (Murphy et al. 2004; Stainforth et al. 2005) is designed specifically to explore the uncertainty arising from parameterization in climate models. The experiment also offers an opportunity to explore a model's sensitivity to parameters. Moreover, ensemble projections from the CMIP3 and CMIP5 experiments provide samples for quantification of modeling uncertainties. The other source of uncertainty that limits the predictability of a climate system is natural climate variability. The relative role of these three uncertainties may vary depending upon lead time and region (Hawkins and Sutton 2009). Hawkins and Sutton (2009) showed that the uncertainty in future projection for the next few decades comes from model uncertainty. Scenario uncertainty becomes the dominant source of uncertainty at the end of the twenty-first century, and the role of internal variability decreases rapidly with lead time. Advancement in climate science will likely reduce the uncertainty associated with lack of knowledge about the climate system. However, the uncertainty stemming from natural variability and the scenarios is likely to remain unaffected in many regions across the globe.

Apart from the uncertainty associated with future projection of climate, there is another source of uncertainty. Downscaling, bias corrections, and hydrological model's uncertainty, which cascade into hydrological impact, have received more attention recently (Bastola et al. 2011a; Chen et al. 2011). Output from GCMs reproduces the global and continental-scale climate fairly well; however, GCMs are inadequate in impact studies because of the differences in the scales at which GCMs provide output and the scale at which impacts are assessed (Maraun et al. 2010; Wilby and Wigley 1997). The gaps between the modeling scales are widely acknowledged and are addressed through statistical/dynamical downscaling and bias correction (Maraun et al. 2010; Maurer et al. 2007). Fowler et al. (2007) and Maraun et al. (2010) provide the most recent comprehensive review on downscaling methods. However, there is no consensus on the methods appropriate for particular downscaling/bias-correction application. This adds a new layer of uncertainty in climate change impact studies. For example, Chen et al. (2011) found significantly large uncertainty associated with the selection of downscaling methods.

Furthermore, conceptual hydrological models, which are widely used to simulate the hydrological impact of future changes, also have uncertainties associated with them. A conceptual model uses relatively simple mathematical equations to aggregate and simplify complex hydrological processes. The predictions from such models contain uncertainties that have largely been neglected until recently (e.g., Bastola et al. 2011a; Kay et al. 2009; Najafi et al. 2010). Jung et al. (2012) provide a brief summary of different climate change impact studies that have included the prediction uncertainties in hydrological models.

## 4.3 APPROACHES TO UNCERTAINTY ANALYSIS

Numerous methods exist that can be used to analyze uncertainty in environmental models (see Refsgaard et al. 2005). Some of the most widely used methods include (a) uncertainty evaluation (e.g., the analytical method, the first-order second moment method, numerical methods using Monte Carlo simulation), (b) sensitivity analysis, and (c) scenario analysis. Analytical or simple propagation rules can be used to quantify uncertainty in output if variables can be derived from the linear combination of other variables. However, if the variables are derived from a functional relationship, then simple propagation rules may not apply and may require methods of moments where uncertainty is the function of input and its sensitivity to output. If the functional relation between input and output is complex and nonlinear, the method based on moment may not suffice and therefore, a numerical method is usually adopted. Numerical methods are based on Monte Carlo simulation, in which the model is run a large number of times. Among the numerous methods, sensitivity analysis is widely used to study the behavior of a model with a range of inputs or parameter, that is, the rate at which output varies with input. Depending on the purpose and scope, sensitivity analysis can be further categorized into three methods: screening analysis, local sensitivity analysis, and global sensitivity analysis. Screening analysis is used to find the most influential factor in the model; local sensitivity analysis is used to analyze the impact of variation in one input factor on output while the other factors remain constant; global sensitivity analysis (Saltelli 2000) is used to study the simultaneous impact of variation in all input factors on system behavior.

Since climate models are very complex and computationally demanding, the implementation of Monte Carlo for the quantification of prediction uncertainties is not practical and requires concerted effort. One of such novel effort that aims at quantifying parametric uncertainty in climate model is PPE experiment. Tebaldi and Knutti (2007) argued that the quantification of all aspects of model uncertainty requires multimodel ensembles. A wide range of GCMs developed by various climate centers is available for simulating Earth's climates (Meehl et al. 2007), and the range of GCM simulations is the basis of most of the methods used to quantify uncertainty in climate projections.

Foley (2010) reviewed uncertainty in climate modeling at the regional level and explored the uncertainty associated with emission scenarios and climate modeling and discussed the implications of uncertainties for regional analysis. Fealy (2013) also discussed the challenges for quantifying various uncertainties that cascade into impact studies at the regional scale.

### 4.3.1 QUANTIFYING UNCERTAINTIES IN CLIMATE MODELS

Two approaches based on multimodel ensembles, namely, the reliability ensemble approach (REA) and Bayesian model averaging (BMA), are increasingly used to quantify uncertainty in future climate projection (details on REA and BMA that can be found elsewhere [e.g., Giorgi and Mearns 2002; Min et al. 2007]). Only brief introductions to these methods are presented here.

*Reliability ensemble averaging* (*REA*): The REA method (Giorgi and Mearns 2002) provides a framework for combining predictions from different climate models. REA uses two criteria: (a) reliability, which reflects the ability of a model to reproduce historical data, and (b) convergence, which reflects the closeness of each model to the ensemble mean. Weighting different models according to reliability is based on the premise that the model that can reproduce present-day climate better is likely to reproduce future climate with a greater level of confidence. Similarly, the notion of convergence as a criterion is to penalize models that produce outputs that are far away from the ensemble mean. In REA, the ensemble mean value of the selected variable is the weighted average value:

$$\bar{T} = \frac{\sum_{i=1}^{n} T_i w_i}{\sum_{i=1}^{n} w_i} \quad \text{where } w_i = R_{p,i}^n R_{c,i}^m; \quad R_{p,i} = \left[\frac{\varepsilon_t}{|Bias_{T,i}|}\right]^m; \quad R_{c,i} = \left[\frac{\varepsilon_t}{|Dist_{T,i}|}\right]^m \quad (4.1)$$

$$\delta_T = \left[ \frac{\sum_{i=1}^{n}(T_i - \bar{T})^2 w_i}{\sum_{i=1}^{n} w_i} \right]^{1/2} \quad (4.2)$$

where

$R_{p,i}$ and $R_{c,i}$ are the functions related to model bias (based on historical data) and convergence

$T_i$ is the variable of interest for the *i*th model

$T$ is the ensemble mean value for the selected variable

*Dist* is the distance between the ensemble mean and the output from the *i*th model

*Bias* is measured as the difference between model and observed value

*m* and *n* are the parameters that allow applying different weight to performance and convergence criteria (typically both are assumed equal to 1)

According to Giorgi and Mearns (2002), $\varepsilon_i$ is the term that reflects the natural variability and is estimated as "the difference between the maximum and minimum values of these 30 yr moving averages" (p. 1144). Giorgi and Mearns (2002) used detrended (remove century-scale trends) Climatic Research Unit (CRU) observations for the twentieth century. The weighted parameter for each model is used to define the model's likelihood. The root mean square error in Equation 4.2 is used to estimate the uncertainty about the mean value (Equation 4.1). Tebaldi et al. (2005) extended the REA method and proposed a Bayesian statistical model that combines information from a multimodal ensemble of AOGCM and observation to determine probability distributions of future temperature change on a regional scale. Several studies used the output archived in CMIP3 to account for uncertainty in GCMs (e.g., Solomon et al. 2007), and several others used the output from PPEs to evaluate the uncertainties arising from GCM model formulation (e.g., Murphy et al. 2007). Similarly, Xu et al. (2010) noted that because model reliability was used as a product of convergence and performance, the REA tends to produce a narrow PDF, as it tends to neglect outliers. The authors proposed upgrades for the REA method in which they define the reliability of a model as a product of five factors: two factors defining models' ability to reproduce mean temperature and precipitation value, two factors defining models' ability in reproducing observed interannual variability for temperature and precipitation, and a fifth factor reflecting the spatial correlations between observed and simulated sea level pressure patterns over the selected domain. Their approach accounts for multiple variables, reproduction of basic circulation, and multiple statistics, an improvement over the original REA method.

*Bayesian model averaging*: Like REA, BMA allows combining predictions from multiple plausible models (Raftery et al., 2005). In BMA, the predictive distribution of the predictand (Equation 4.3) is calculated as the average of the posterior predictive distribution of the quantity derived from each individual model weighted by the corresponding posterior model probability:

$$p(\Delta \mid f_1,\ldots,f_K, D) = \sum_{k=1}^{K} p(\Delta \mid f_k, D) p(f_k \mid D) \qquad (4.3)$$

The posterior model probability, $p(f_k|D)$, of model $f_k$ given the data, is given by the following equation:

$$p(f_k \mid D) \propto P(D \mid f_k) P(f_k) \qquad (4.4)$$

where the constant of proportionality is chosen so that the posterior model probabilities add up to one. The prior probability, $P(f_k)$, in Equation 4.4 is the likelihood of $f_k$ before reevaluation. Therefore, a model that reproduces the past behavior will provide a greater confidence in its ability to simulate future behavior. Note that without any prior knowledge of model preference, the prior probability is assumed to have a uniform distribution among the $N$ models. The quantity $P(D|f_k)$ is the integrated likelihood of model $f_k$.

The posterior mean and variance of $\Delta$ are as follows:

$$E[\Delta \mid f_1,\ldots,f_k, D)] = \sum_{k=1}^{K} w_k \hat{\Delta}_k \qquad (4.5)$$

$$Var[\Delta \mid f_1,\ldots,f_k, D)] = \sum_{k=1}^{K} (Var(\Delta \mid D, f_k) + \hat{\Delta}_k) w_k - E(\Delta \mid D)^2 \qquad (4.6)$$

where $\hat{\Delta}_k = E(\Delta|D, f_k)$. Note that weight $w_k$ has a value between 0 and 1. At each time step, the chosen PDF is centered on the individual forecasts with an associated variance. The BMA parameters, that is, BMA weights and variances, can be obtained from historical data by using the expectation–maximization algorithm, the Bayes factor (e.g., Duan et al. 2007; Min et al. 2007), or Markov chain Monte Carlo (MCMC) sampling (e.g., Vrugt et al. 2008).

Min et al. (2007) used BMA to produce a probabilistic climate change projection for temperature using simulation with 21 climate models (SRES A1B scenario) that participated in the CMIP3 experiment. The authors calculated BMA weight parameter using the Bayes factor and the expectation–maximization method using twentieth-century data (50 years). The authors argue that probabilistic future projection of climate using BMA is feasible, provided the model's weighted factors are robust, and may be superior to raw ensembles in terms of information content.

## 4.4. UNCERTAINTY IN HYDROLOGICAL IMPACTS

Apart from uncertainties in climate models, additional source of uncertainty is introduced when hydrological impact of future changes is assessed. The additional layer of uncertainty arises from (a) the differences in the scale at which climate models are run and the scale at which impact models are run, (b) the existence of biases in the output of climate models, and (c) structural and parametric uncertainty in impact models. Over the past decade, needs to bridge the gap that exist between climate and hydrological models have resulted in the exploration of wide range of downscaling and bias correction methods.

### 4.4.1 Downscaling Climate Model Output

GCMs can reproduce global and continental-scale climate reasonably well, but they still lack the resolution that is required for hydrological simulation. Scores of methods have surfaced in the attempt to address the gap between climate and impact models. These methods range from empirical statistical downscaling to complex dynamic downscaling with RCMs. RCMs solve physically based equations at the regional scale. In the last decade, there has been a tremendous advancement in RCMs and their ability to reproduce the present-day climate. However, the systematic errors in GCMs that provide boundary conditions to RCM, the high computational cost, and the need for further downscaling for impact studies (Wilby and Wigley 1997) are among the many factors that hinder the use of RCMs as a downscaling method. Alternatively, downscaling methods based on the statistical relationship between large-scale climate variables and regional variables can also be used. Such empirical methods are based on the premise that statistical relationships between the predictors predictand remain valid for a future period. However, statistical downscaling can provide information only at places where observations are available. The outcome from a number of studies that compared both downscaling methods showed that their performances are similar. Wilby et al. (2000) observed similar performance with both the statistical and the dynamical methods. Similarly, Wood et al. (2004), who compared six different downscaling methods in terms of their ability to reproduce precipitation and temperature, primary input to a hydrological model, did not observe any additional skill for dynamical downscaling methods. Wilby and Wigley (1997) provided the relative advantages and disadvantages of the statistical and dynamical downscaling techniques. According to the authors, the dynamic downscaling method, which is based on a physically consistent process, can produce fine resolution data but its skill depends upon the bias in GCM boundary conditions and the strength of regional forcing. Also, its use is limited by the computational cost.

Alternatively, stochastic WGENs can be used to produce future climate scenarios. They can produce synthetic daily time series of meteorological variables with statistical characteristics similar to those of historical data series (e.g., Racsko et al. 1991; Richardson 1981). A stochastic WGEN allows the spatial and temporal extrapolation of observed weather data needed for risk assessment of climate change. Furthermore, their parameters can easily be scaled to match future projection. They have found widespread application as a cheap alternative to dynamic downscaling in constructing regionalized scenarios based on GCM-simulated or subjectively introduced changes in climate (e.g., Semenov et al. 1998; Wilks 1992).

Downscaling methods also add a layer of uncertainty in climate change impact studies. Khan et al. (2006) used three statistical downscaling methods, Statistical DownScaling Models (SDSM), LARG-WG, and Artificial Neural Network, and compared their performance in terms of reproducing observations, ranking SDSM as the best for the authors' domain. Wilby and Harris (2006) agreed that uncertainty from downscaling methods is significant. Similarly, Chen et al. (2011) investigated six downscaling methods, including dynamical and statistical downscaling, and quantified uncertainty in climate change impact studies arising from the choice of downscaling method. The authors observed apparent differences in projections derived from different downscaling methods and cautioned users to interpret the impact simulated with a single downscaling method.

### 4.4.2 Bias Correction of Climate Model Output

Both regional and global models are known to have biases (Christensen et al. 2008; Stefanova et al. 2012) that limit their direct use in impact studies. These biases may amplify when they are propagated through the impact models, resulting in large simulation biases. The systematic biases in climate models can be removed using bias-correction methods, which have recently received much attention in climate change impact literature (Hagemann et al. 2004; Stoll et al. 2011). Bias-correction methods include correction to the mean standardization (Wilby et al. 2004) and correction of distribution, for example, quantile-based mapping (Li et al. 2010; Wood et al. 2004). The quantile mapping (QM) method has recently found widespread application in impact studies. In QM, the systematic error is corrected for each variable and month separately. First, the cumulative distribution function (CDF) of observed and regional reanalysis datasets is derived. Then the CDF is applied to variables to correct its mean and distribution using the following equation:

$$\hat{x}_{i,t}^m = F_{\text{obs}}^{-1}(F_{\text{mod}}(x_{i,t}^m)) \quad (4.7)$$

where
  $\hat{x}_{i,t}^m$ and $(x_{i,t}^m)$ are the $t$th corrected and the uncorrected estimate of a variable $i$
  $t$ is the time step
  $m$ is the month
  $F_{\text{obs}}(\cdot)$ and $F_{\text{mod}}(\cdot)$ are the empirical CDFs of the observed (O) and the modeled (M) datasets

The QM method does not properly account for extreme values that lie outside the observations. Furthermore, it presumes that the relationships between model and observed variable remain valid in the future. The QM-based approach has been successfully implemented in hydrological applications (Bastola and Misra 2013; Wood et al. 2004). This method was found, however, to produce bias, as it does not preserve the relationship between precipitation and temperature (Zhang and Georgakakos 2011).

Li et al. (2010) modified a QM-based bias-correction method (Equation 4.7), termed the equidistant CDF matching method, for correcting biases in monthly output from general circulation models. Instead of using CDFs for the model and observations for the historical period, the authors used the CDFs for future projection. For a given percentile, the authors assumed *the difference between the model and observed value during the training period also applies to the future period, which means the adjustment function remains the same. However, the difference between the CDFs for the future and historic periods is also taken into account* (p. 6). Teutschbein and Seibert (2012) used differential split sample testing to test bias-correction methods under nonstationary conditions. The authors used five commonly used bias-correction methods ranging from a very simple change factor to a more complex QM-based method and observed that the distributional mapping method performed well in the validation period as compared to the simple delta-change approach and linear scaling.

### 4.4.3 Uncertainty in Hydrological Models

A hydrological model uses a set of interrelated mathematical equations to aggregate spatially distributed and complex hydrological processes. This aggregation of processes, simplifications, lack of knowledge, and randomness in nature often results in prediction uncertainties. Melching (2001) reviewed various methods that have been used to account for uncertainty in hydrological modeling. Refsgaard et al. (2005) discussed different strategies and methods to account for uncertainty. The generalized likelihood uncertainty estimation (GLUE) method (Beven and Binley 1992) and BMA are the two strategies that have been widely applied to quantify hydrological model uncertainty (Beven and Binley 1992; Freer et al. 1996; Vrugt et al. 2008). The GLUE is the most widely used and debated method in hydrological modeling literature. The method is based on the premise that for a hydrological model, a large number of model parameter values may represent the process equally well. The GLUE is based on Monte Carlo simulation in which a hydrological model is run with a large number of model parameters sampled from its prior distribution. Contrary to the optimization-based approach, the GLUE embraces the fact that a range of values of model parameters can result in acceptable simulation and rejects the notion of a single best value of model parameters. The application of GLUE starts with the definition of the prior distribution or range of model parameters. Then, the model is run with a large number of model parameters sampled from its prior probability distribution. The next step involves differentiating model parameters, from a large sample, that represent the system in an acceptable way. For this, the likelihood measures that reflect the model's relative ability to reproduce observation and a threshold value that helps in defining the border between acceptable and unacceptable solutions must be defined. Subsequently, all behavioral simulations are ranked and the likelihood weighted to produce a prediction range (see Beven and Freer 2001). The GLUE framework not only allows accounting for plausible parameters but also allows combining competing models. Bastola et al. (2011a,b) used GLUE to account for uncertainty in both model parameterization and model structure by using a suite of hydrological models and their behavioral simulators.

The BMA is another statistical postprocessing method that can be used to quantify prediction uncertainties (Section 2.1). It allows construction of probabilistic prediction from a set of diverse calibrated hydrological models (Ajami et al. 2006; Bastola et al. 2011a; Duan et al. 2007).

In the case study presented in Section 4.4, the future projection of hydrological data, for example, streamflow, is based on four lumped conceptual models: Hydrological Model (HyMOD) (see Wagener et al. 2001), Nedbør-Afstrømnings-Model (NAM) (see Madsen 2000), tank model (Sugawara 1995), and Topography-based model (TOPMODEL) (Beven et al. 1995). The HyMOD uses probability distribution, and the TOPMODEL uses distribution of topographic index to characterize spatial variability in soil moisture. The tank model and NAM models both assume constant soil moisture storage. All four models use a single linear reservoir to model groundwater. These models have been used in numerous applications and their potential for application to simulate flow under changed climate has been discussed previously (e.g., Andersen et al. 2006; Najafi et al. 2010; Tanakamaru and Kadoya 1993).

## 4.5 CASE STUDIES

### 4.5.1 Quantifying Uncertainty in Future Simulations: Moy River Basin, Republic of Ireland (Bastola et al. 2011a)

*Context*: The hydrological simulation result for the Moy River basin at Rahans, Republic of Ireland, is presented to demonstrate the role of hydrological model uncertainty (parameter and structural uncertainty) in climate change impact studies. The Moy River basin has a total of 1803 km$^2$. The basin receives nearly 1425 mm of rainfall annually; it discharges 57.9 cumecs of runoff. Loam is the dominant soil texture and peat bogs are the dominant land use type.

*Methods*: Scenario-led approach, the most widely used one to assess the hydrological impact of climate change, was used in this study. For hydrological simulation, six downscaled climate scenarios corresponding to two medium SRES, A2 and B2, and three GCMs (Fealy and Sweeney 2007, 2008) were used. The selection of only three GCMs does not reflect the fuller uncertainty envelope. However, the GCMs selected for this study represent a diverse estimate of climate sensitivity. Figure 4.1 shows the schematic of the four top-down-based experiments implemented to quantify four sources of uncertainty:

1. *Hydro*: This experiment aims to quantify the uncertainty in future simulations due to hydrological model structure and their parameters for each GCM and scenario separately.
2. *Scenario+Hydro*: This experiment aims to quantify the uncertainty in future simulations due to the selection of the emission scenario and hydrological model for each GCM separately.
3. *GCM+Hydro*: This experiment aims to quantify the uncertainty in future simulations due to the selection of climate models and hydrological models for each scenario separately.
4. *Total*: This experiment aims to define the total uncertainty envelope in future simulations of stream flow.

# Uncertainty in Climate Change Studies

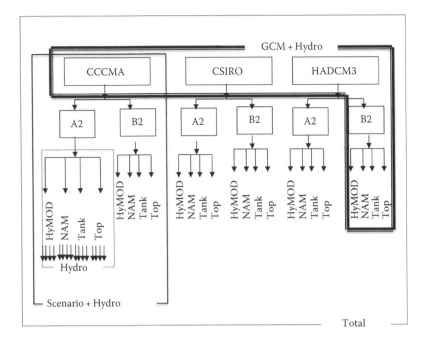

**FIGURE 4.1** Schematic of the experiment implemented to quantify different sources of uncertainty. (From Bastola, S. et al., *Adv. Water Resour.*, 34(5), 562, 2011a.)

These four experiments are conducted using both the GLUE and the BMA methods.

*Results*: The uncertainty in projected streamflow, expressed in terms of average width of the prediction interval (% of long-term average flow) arising from uncertainty in hydrological model (parameterization and model selection), GCM, and scenario is shown in Table 4.1. The prediction interval arising from model parameters alone is almost two-thirds of the average long-term flow. With scenarios, the combined hydrological and scenario uncertainty is nearly 90% of the average flow. With GCM, the combined hydrological and GCM uncertainty nearly equals the average flow. It equals to nearly 114% when hydrological, scenario, and GCM uncertainty is accounted for. The seasonal flow, along with associated uncertainty estimates from both BMA and GLUE for the Moy River basin, is shown in Figure 4.2. The uncertainty envelope estimated with BMA, however, is marginally wider than that estimated with GLUE framework. In this case, full range of emission scenarios and GCM sensitivities are not sampled, and therefore, results are indicative only.

## 4.5.2 Probabilistic Approach for Impact Assessment in Bastola et al. (2012)

*Context*: The probabilistic approach for the assessment of the hydrological impact is on the basis of the result from the Moy River basin as introduced in the previous section. For this study, monthly output from CMIP3 experiment is utilized to derive locally relevant climate scenarios.

## TABLE 4.1
### Average Width of the Prediction Interval (%) for 2070s for Each of the Four Experimental Designs

| S.No. | GCM | SRES Scenario | Hydro 1990s | Hydro | Hydro and Scenario | Hydro+GCM (A2) | Hydro+GCM (B2) | GLUE (Total) | BMA (Total) |
|---|---|---|---|---|---|---|---|---|---|
| 1 | HADCM3 | A2 |  | 68.2 | 90.5 |  |  |  |  |
| 2 | HADCM3 | B2 |  | 66.1 |  |  |  |  |  |
| 3 | CCCMA | A2 | 81.4 | 63.4 |  | 102.6 |  | 114.7 | 122.7 |
| 4 | CCCMA | B2 |  | 66.3 | 77.7 |  | 103.9 |  |  |
| 5 | CSIRO | A2 |  | 58.7 |  |  |  |  |  |
| 6 | CSIRO | B2 |  | 56.9 | 91.1 |  |  |  |  |

*Note:* Uncertainty in the hydrological models (Hydro), uncertainty in the selection of scenario and Hydro (Hydro and SRES), uncertainty in the selection of GCM forced with A2 (Hydro and GCM [A2]), Hydro and GCM [B2]), uncertainty in the selection of GCM and selection of scenario, and uncertainty in hydrological models (total). The prediction uncertainty estimated from BMA is also shown.

# Uncertainty in Climate Change Studies

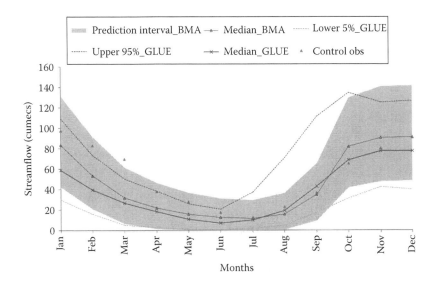

**FIGURE 4.2** The 90% prediction interval (shaded region) simulated by the behavioral parameter sets of four hydrological models forced with projected climate scenarios from three climate models and two scenarios for the Moy River basin, Republic of Ireland (based on simulation from 2070 to 2079).

*Methods*: The schematic of the method adopted is shown in Figure 4.3. A probabilistic approach was considered and outputs from a suite of climate models that participated in CMIP3 experiments were used. GCM outputs were downscaled using a computationally cheap scenario generator based on change factors (CF) derived from a suite of GCMs and a WGEN. The application of CF first requires the estimation of baseline climatology. Then the CF for temperature and precipitation is calculated. CF in temperature is defined as the difference in temperature between control and future period; CF in precipitation is defined as the percentage of change in precipitation between the control and the future periods.

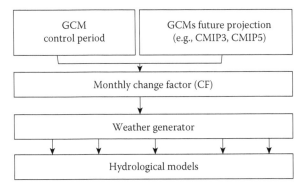

**FIGURE 4.3** Schematic of the probabilistic method (see section 4.5.2) used to account for modeling uncertainties in climate change impact studies by propagating climate scenarios through hydrological models.

From the CFs derived from 17 GCMs, a minimum, maximum, first quartile, third quartile, and median for both temperature and precipitation were defined. Then using the distribution of CFs, CFs were sampled such that their number in each specified interval was proportional to the area enclosed by its probability distribution function. This was done to improve the representation of sampling density. Samples from the distribution of CFs were used to scale the parameter of the WGENs to produce regionalized future scenarios. The WGEN of Richardson and Wright (1984), which uses a first-order Markov to model dry and wet days, was used. The distribution of rainfall amounts in WGEN is modeled using a two-parameter gamma distribution; the parameters for the selected location are then determined from the historical data. The potential evapotranspiration (PET) was modeled using the Hargreaves method (Hargreaves et al. 1985)—a radiation-based empirical method—that uses solar radiation and minimum and maximum temperatures.

Regionalized projections are then propagated through a suite of calibrated hydrological models. As the model is calibrated using 20 years of data, it was assumed that the calibrated parameters remain valid for future simulation. The simulation was based on a GLUE framework that allows combining output from different plausible hydrological models.

*Results*: The average width of the uncertainty envelope (A2 SRES), derived on the basis of upper 95% and lower 5% flow, is nearly 115% of the long-term average flow (Figure 4.4). Figure 4.5 shows the PDF for seasonal mean flow estimated for the 2070s. The density in PDF is estimated using a number of simulated flows lying within the chosen interval. Figures 4.4 and 4.5 show that the amount of uncertainty cascaded into the hydrological impact is large and subsequently reflects the challenges in using them to formulate and design adaptation measures.

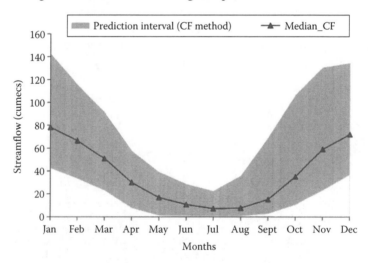

**FIGURE 4.4** Ninety percent prediction interval simulated by behavioral parameter sets of four hydrological models forced with projected climate scenarios from a suite of climate models and two scenarios for the 2080s (2070–2099) for the Moy River basin, Republic of Ireland. Locally relevant scenarios are generated using a change factor method.

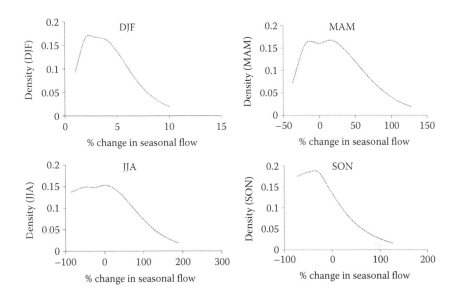

**FIGURE 4.5** Probability distribution of mean seasonal streamflow (Moy River) for the 2080s (2071–2099) estimated using the behavioral parameters of four hydrological models forced with daily probabilistic scenarios derived using 17 GCMs and a WGEN.

### 4.5.3 Hydrological Impacts from CMIP3 and CMIP5 Experiments: Paint Rock River, Alabama, United States (Bastola 2013)

*Context*: Hydrological impacts simulated in the preceding two case studies are based on the CMIP3 experiment, which does not account for human response in mitigating the future consequences of climate change through technological innovation, change in behavior, and policies. CMIP3 is run with only SRES, which are no-climate-policy scenarios. The RCP scenarios were developed to cater to the need for scenarios that account for a human response to future risk. Moreover, the climate models used in the CMIP3 experiment have evolved over time, especially in terms of process representation and resolution. Therefore, in this case study, the focus is shifted toward comparing the hydrological impact derived from the CMIP5 experiment with that derived from the CMIP3 experiment. The hydrological simulation of the Paint Rock River (Woodville, Alabama: USGS ID 3574500) run with output from a suite of climate models from the CMIP3 and the CMIP5 experiments is presented in this case study. The Paint Rock River basin receives nearly 1467 mm of rainfall annually and drains water from nearly 828.9 km$^2$.

*Methods*: The hydrological model tank, HyMOD, and NAM were calibrated using hydrometeorological data from the model parameter estimation experiment (MOPEX) database (Schaake et al. 2006) and then run with plausible future scenarios. Hundred plausible scenarios were generated using the CF method derived from the CMIP3 and CMIP5 experiments and the WGEN (as in case study 4.5.2). Model output from the CMIP3, forced with three SRES (A1B, A2, and B1), and the CMIP5, forced with four RCP scenarios (RCP2.6, RCP4.5, RCP6.0, and RCP8.5), were used to construct the probabilistic CFs (see Section 4.2). Subsequently, the CFs was then applied to the parameters of the WGEN to produce multiple realizations of plausible future scenarios. The hydrological simulation was then based on these projected scenarios for the study area.

*Results*: The CFs for precipitation and temperature (CFP and CFT) relative to the historical period (1961–1990) are derived for the 2.5° grid cells centered at 31.25°N and 276.50°E. The CFP and CFT for 2061–2080 derived from the CMIP3 and CMIP5 models show a rise in temperature (Figure 4.6). However, disagreement in sign and strength is more apparent for precipitation. Temperature and rainfall inter-model variation for both the CMIP3 and the CMIP5 experiments is high; the range of temperature and precipitation estimated from the CMIP3 models is relatively larger than that estimated from the CMIP5 models. The ensembles' mean hydrological responses (seasonal) run with both SRES and RCP scenarios are similar (Figure 4.7). Nevertheless, the uncertainty about the mean is high (see Table 4.2). In Table 4.2, the uncertainty, which is defined as the width of the envelope of simulated flow for all the three quartiles, is expressed in terms of fraction of long-term average flow. The closeness of the interquartile range among different scenarios reflects that the uncertainties in GCMs are relatively higher than the uncertainties in the emission scenarios.

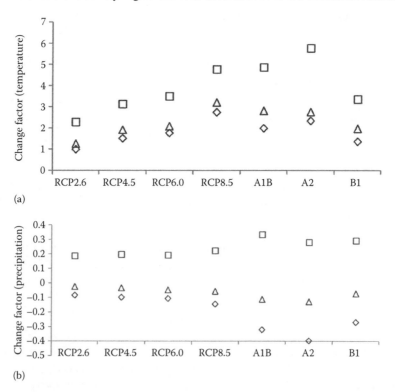

**FIGURE 4.6** The 5th, 95th, and median percentiles of change factor for temperature and precipitation derived from two different experiments (CMIP3 and CMIP5) and seven different scenarios: (a) temperature and (b) precipitation.

# Uncertainty in Climate Change Studies

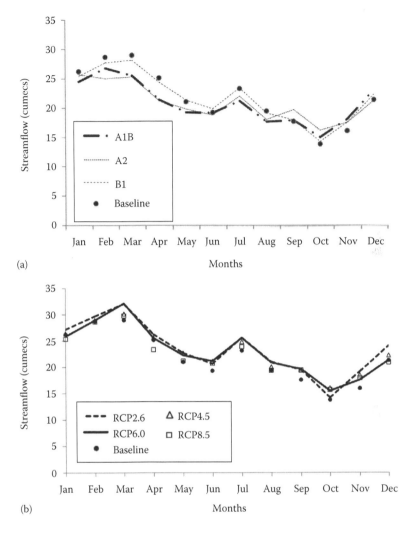

**FIGURE 4.7** Projected ensemble mean simulated flows for the Paint Rock River basin (Alabama) with the CMIP3 and CMIP5 experiments: (a) SRES and (b) RCP scenarios.

Figure 4.8 shows the typical example of PDF of seasonal (June–August [JJA]) average streamflow for 2061–2080. From the high-end to the low-end scenario, a marginal shift in the PDF toward the right is observed for both experiments. The highly dispersed probability distribution of streamflow indicates that uncertainty arising from GCMs and hydrological models is high. Moreover, a marginal shift in the location of the PDF simulated with different scenarios reflects that uncertainty associated with multiple GCMs is high compared to the uncertainty associated with future emission scenarios. Even with a new set of climate change experiments, the uncertainties in hydrological simulation have not decreased as expected, supporting the argument that there is a to shift focuses of impact modeling toward decision appraisal rather than the top-down approach.

## TABLE 4.2
### Uncertainties in Future Streamflow Expressed as Fraction of Baseline

| | | | Future Streamflow (Fraction of Baseline) | | |
|---|---|---|---|---|---|
| S. No. | Experiment | Scenario | First Quartile | Median | Third Quartile |
| 1 | CMIP5 | RCP2.6 | 0.48 | 1.1 | 1.89 |
| 2 | CMIP5 | RCP4.5 | 0.48 | 1.07 | 1.84 |
| 3 | CMIP5 | RCP6.0 | 0.47 | 1.1 | 1.92 |
| 4 | CMIP5 | RCP8.5 | 0.45 | 1.04 | 1.8 |
| 5 | CMIP3 | A1B | 0.35 | 0.94 | 1.65 |
| 6 | CMIP3 | A2 | 0.38 | 0.95 | 1.69 |
| 7 | CMIP3 | B1 | 0.43 | 1.01 | 1.75 |

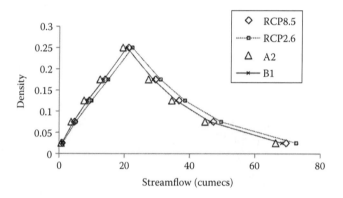

**FIGURE 4.8** Probability distribution of projected mean seasonal streamflow (JJA) simulated with low- and high-end scenarios of CMIP3 and CMIP5 models for Paint Rock River basin (Alabama).

### 4.5.4 Climate Change Sensitivity of the Moy River Basin, Republic of Ireland (Bastola et al. 2011b)

*Context and methods*: It is apparent from the preceding case studies that the top-down approach results in huge uncertainty, which is difficult to use in designing a robust adaptation strategy. Therefore, in this case study, the same problem is revisited using a sensitivity-based approach. In contrast to the previously discussed case studies, which focus on the hydrological impact on seasonal flow, this case study focuses on the fluvial flood risk. The impact of climate change on flood frequency is defined as the percentage of change in the flood peak value corresponding to a given return period. This case study is based on simulation of the Moy River basin using the four calibrated hydrological models (NAM, HyMOD, tank, and TOPMODEL).

The sensitivity of hydrological response to the range of future changes is derived according to CFs, estimated from the 17 GCMs (run with the A2 scenario) that participated in the CMIP3 experiment. As in the earlier mentioned case studies, the generation of regional climate scenarios for hydrological simulation is based on the CF method and a WGEN.

The focus in this case study is on sensitivity analysis, that is, quantifying the sensitivity of flood quantiles to monthly CFs (precipitation and temperature). As CFs are derived separately for each month, the dimension of the sensitivity analysis was reduced by applying harmonic analysis to model the monthly CFs and to synthesize and smooth the larger interannual variations, reducing the required number of parameters to three (see the following equation):

$$\mu_t = \bar{\mu} + A\cos(2\pi t/P - \Phi) \tag{4.8}$$

where

$\mu_t$ is the value of the series at time $t$
$\bar{\mu}$ is the arithmetic mean
$A$ and $\Phi$ are the amplitude and phases (in radian)
$P$ is the period of observation

The phase angle $\Phi$ indicates the time of the year the maximum of a given harmonic occurs and was converted to months. For the Moy River basin, $\Phi$ was fixed to the month of July. The sensitivity domain of the mean and amplitude parameter was then derived according to the range of CFs. A combination of mean and amplitude was used to estimate the monthly CFs, which were subsequently used to generate daily scenarios for hydrological simulations. Using the simulated flow, the time series of the annual maximum flow series was constructed, which was then fit to generalized extreme value distribution using probability-weighted moments, a method equivalent to L-moments. Using the different combinations of mean and amplitude parameter, which lie within the sensitivity domain derived from 17 GCMs, the response surface defining the sensitivity of mean and amplitude parameter on flood frequency of different return period is constructed. The schematic of the method is shown in Figure 4.9.

The sensitivity domain of the mean and amplitude parameter is derived from 17 GCMs run with the A2 SRES. To include future unexpected scenarios, which cannot be properly accounted for in the present scenarios, the sensitivity domain derived from the GCMs is increased at both ends, that is, the sensitivity domain is 1.5 times greater than the range derived for the modeled CFs (Table 4.3).

A full factorial experiment, whose design consists of two parameters, the mean and amplitude parameter that characterizes the monthly CFs, each with 10 discrete equally spaced values, was conducted. The simulated time series of flow were used to estimate 20,000 sets of annual maximum series (AMS). These AMSs are then subsequently fitted to the GEV distribution using the method of probability-weighted moments.

*Results*: Results of the sensitivity analysis of precipitation scenarios are summarized using the 3D contour plot (Figure 4.10), which shows the percentage changes in the 95th percentile flow of 100- and 5-year return periods (changed with respect to the 95th percentile flow of the same return period estimated for the present climatic condition). Such a response surface can be used to assess the robustness and to

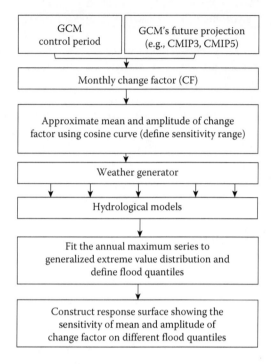

**FIGURE 4.9** Schematic of the method used to account for modeling uncertainties in climate change impact studies by propagating climate scenarios through hydrological models.

quantify the residual risk associated with a policy decision. For example, the Office of Public Works (OPW), the national body responsible for flood risk management in the Republic of Ireland, has advised an allowance of +20% of peak flows under a midrange future scenario and +30% as a high-end future scenario (OPW 2001). It is essential to test the robustness of such safety margins using sensitivity analysis.

**TABLE 4.3**
**Range of Change Factors Derived from the Differences in Change Factors Estimated from Different GCMs and the Modeled Range Used for the Sensitivity Testing**

| | | Parameter of Cosine Curve Characterizing the Monthly Change Factor for Precipitation | | | | Annual Average Changes in Temperature | | | |
|---|---|---|---|---|---|---|---|---|---|
| | | Mean | | Amplitude | | Mean | | Coeff. of Variation | |
| S.No. | Scheme | Min | Max | Min | Max | Percentile (5) | Percentile (95) | Percentile (5) | Percentile (95) |
| 1 | Future period | −0.08 | 0.08 | 0.10 | 0.35 | 0.95 | 2.74 | −0.27 | 0.09 |
| 2 | Sensitivity | −0.13 | 0.13 | 0.00 | 0.50 | 0.95 | 2.74 | −0.27 | 0.09 |

# Uncertainty in Climate Change Studies

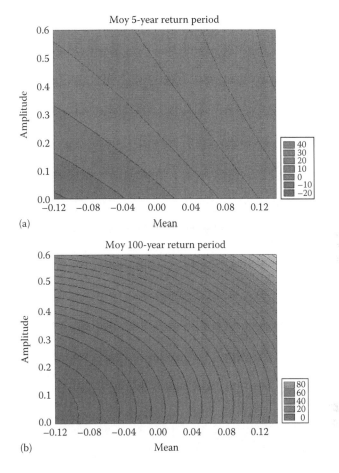

**FIGURE 4.10** Three-dimensional contour plot showing percentage changes in the 5- and 100-year return period peak flows for the Moy River basin under the range of scenarios constructed from the change factors synthesized by amplitude (0–0.6) and mean (−0.125–0.125) of the cosine curve: (a) 5 year return period and (b) 100 year return period.

Figure 4.10a and b shows that the residual risk to a policy decision that allows a 20% increase in safety margins to accommodate future fluvial risk is significant. Furthermore, residual risks are apparently higher for low-frequency events than for high-frequency events (floods with a high recurrence period, e.g., a 5-year return period event) indicating that the risk of exceedance of design allowances is greater with considerable implications for critical infrastructure.

## 4.6 CONCLUSIONS

Conceptual hydrological models forced with projections from GCMs are widely used to simulate the hydrological impact of climate change. Despite being straightforward, such a top-down approach is associated with a range of uncertainties: uncertainty in scenarios, climate models, downscaling, and bias-correction methods, as well as uncertainty in impact models. If these associated uncertainties are robustly quantified, then they can be meaningfully used in managing future risk. Deterministic scenario-based approaches, probabilistic-based approaches, and sensitivity-based approaches are widely used in impact studies. All of them are illustrated in this chapter with a case study involving the propagation of output from a range of climate models through a suite of hydrological models.

Quantification and investigation of uncertainty are essential and require a large sample of climate models. This may hinder the implementation of a more rigorous downscaling method that focuses predominately on reproduction of spatial and temporal variability of rainfall. This hindrance is in part due to the high computational cost involved in such an attempt. Therefore, a method based on a change factor can be alternatively used to generate regionalized climate scenarios as it allows characterization of uncertainty in climate models.

The projection of hydrological variables with the CMIP3 and CMIP5 experiments shows that the projected streamflow contains significant uncertainty. From the case studies, it can be concluded that the range of uncertainty cascaded into the impact studies is the biggest challenge for using the top-down approach in designing and planning robust adaptation measures for future risk. The major sources of uncertainty arise from the model, the scenario, and the natural variability in a projection whose relative role varies from region to region and depends upon the lead time. Furthermore, it is also evident from the result that the role of uncertainty in hydrological models can be high and should be properly accounted for.

Considerable uncertainties are associated with the top-down approach; therefore, the sensitivity-based approach was revisited to test widely adopted design allowances to increase the resilience of critical infrastructure to future risk. Because of the high uncertainty and challenges in using the future impact in designing adaptation strategy, the use of climate model output may be more meaningfully used for the appraisal of policy decision rather than assessing the impact based on a top-down approach.

## ACKNOWLEDGMENTS

I acknowledge Kathy Fearon of the Center for Ocean-Atmospheric Prediction Studies (COAPS) for the help with editing the chapter. I thank Prof. J. Sweeney, H. Ishidaira, C. Murphy, and V. Misra for their continual support and constructive suggestions.

## REFERENCES

Ajami, N.K., Duan, Q., Gao, X., and Sorooshian, S. 2006. Multimodel combination techniques for analysis of hydrological simulations: Application to Distributed Model Intercomparison Project results. *Journal of Hydrometeorology* 7: 755–768.

Andersen, H.E., Kronvang, B., Larsen, S.E., Hoffmann, C.C., Jensen, T.S., and Rasmussen, E.K. 2006. Climate-change impacts on hydrology and nutrients in a Danish lowland river basin. *Science of Total Environment* 365: 223–237.

Bastola, S. 2013. Hydrologic impact of climate change in watersheds of southeast US using multimodel ensembles from CMIP3 and CMIP5 models. *Regional Environmental Change* 13: 131–139.

Bastola, S. and Misra, V. 2013. Evaluation of dynamically downscaled reanalysis precipitation data for hydrological application. *Hydrological Processes* 28(4): 1989–2002.

Bastola, S., Murphy, C., and Fealy, R. 2012. Generating probabilistic estimates of hydrological response using a weather generator and probabilistic climate change scenarios. *Hydrological Processes* 26(15): 2307–2321.

Bastola, S., Murphy, C., and Sweeney, J. 2011a. Role of hydrological model uncertainty in climate change impact studies. *Advance in Water Resources* 34(5): 562–576.

Bastola, S., Murphy, C., and Sweeney, J. 2011b. The sensitivity of fluvial flood risk in Irish catchments to the range of IPCC AR4 climate change scenarios. *Science of Total Environment* 409: 5403–5415.

Beven, K. and Binley, A.M. 1992. The future of distributed models: Model calibration and uncertainty prediction. *Hydrological Processes* 6: 279–298.

Beven, K. and Freer, J. 2001. Equifinality, data assimilation, and uncertainty estimation in mechanistic modelling of complex environmental systems. *Journal of Hydrology* 249: 11–29.

Beven, K., Lamb, R., Quinn, P., Romanowicz, R., and Freer, J. 1995. TOPMODEL. In: Singh, V.P. (ed.), *Computer Models of Watershed Hydrology*, pp. 627–668. Highlands Ranch, CO: Water Resource Publication.

Brown, C., Werick, W., Leger, W., and Fay, D. 2011. A decision-analytic approach to managing climate risks: Application to the upper great lakes. *Journal of the American Water Resources Association* 47: 524–534.

Buytaert, W., Celleri, R., and Timbe, L. 2009. Predicting climate change impacts on water resources in the tropical Andes: The effects of GCM uncertainty. *Geophysical Research Letter* 36: L07406. doi:10.1029/2008GL037048.

Chen, J., Brissette, F.P., and Leconte, R. 2011. Uncertainty of downscaling method in quantifying the impact of climate change on hydrology. *Journal of Hydrology* 401: 190–202.

Christensen, J.H., Boberg, F., Christensen, O.B., and LucasPicher, P. 2008. On the need for bias correction of regional climate change projections of temperature and precipitation. *Geophysical Research Letter* 35: L20709. doi:10.1029/2008GL035694.

Christensen, N.S., Wood, A.W., Voisin, N., Lettenmaier, D.P., and Palmer, R.N. 2004. Effects of climate change on the hydrology and water resources of the Colorado River basin. *Climatic Change* 62(1–3): 337–363.

Dessai, S. and Hulme, M. 2004. Does climate adaptation policy need probabilities? *Climate Policy* 4: 107–128.

Duan, Q., Ajami, N.K., Gao, X., and Sorooshian, S. 2007. Multi-model ensemble hydrologic prediction using Bayesian model averaging. *Advance in Water Resources* 30(5): 1371–1386.

Fealy, R. 2013. Deriving probabilistic based climate scenarios using pattern scaling and statistically downscaled data: A case study application from Ireland. *Progress in Physical Geography* 30(10): 1–28.

Fealy, R. and Sweeney, J. 2007. Statistical downscaling of precipitation for a selection of sites in Ireland employing a generalized linear modeling approach. *International Journal of Climatology* 27: 2089–2094.

Fealy, R. and Sweeney, J. 2008. Statistical downscaling of temperature, radiation and potential evapotranspiration to produce a multiple GCM ensemble mean for a selection of sites in Ireland. *Irish Geography* 41(1): 1–27.

Foley, A.M. 2010. Uncertainty in regional climate modelling: A review. *Progress in Physical Geography* 34(5): 647–670.

Fowler, H.J., Blenkinsopa, S., and Tebaldi, C. 2007. Review: Linking climate change modelling to impacts studies: Recent advances in downscaling techniques for hydrological modelling. *International Journal of Climatology* 27: 1547–1578.

Freer, J., Beven, K., and Ambroise, B. 1996. Bayesian uncertainty in runoff prediction and the value of data: An application of the GLUE approach. *Water Resources Research* 32(7): 2161–2173.

Giorgi, F. and Mearns, L.O. 2002. Calculation of average, uncertainty range, and reliability of regional climate changes from AOGCM simulations via the "Reliability Ensemble Averaging" (REA) method. *Journal of Climate* 15: 1141–1158.

Hagemann, S., Machenhauer, B., Jones, R., Christensen, O.B., Déqué, M., Jacob, D., and Vidale, P.L. 2004. Evaluation of water and energy budgets in regional climate models applied over Europe. *Climate Dynamics* 23: 547–567.

Hargreaves, G.L., Hargreaves, G.H., and Riley, J.P. 1985. Irrigation water requirement for Senegal River Basin. *Journal of Irrigation and Drainage Engineering, ASCE* 111(3): 265–275.

Hawkins, E.D. and Sutton, R. 2009. The potential to narrow uncertainty in regional climate predictions. *Bulletin of the American Meteorological Society* 90: 1095–1107.

IPCC. 2007. Climate change 2007. In: Solomon, S., Qin, D., Manning, M., Chen, Z., Marquis, M., Averyt, K.B., Tignor, M., and Miller, H.L. (eds.), *The Physical Science Basis. Contribution of Working Group I to the Fourth Assessment Report of the Intergovernmental Panel on Climate Change*. Cambridge, U.K.: Cambridge University Press, 996pp.

Jung, I.W., Moradkhani, H., and Chang, H. 2012. Uncertainty assessment of climate change impact for hydrologically distinct river basins. *Journal of Hydrology* 466–467(12): 73–87.

Kay, A.L., Davies, H.N., Bell, V.A., and Jones, R.G. 2009. Comparison of uncertainty sources for climate change impacts: flood frequency in England. Climatic Change, 92(1–2): 41–63.

Khan, M.S., Coulibaly, P., and Dibike, Y. 2006. Uncertainty analysis of statistical downscaling methods. *Journal of Hydrology* 319: 357–382.

Klir, G. and Wierman, M. 1999. *Uncertainty-Based Information: Elements of Generalized Information Theory*, 2nd edn. Heidelberg, Germany: Physica-Verlag.

Lempert, R., Nakicenovic, N., Sarewitz, D., and Schlesinger, M. 2004. Characterizing climate-change uncertainties for decision-makers. *Climate Change* 65: 1–9.

Li, H., Sheffield, J., and Wood, E.F. 2010. Bias correction of monthly precipitation and temperature fields from Intergovernmental Panel on Climate Change AR4 models using equidistant quantile matching. *Journal of Geophysical Research* 115: D10101. doi:10.1029/2009JD012882.

Madsen, H. 2000. Automatic calibration of a conceptual rainfall-runoff model using multiple objectives. *Journal of Hydrology* 235: 276–288.

Maraun, D. et al. 2010. Precipitation downscaling under climate change: Recent developments to bridge the gap between dynamical models and the end user. *Reviews of Geophysics* 48: RG3003. doi:10.1029/2009RG000314.

Maurer, E.P., Brekke, L., Pruitt, T., and Duffy, P.B. 2007. Fine-resolution climate projections enhance regional climate change impact studies. *Eos, Transactions American Geophysical Union* 88(47): 504. doi:10.1029/2007EO470006.

Meehl, G., Covey, C., Delworth, T. et al. 2007. The WCRP CMIP3 multimodel dataset. *Bulletin of the American Meteorological Society* 88: 1383–1394.

Melching, C. 2001. Reliability estimation V.P. Singh (Ed.), Computer models of watershed hydrology, Water Resource Publications, Colorado, USA (1995), pp. 69–118.

Min, S.K., Simonis, D., and Hense, A. 2007. Probabilistic climate change predictions applying Bayesian model averaging. *Philosophical Transactions of the Royal Society A* 365: 2103–2116.

Moss, R.H. et al. 2010. The next generation of scenarios for climate change research and assessment. *Nature* 463: 747–756.

Murphy, J.M., Booth, B.B.B., Collins, M., Harris, G.R., Sexton, D.M.H., and Webb, M.J. 2007. A methodology for probabilistic predictions of regional climate change from perturbed physics ensembles. *Philosophical Transactions of the Royal Society A* 365: 1993–2028.

Murphy, J.M. et al. 2004. Quantifying uncertainties in climate change from a large ensemble of general circulation model predictions. *Nature* 430: 768–772.

Najafi, M.R., Moradkhani, H., and Jung, W.I. 2010. Combined effect of global climate projection and hydrologic model uncertainties on the future changes of streamflow. *ASCE Conference Proceedings* 371: 10. doi:10.1061/41114(371)10.

Nakicenovic, N. et al. 2000. *Special Report on Emissions Scenarios: A Special Report of Working Group III of the Intergovernmental Panel on Climate Change.* Cambridge, U.K.: Cambridge University Press, 599pp. Available online at: http://www.grida.no/climate/ipcc/emission/index.htm.

New, M.G. and Hulme, M. 2000. Representing uncertainties in climate change scenarios: A Monte Carlo approach. *Integrated Assessment* 1: 203–213.

OPW. 2001. Assessment of potential future scenarios for flood risk management. August 14, 2001. http://www.opw.ie/en/FloodRiskManagement/TechnicalInformation/OperationalGuidance/ClimateChange/2009, March 10, 2011.

Prudhomme, C., Wilby, R.L., Crooks, S., Kay, A.L., and Reynard, N.S. 2010. Scenario neutral approach to climate change impact studies: Application to flood risk. *Journal of Hydrology* 390: 198–209.

Racsko, P., Szeidl, L., and Semenov, M. 1991. A serial approach to local stochastic weather models. *Ecological Modelling* 57: 27–41.

Raftery, A.E., Gneiting, T., Balabdaoui, F., and Polakowski, M. 2005. Using Bayesian model averaging to calibrate forecast ensembles. *Monthly Weather Review* 133: 1155–1174.

Refsgaard, J.C., Henriksen, H.J., Harrar, W.G., Scholten, H., and Kassahun, A. 2005. Quality assurance in model based water management—Review of existing practices and outline of new approaches. *Environmental Modelling & Software* 20(10): 1201–1215.

Richardson, C.W. and Wright, D.A. 1984. *WGEN: A Model for Generating Daily Weather Variables.* Springfield, VA: United States Department of Agriculture, Agricultural Research Service, ARS-8, p. 83.

Rowell, D.P. 2006. A demonstration of the uncertainty in projections of UK climate change resulting from regional model formulation. *Climatic Change* 79: 243–257.

Saltelli, A. 2000. What is sensitivity analysis? In: Saltelli, A., Chan, K., and Scott, E.M. (eds.), *Sensitivity Analysis, Wiley Series in Probability and Statistics*, pp. 3–13. Chichester, England: Wiley.

Schaake, J., Cong, S., and Duan, Q. 2006. US MOPEX datasets. IAHS Publication Series. https://e-reports-ext.llnl.gov/pdf/333681.pdf. December 1, 2011.

Semenov, M.A., Brooks, R.J., Barrow, E.M., and Richardson, C.W. 1998. Comparison of the WGEN and LARS-WG stochastic weather generators in diverse climates. *Climate Research* 10: 95–107.

Solomon, S., Qin, D., Manning, M. et al. 2007. *Climate Change 2007: The Physical Science Basis.* Cambridge, U.K.: Cambridge University Press.

Stainforth, D.A., Aina, T., Christensen, C. et al. 2005. Uncertainty in predictions of the climate response to rising levels of greenhouse gases. *Nature* 433: 403–406.

Stainforth, D.A., Downing, T.E., Washington, R., Lopez, A. and New, M. 2007. Issues in the interpretation of climate model ensembles to inform decisions. Phil. Trans. R. Soc. A 365: 2163–2177. (doi:10.1098/rsta.2007.2073).

Stefanova, L., Misra, V., Chan, S., Griffin, M., O'Brien, J.J., and Smith, III, T.J. 2012. A proxy for high-resolution regional reanalysis for the Southeast United States: Assessment of precipitation variability in dynamically downscaled reanalyses. *Climate Dynamics* 38: 2449–2446.

Stewart, T. 2000. Uncertainty, judgment, and error in prediction. In: Sarewitz, D., Pielke, R., and Byerly, R. (eds.), *Prediction: Science, Decision Making, and the Future of Nature*, pp. 41–57. Washington, DC: Island Press.

Stoll, S., Hendricks Franssen, H.J., Butts, M., and Kinzelbach, W. 2011. Analysis of the impact of climate change on groundwater related hydrological fluxes: A multi-model approach including different downscaling methods. *Hydrology and Earth System Sciences* 15: 21–38. doi:10.5194/hess-15-21-2011.

Sugawara, M. 1995. Tankmodel. In: Singh, V.P. (ed.), *Computer Models of Watershed Hydrology*, pp. 165–214. Littleton, CO: Water Resources Publication.

Tanakamaru, H. and Kadoya, M. 1993. Effects of climate change on the regional hydrological cycle of Japan. In: *Proceedings of the Yokohama Symposium*, Yokohama, Japan, July 13–16, 1993, IAHS Publ. No. 212.

Taylor, K.E., Stouffer, R.J., and Meehl, G.A. 2012. An overview of CMIP5 and the experiment design. *Bulletin of the American Meteorological Society* 93: 485–498.

Tebaldi, C. and Knutti, R. 2007. The use of the multi-model ensemble in probabilistic climate projections. *Philosophical Transactions of the Royal Society A* 365(1857): 2053–2075.

Tebaldi, C., Smith, R., Nychka, D., and Mearns, L. 2005. Quantifying uncertainty in projections of regional climate change: A Bayesian approach to the analysis of multi-model ensembles. *Journal of Climate* 18: 1524–1540.

Teutschbein, C. and Seibert, J. 2012. Is bias correction of Regional Climate Model (RCM) simulations possible for non-stationary conditions? *Hydrology and Earth System Sciences Discussions* 9: 12765–12795.

Vrugt, J.A., Diks, C.G.H., and Clark, M.P. 2008. Ensemble Bayesian model averaging using Markov chain Monte Carlo sampling. *Environmental Fluid Dynamics* 8: 579–595.

Wagener, T., Boyle, D.P., Lees, M.J., Wheater, H.S., Gupta, H.V., and Sorooshian, S. 2001. A framework for development and application of hydrological models. *Hydrology and Earth System Sciences* 5(1): 13–26.

Wilby, R.L., Charles, S.P., Zorita, E., Timbal, B., Whetton, P., and Mearns, L.O. 2004. The guidelines for use of climate scenarios developed from statistical downscaling methods. Supporting Material of the Intergovernmental Panel on Climate Change (IPCC), Prepared on Behalf of Task Group on Data and Scenario Support for Impacts and Climate Analysis (TGICA).

Wilby, R.L. and Dessai, S. 2010. Robust adaptation to climate change. *Weather* 65: 180–185. doi:10.1002/wea.543.

Wilby, R.L. and Harris, I. 2006. A framework for assessing uncertainties in climate change impacts: Low flow scenarios for the River Thames, UK. *Water Resources Research* 42: W02419. doi:10.1029/2005WR004065.

Wilby, R.L., Hay, L.E., Gutowski, Jr., W.J. et al. 2000. Hydrological responses to dynamically and statistically downscaled climate model output. *Geophysical Research Letter* 27(8): 1199–1202.

Wilby, R.L. and Wigley, T.M.L. 1997. Downscaling general circulation model output: A review of methods and limitations. *Progress in Physical Geography* 21: 530–548.

Wilks, D.S. 1992. Adapting stochastic weather generation algorithms for climate changes studies. *Climatic Change* 22: 67–84.

Wood, A.W., Leung, L.R., Sridhar, V., and Lettenmaier, D.P. 2004. Hydrologic implications of dynamical and statistical approaches to downscaling climate model outputs. *Climatic Change* 62: 189–216.

Xu, Y., Gao, X.J., Giorgi, F. et al. 2010. Upgrades to the REA method for producing probabilistic climate change predictions. *Climate Research* 41: 61–81.

Young, R. 2001. *Uncertainty and the Environment*. Cheltenham, U.K.: Edward Elgar, 249pp.

Zhang, F. and Georgakakos, A. 2011. Joint variable spatial downscaling. *Climatic Change* 111(3–4): 945–972.

# 5 Climate Change Impacts on Water Resources and Selected Water Use Sectors

*Mukand S. Babel, Anshul Agarwal, and Victor R. Shinde*

## CONTENTS

| | | |
|---|---|---|
| 5.1 | Introduction | 110 |
| 5.2 | Methods to Assess Impacts of Climate Change | 111 |
| | 5.2.1 Water Resources | 113 |
| | 5.2.2 Water Use Sectors | 114 |
| |     5.2.2.1 Irrigation | 114 |
| |     5.2.2.2 Hydropower | 114 |
| 5.3 | Climate Change Impacts on Water Resources | 115 |
| | 5.3.1 Water Availability | 115 |
| |     5.3.1.1 Surface Water | 115 |
| |     5.3.1.2 Groundwater | 117 |
| | 5.3.2 Sea-Level Rise | 119 |
| | 5.3.3 Floods and Droughts | 121 |
| |     5.3.3.1 Floods | 121 |
| |     5.3.3.2 Drought | 127 |
| | 5.3.4 Water Quality | 127 |
| |     5.3.4.1 Surface Water Quality | 127 |
| |     5.3.4.2 Groundwater Quality | 129 |
| 5.4 | Climate Change Impacts on Water Use Sectors | 129 |
| | 5.4.1 Agriculture | 129 |
| | 5.4.2 Industry (Hydropower) | 132 |
| | 5.4.3 Water Supply and Sanitation | 134 |
| 5.5 | Examples to Evaluate Climate Change Impacts | 135 |
| | 5.5.1 Water Resources: Case of the Bagmati River Basin, Nepal | 135 |
| |     5.5.1.1 Introduction | 135 |
| |     5.5.1.2 Methodology | 138 |
| |     5.5.1.3 Results and Discussion | 138 |
| |     5.5.1.4 Conclusions | 141 |

> 5.5.2 Water Use Sector: Case of Rice Cultivation in Northeast Thailand....142
>         5.5.2.1 Introduction..................................................................... 142
>         5.5.2.2 Methodology .................................................................. 144
>         5.5.2.3 Results and Discussion.................................................... 144
>         5.5.2.4 Conclusions .................................................................... 147

5.6 Summary ................................................................................................. 147
References........................................................................................................... 148

## 5.1 INTRODUCTION

Climate change is now largely accepted as a real, pressing, and truly global problem, and scientific evidence for global warming is now considered irrevocable (Allison et al. 2009). Understanding the potential impacts of current and future climate conditions on hydrological processes is gaining more impetus in the present day because of the social and political implications of water. The Fourth Assessment Report of the Intergovernmental Panel on Climate Change (IPCC 2007) has addressed many previous concerns pertaining to the credibility of climate change in scientific and policy discussions. There is an increasing consensus among the scientific community that climate change will have a significant effect on water resources (Bates et al. 2008; Cromwell et al. 2009; Xu et al. 2007; etc.). Listed hereafter is a summary of the various potential impacts of climate change on water supply systems (Arnell and Delaney 2006; Bates et al. 2008):

- It is expected to cause an increase in volume of precipitation and average runoff in high latitudes and part of the tropics and a decreased volume in some subtropical and lower latitude regions.
- It is likely to alter the reliability of raw water sources and supply infrastructure (e.g., dams, reservoirs) by changing the magnitude and frequency of flows.
- It may alter the demand of water and the ability to meet these demands, particularly at times of peak demand.
- By 2050s, the area of land subjected to water stress would be two times the area that is not stressed. For global assessment, water stress occurs when the per capita availability of water is less than 1000 $m^3$/year or when the ratio of water withdrawals to long-term annual runoff exceeds 0.4.
- It is likely to cause an increase in the number of extreme events like floods and droughts.
- It may alter the raw water quality because of rise in water temperatures and thereby the ability to treat raw water to potable standards.
- Current management practices worldwide may not be able to cope up with the ill effects of climate change.

Climate change is of particular relevance to policymaking because rise in the average global temperature is expected to change the hydrological cycle, which may have multiple impacts on natural resources. One of the most crucial effects of climate change would be the changes in local and regional water availability. Even in the present times, there are countries that face high hydrological variability, and climate change will only aggravate the problem. For countries that currently have reliable water supplies, climate change may reintroduce water security challenges. Countries in the developing world are more prone to the adverse effects of climate change: particularly droughts and/or floods. Currently, 1.6 billion people live in countries and regions with absolute water scarcity, and the number is expected to rise to 2.8 billion people by 2025 (World Bank 2013). Climate change could profoundly alter future patterns of both water availability and use, thereby increasing water stress globally.

It is important to note that impact of climate change on water resources depends not only on direct climatic drivers (e.g., changes in the volume, timing, and quality of streamflow) but also on nonclimate drivers such as urbanization and pollution, which can influence systems directly and indirectly through their effects on climate variables. Table 5.1 presents a list of some of these nonclimatic drivers and their potential effects on climate. This chapter focuses only on the climate drivers and their impacts on water resources and water use sectors.

## 5.2 METHODS TO ASSESS IMPACTS OF CLIMATE CHANGE

As climate change becomes more evident, there is a requirement to analyze its impacts on water resources and water use sectors. There are various methods of assessing climate change impacts, and the use of a method depends on factors such as the level of detail required, the geographical coverage, and availability of observed data. For example, the level of detail required for a global assessment differs from that needed for basin-level assessments. Basin-level assessment involves downscaling of climatic information from Global Climate Models (GCMs) and detailed hydrological modeling (Hamududu and Killingtveit 2012).

IPCC has developed certain potential scenarios as alternative images of how the future might unfold. The evaluation of future GHG emissions is the product of very complex dynamic systems determined by driving forces such as demographic growth, socioeconomic development, and technological changes (Anandhi et al. 2008). The IPCC scenarios are an appropriate tool to analyze how driving forces may influence future emission outcomes and to assess the associated impacts. They assist in climate change analysis, including climate modeling and the assessment of impacts, adaptation, and mitigation. The consequences of these developments on climate are estimated by the general circulation models (GCMs) (IPCC 2007). GCMs are an important tool to produce the virtual estimate of climate change in the future. However, GCM information remains relatively coarse in resolution and is unable to resolve subgrid scale features such as topography, clouds, and land use. Further, GCMs cannot resolve the circulation patterns leading to extreme events. This represents a considerable problem for direct use in hydrological modeling as well for the impact assessment of climate change on hydrological dynamics in river systems (Christensen and Christensen 2007; Tisseuil et al. 2010). Thus, GCMs can best be used to suggest the likely direction and rate of change of future climate.

Climate change impact studies at regional level require point climate observations and are highly sensitive to fine-scale climate variations that are parameterized in coarse-scale models. This is especially true for regions of complex topography and

## TABLE 5.1
### Nonclimatic Drivers and Their Effects on Climate

| Nonclimate Driver | Examples | Direct Effects on Systems | Indirect Effects on Climate |
|---|---|---|---|
| Geological processes | Volcanic activity, earthquakes, tsunamis | Lava flow, mudflows (lahars), ashfall, shock waves, coastal erosion, enhanced surface and basal melting of glaciers, rockfall and ice avalanches | Cooling from stratospheric aerosols, change in albedo |
| Land-use change | Conversion of forest to agriculture | Declines in wildlife habitat, biodiversity loss, increased soil erosion, nitrification | Change in albedo, lower evapotranspiration, altered water and heat balances |
| | Urbanization and transportation | Ecosystem fragmentation, deterioration of air quality, increased runoff and water pollution | Change in albedo, urban heat island, local precipitation reduction, downwind precipitation increase, lower evaporation |
| | Afforestation | Restoration or establishment of tree cover | Change in albedo, altered water and energy balances, potential carbon sequestration |
| Land-cover modification | Ecosystem degradation (desertification) | Reduction in ecosystem services, reduction in biomass, biodiversity loss | Changes in microclimate |
| Invasive species | Tamarisk (United States), Alaska lupin (Iceland) | Reduction of biodiversity, salinization | Change in water balance |
| Pollution | Tropospheric ozone, toxic waste, oil spills, exhaust, pesticides, increased soot emissions | Reduction in breeding success and biodiversity, species mortality, health impairment, enhanced melting of snow and ice | Direct and indirect aerosol effects on temperature, albedo, and precipitation |

*Source:* IPCC, *Climate Change 2007: The Physical Science Basis*, Contribution of WG I to AR4 of the IPCC, Cambridge University Press, Cambridge, U.K., 2007.

coastal or island locations and in regions of highly heterogeneous land cover (Wilby et al. 2004). The quality of GCM output precludes their direct use for hydrological impact studies. They cannot resolve the important processes relating subgrid scale and topographic effects that are of significance to many impact studies (Moriondo and Bindi 2006; Stehlik and Bardossy 2002). In order to anticipate the consequences of climate change on water resources, reliable regional climate scenarios are needed

(Boe et al. 2007). Two fundamental approaches exist to bridge the gap between large- and local-scale climate data: dynamical downscaling and statistical downscaling (Ines and Hansen 2006; Maraun et al. 2010).

Dynamical downscaling refers to the use of regional climate models (RCMs). These use lateral boundary conditions from GCMs to produce higher-resolution outputs (Fowler et al. 2007). RCMs provide future climate data at 12–60 km resolution by remodeling GCM outputs. RCMs require a considerable processing capacity, time, and storage for obtaining a single scenario-by-period output, thus making it barely feasible to get RCM outputs for most assessment studies (Ramirez and Jarvis 2010). Statistical downscaling provides an easy to apply and much rapid method for developing high-resolution climate data for climate change impact assessment studies. However, it has been criticized by climatologists, since it tends to reduce variances (and thus alter uncertainties) and to cause a wrong sensation of more accuracy, when actually it only provides a smoothed surface of future climates (Fowler et al. 2007).

Statistical downscaling involves the use of empirical relationships between coarse-scale GCM output and higher-resolution observations. These relationships are developed using climate model output from the twentieth century and comparing it with observations. Statistical downscaling relies on the assumption that the relation between model output in the twentieth century and observations will hold in the twenty-first century (EPRI 2009). Statistical downscaling methods, as reviewed by Wilby et al. (2004), Fowler et al. (2007), and Hessami et al. (2008), are divided into three general categories: regression-based methods, weather pattern approaches, and stochastic weather generators. Each group covers a range of methods, all relying on the fundamental concept that regional climates are largely a function of the large-scale atmospheric state. This relationship may be expressed as a stochastic or deterministic function between predictors and predictands. The nature of local climate predictands determines the choice of statistical methods. For example, daily precipitation that is highly heterogeneous and discontinuous in space and time requires complicated nonlinear approach or transformation of raw data, while monthly temperature requires multiple regression approach as they show linear relationship (Wilby et al. 2004).

### 5.2.1 Water Resources

The assessment of water resources availability under different climatic conditions is performed by hydrological modeling. Hydrological models use climatic variables like precipitation, temperature, and catchment topography and land-use characteristics to simulate the runoff. The hydrological models can be separated broadly into two categories: physically based distributed-parameter models and simple models. The choice of a model for a particular study depends on many factors, among which the purposes of study and data availability have been the dominant ones (Jiang et al. 2007).

Physically based distributed-parameter models are complex in terms of structure and input requirements and can be expected to provide adequate results for a wide range of applications. On the other hand, simpler models that have a smaller range of applications can yield adequate results at greatly reduced cost, provided that the objective function is suitable. Thus, choosing a suitable model is equivalent to distinguishing the situation between when simple models can be used and when complex model must be used. For example, for assessing water resources management on a regional scale, monthly rainfall–runoff (water balance) models were found useful for identifying hydrological consequences of changes in temperature, precipitation, and other climate variables. For example, Jiang et al. (2007) investigated potential impacts of climate change on the water availability in the Dongjiang basin, south China, using six monthly water balance models, namely, the Thornthwaite–Mather, Vrije Universitet Brussel, Xinanjiang, Guo, WatBal, and Schaake models.

For detailed assessments of surface flow and other water balance components, conceptual lumped-parameter models are used. Many researchers have used these models for studying the impact of climate change. For example, Wang et al. (2006) used HBV of the Swedish Meteorological and Hydrological Institute to analyze the impact of the climate change on discharge of Suir River Catchment, Ireland. Boyer et al. (2010) used HSAMI model to analyze the impact of climate change on the hydrology of St. Lawrence tributaries. Thodsen (2007) used NAM model to analyze the influence of climate change on streamflow in Danish rivers. Jones et al. (2006) estimated the hydrological sensitivity, measured as the percentage change in mean annual runoff, of two lumped-parameter rainfall–runoff models, SIMHYD and AWBM, and an empirical model, Zhang01, to analyze changes in rainfall and potential evaporation. For simulation of spatial patterns of hydrological response within a basin, process-based distributed-parameter models are needed. The fully distributed models require extensive data and analysis of various processes.

### 5.2.2 WATER USE SECTORS

#### 5.2.2.1 Irrigation

Water is a key driver of agricultural production and its most precious input. Irrigation water has enabled farmers to increase crop yields by reducing their dependence on rainfall patterns (Fischer et al. 2007). Irrigation water requirements vary according to the balance between precipitation and evapotranspiration and the resultant fluctuations in soil moisture status. Because global warming will influence temperature and rainfall patterns, there will be direct impacts on soil moisture (De Silva et al. 2007).

Save et al. (2012) used SWAT model to analyze the potential changes in irrigation requirements of maize, apple trees, and alfalfa. De Silva et al. (2007) used CROPWAT model to assess the impacts of climate change on paddy irrigation water requirements in Sri Lanka. DSSAT and Aquacrop are the other useful modeling tools used to analyze the impacts of climate change on irrigation water requirements.

#### 5.2.2.2 Hydropower

The hydropower is sensitive to the amount, timing, and pattern of rainfall as well as the temperature (IPCC 2007). Climate change may also increase the variability of river runoff. Changes in the quantity and timing of river runoff, together with increased reservoir evaporation, may affect system operations. The analysis of potential climate change impacts on hydropower requires setting up an integrated simulation tool to simulate the behavior of the system for different climatic conditions. The simulation tool includes different types of models: a hydrological model, a water management model, and a model for assessment of hydropower under a given climate scenario (Schaefli et al. 2007). Various methods have been developed by researchers to assess the impact of climate change on hydropower. For example, Vicuna et al. (2008) developed a deterministic linear programming model, and Madani et al. (2010) developed an energy-based hydropower optimization model to estimate the hydropower in future periods.

## 5.3 CLIMATE CHANGE IMPACTS ON WATER RESOURCES

### 5.3.1 Water Availability

#### 5.3.1.1 Surface Water

Climate change has the potential to substantially alter river flow regimes and thereby surface water availability, as indicated by a number of modeling studies undertaken in many different environments (e.g., Arnell and Gosling 2013; Milly et al. 2005; World Bank 2013; etc.). Bates et al. (2008) suggest that globally there has been a discernible and contrasting change in the pattern of runoff: the regions lying in the higher latitudes have been experiencing an increase, while parts of west Africa, southern Europe, and southern Latin America have had a decrease. To project the impacts of climate change on future runoff of one of the most widely cited studies, Milly et al. (2005) used 12 different GCMs to estimate the mean runoff change until 2050 for the A1B scenario. As seen in Figure 5.1, the total annual river runoff globally is generally projected to increase between 10% and 40%, although there is significant decrease in midlatitudes and some parts of the dry tropics. However, whether or not all this change could be attributed to climate

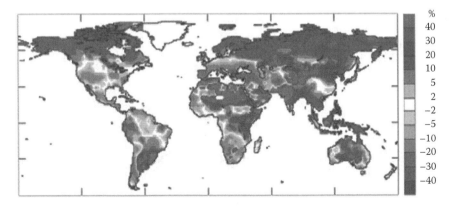

**FIGURE 5.1** Changes in runoff (in percent) for the period 2050s, compared to the 1900–1970 period. (From Milly, P.C.D. et al., *Nature*, 438, 347, 2005.)

is debatable given the potential influence of nonclimatic factors such as land-use changes and as mentioned in Table 5.1.

In a more recent study, Arnell and Gosling (2013) carried out a similar study with climate scenarios constructed using pattern scaling from 21 Coupled Model Intercomparison Project Phase 3 (CMIP3) climate models to project the runoff in 2050. Results of one of the models (HadCM3) indicated that the average annual runoff increases significantly over 47% of the land surface and decreases over 36%; only 17% sees no significant change. Like Milly et al. (2005), considerable variability between regions was projected.

From a regional point of view, a study carried out by Arnell (2004) projects considerable decrease in runoff in the north and south of the African continent by 2055. Contrasting projections were made for east Africa and parts of semiarid sub-Saharan Africa where the runoff is projected to increase.

In Asia (Bates et al. 2008), climate change and multiple socioeconomic stresses are expected to exacerbate the already disconsolate water scarcity situation. Climate change is expected to impact both seasonality and amount of river flows. For example, while the maximum monthly flow of the Mekong is projected to increase by 35%–41% in the basin (with the lower value estimated for the years 2010–2038 and the higher value for the years 2070–2099), compared with 1961–1990 levels, the minimum monthly flows are estimated to decline by 17%–24%, suggesting increased flooding risks during the wet season and a greater possibility of water shortages in the dry season.

In southern and eastern Australia and eastern New Zealand, the existing water security problems are very likely to intensify by 2030. In the Murray–Darling basin, the largest in Australia, the annual streamflow is projected to fall 10%–25% by 2050 and 16%–48% by 2100 (Beare and Heaney 2002).

For Europe, Alcamo et al. (2007) estimated that annual average runoff will increase in the north (north of 47°N) by approximately 5%–15% up to the 2020s and by 9%–22% up to the 2070s, for the A2 and B2 scenarios, while the runoff in southern Europe (south of 47°N) will decrease by 0%–23% up to the 2020s and by 6%–36% up to the 2070s.

For South America, Bates et al. (2008) report that most GCM projections indicate larger (positive or negative) rainfall anomalies for the tropical region and smaller ones for the extratropical part of South America. Further, extreme dry seasons are projected to become more frequent in Central America throughout the year.

For North America, generally the annual mean precipitation is projected to decrease in the southwestern United States but increase over most of the remainder of North America up to 2100. Changes in the magnitude and frequency of extreme precipitation events are likely to be more significant than changes in the average precipitation. This will have a major impact on runoff and river flows, thereby affecting seasonal water availability. Areas in the higher latitudes (e.g., Canada) are projected to receive increased precipitation (Bates et al. 2008). Because runoff is a function of precipitation, it is expected that runoff will also increase.

### 5.3.1.2 Groundwater

Groundwater is an important source of water in many parts of the world, and for centuries, it has been considered a reliable source of water supply for the human society. However, the overexploitation of this resource has cast serious aspersions on its sustainable use especially because a majority of the groundwater resources are nonrenewable on meaningful time scales. Climate change effects—reduced precipitation and increased evapotranspiration—will reduce recharge and possibly increase groundwater withdrawal rates (Treidel et al. 2012). More importantly because of variations in the volume of snowmelt and distribution of rainfall, the timing of recharge will be affected: typically with a shift in seasonal mean and annual groundwater levels (Hiscock et al. 2012). The FAO (2011) describes some obvious climate-related impacts in general terms, listed hereafter:

- If flooding increases, aquifer recharge will increase, except in continental outcrop areas.
- If drought frequency, duration, and severity increase, the cycle time will lengthen and abstraction will require better balance, with less in sequences of wet years and more in dry years.
- If snowmelt increases, aquifer recharge rates should increase, but this is dependent on permafrost behavior and recharge patterns, which largely remains unknown.

Klove et al. (2013) provide an interesting indication of the impacts of climate change on groundwater levels and flow paths for terrestrial and aquatic ecosystems (TGDE and AGDE, respectively) as shown in Figure 5.2. From a regional availability perspective, a study carried out by a franchise of the World Bank, Alavian et al. (2009) indicates that climate change is likely to reduce groundwater availability (recharge) in Africa and Latin America and the Caribbean, whereas there will be an increase in east Asia and the Pacific and Europe and Central Asia.

Table 5.2 presents the expected groundwater use and effects of climate change on the recharge capacity for the major regions in the world for the year 2050. Worryingly, predictions for the Middle East and north Africa are uncertain given the high-current usage of groundwater.

Using the HadCM3 model, Ranjan et al. (2006) made projections of future groundwater availability in the most water resources-stressed regions in the world, for A2 and B2 scenarios, as described in Table 5.3.

Apart from population and total groundwater availability statistics, Table 5.3 also shows the per capita resource availability per unit aquifer thickness (1 m) for 2010 and 2100. Except for north Africa, both groundwater availability and per capita availability are expected to reduce in all the other regions, under both scenarios. Interestingly, while for the Mediterranean and south Asia groundwater availability is less under the B2 scenario, the per capita availability is less for the A2 scenario. This is because of the significant difference in population densities for the two regions under the two scenarios, emphasizing on the notion that the impacts of nonclimatic drivers are equally significant.

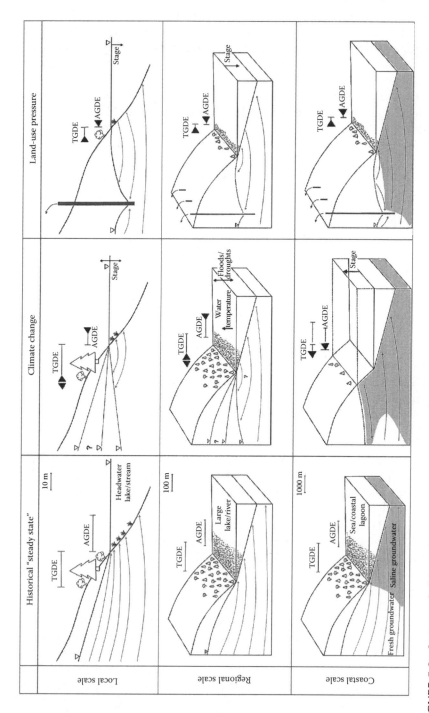

**FIGURE 5.2** Impacts of climate change and land-use pressures on groundwater levels and flow paths in terrestrial (TGDE) and aquatic (AGDE) groundwater-dependent ecosystems at different scales of water bodies. (From Klove, B. et al., *J. Hydrol.*, in press.)

**TABLE 5.2**
**Region-Wise Vulnerability of Groundwater to Climate Change**

| Region | Utilization of Groundwater | Climate Change Impact on Recharge |
|---|---|---|
| East Asia and the Pacific | Moderate | Increase |
| Europe and central Asia | Low | Increase |
| Latin America and the Caribbean | Moderate | Reduction |
| Middle East and north Africa | High | Uncertain |
| South Asia | Moderate | Negligible |
| Africa | Moderate | Reduction |

*Source:* Adapted from Alavian, V. et al., *Water and Climate Change: Understanding the Risks and Making Climate-Smart Investment Decisions*, International Bank for Reconstruction and Development/The World Bank, Washington, DC, 2009.

### 5.3.2 Sea-Level Rise

Increase in sea level has serious implications for both human security (increased flood risks, degraded groundwater quality, etc.) and ecosystems (impacts on mangrove forests and coral reefs, etc.), especially so in coastal regions. Because 60% of the world's 39 metropolises with a population of over 5 million are located within 100 km of the coast, including 12 of the world's 16 cities with populations greater than 10 million (IPCC 2007), the expected effects of sea-level rise are particularly crucial. Coastal cities in developing regions are particularly vulnerable to sea-level rise because of high population densities and often inadequate urban planning and the added burden due to urban migration. Of the impacts projected for 31 developing countries, only 10 cities account for two-thirds of the total exposure to extreme floods, for which rise in sea levels is an integral driver. Highly vulnerable cities are to be found in Mozambique, Madagascar, Mexico, Venezuela, India, Bangladesh, Indonesia, the Philippines, and Vietnam (Brecht et al. 2012).

There has always been a steady increase in the global sea level, but because of accelerated glacier melting in Greenland and the Antarctic, the rise has been quite rapid in the last decade and is projected to rise at a greater rate in the twenty-first century. With an average rise of 4 mm/year, the global sea level will reach 0.22–0.44 m above 1990 levels by the mid-2090s under the A1B scenario (IPCC 2007), although significant uncertainty remains as to the rate and scale of future sea-level rise. Satellite data, climate models, and hydrographic observations indicate that sea-level rise is not uniform around the world. This spatial variability of the rates of sea-level rise can be attributed to nonuniform changes in temperature and salinity, which bring about changes in the ocean circulation. It is expected that sea-level change in the future will also not be geographically uniform. Thermal expansion because of rapidly melting ice will account for more than half of the average rise in sea level in the next few decades. However, whether or not accelerated ice flow and subsequent melting will continue, as has been observed in recent years, is still a matter of uncertainty. If so, the sea level will rise further but

**TABLE 5.3**
**Groundwater Availability in Water Resources-Stressed Regions**

| Emission Scenario | Region | 2010 Population Density | 2010 Fresh Groundwater Availability (m³/km²) | 2010 Per Capita Resources (m³/cap/Year) | 2100 Population Density | 2100 Fresh Groundwater Availability (m³/km²) | 2100 Per Capita Resources (m³/cap/Year) |
|---|---|---|---|---|---|---|---|
| A2 | Central America | 120 | 64,000 | 533.3 | 200 | 25,000 | 125 |
|  | Mediterranean | 82 | 195,000 | 2378 | 150 | 178,000 | 1186.7 |
|  | North Africa | 11 | 12,000 | 1090.9 | 18 | 15,000 | 833.3 |
|  | South Africa | 18 | 67,000 | 3722.2 | 26 | 39,000 | 1500 |
|  | South Asia | 250 | 117,000 | 468 | 450 | 52,000 | 115.6 |
| B2 | Central America | 116 | 79,000 | 681 | 140 | 45,000 | 321.4 |
|  | Mediterranean | 80 | 180,000 | 2250 | 112 | 175,000 | 1562.5 |
|  | North Africa | 10 | 9,000 | 900 | 16 | 15,000 | 937.5 |
|  | South Africa | 12 | 76,000 | 6333.3 | 19 | 45,000 | 2368.4 |
|  | South Asia | 200 | 110,000 | 550 | 300 | 35,000 | 116.7 |

*Source:* Ranjan, P. et al., *Global Environ. Change*, 16, 388, 2006.

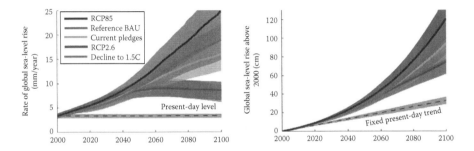

**FIGURE 5.3** Rate of global sea-level rise and global sea-level rise. *Notes*: (1) Lines show "best-estimate" median projections for each emission scenario, while shaded areas indicate the 66% uncertainty range. (2) RCP8.5: a no-climate-policy baseline with high greenhouse gas emissions, referred to at 4°C world by the World Bank. (3) RCP2.6: a scenario that is representative of the literature on mitigation aiming to limit the increase of global temperature to 2°C. (From World Bank, *Turn Down the Heat: Climate Extremes, Regional Impacts, and the Case for Resilience*, World Bank, Washington, DC, 2013.)

quantitative projections of the magnitude are difficult to make because of limited understanding of the relevant processes.

The World Bank (2013) reports that as much as 100 cm sea-level rise may occur if emission increases continue and raise the global average temperature to 4°C by 2100 and higher levels thereafter. Figure 5.3 shows the World Bank projections for sea-level rise for various scenarios.

### 5.3.3 Floods and Droughts

Floods and droughts cause significant damages every year and are responsible for a large fraction of water-related disasters. While droughts are a creeping disaster, in which the effects are felt over a longer duration of time, flooding phenomenon is usually more rapid in nature especially in urban areas because of the imperious nature of the ground surface. The IPCC (2012) projects that the frequency of heavy precipitation or the proportion of total rainfall from heavy falls will increase in the twenty-first century over many areas of the globe (Figure 5.4). The increase will be more intense in the high latitudes and tropical regions and in winter in the northern midlatitudes. Additionally, the maximum daily temperatures are projected to increase globally, while extremes in low temperatures will reduce (see Figure 5.5). Based on the A1B and A2 emission scenarios, the IPCC (2012) suggests that a 1-in-20 year hottest day is likely to become a 1-in-2 year event by the end of the twenty-first century in most regions, except in the high latitudes of the Northern Hemisphere, where it is likely to become a 1-in-5 year event.

#### 5.3.3.1 Floods

Although the risk of flooding is a global concern, coastal and deltaic regions are particularly vulnerable because of the high numbers of exposed people. Climatic change exacerbates the risk of flooding through extreme precipitation events, higher peak river flows, accelerated glacial melt, increased intensity of the most extreme tropical cyclones, and sea-level rise (Eriksson et al. 2009; Mirza 2010). These changes are

122                                                                 Climate Change and Water Resources

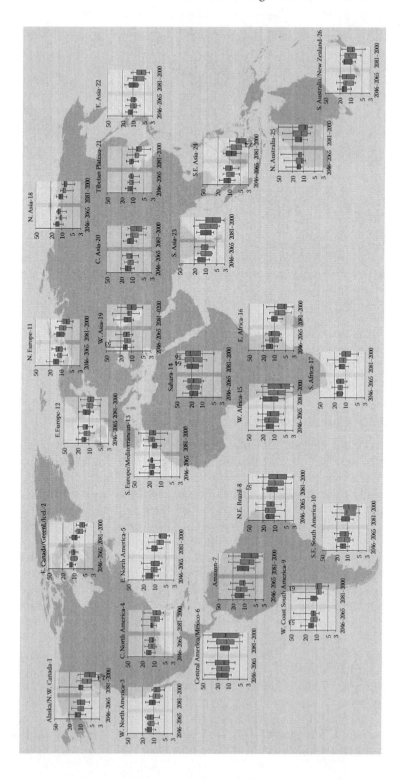

Climate Change Impacts 123

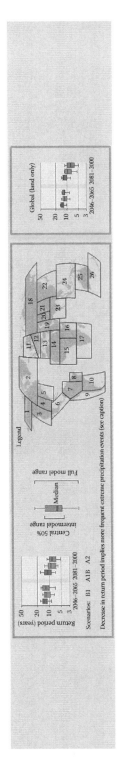

**FIGURE 5.4** Projected return periods for a daily precipitation event that was exceeded in the late twentieth century on average once during a 20-year period (1981–2000). A decrease in return period implies more frequent extreme precipitation events (i.e., less time between events on average). The box plots show results for regionally averaged projections for two time horizons, 2046–2065 and 2081–2100, as compared to the late twentieth century, and for three different SRES emissions scenarios (B1, A1B, and A2). Results are based on 14 GCMs contributing to the CMIP3. The level of agreement among the models is indicated by the size of the colored boxes (in which 50% of the model projections are contained) and the length of the whiskers (indicating the maximum and minimum projections from all models). Values are computed for land points only. The "globe" inset box displays the values computed using all land grid points. (From IPCC, *Managing the risks of extreme events and disasters to advance climate change adaptation*, in Field, C.B., Barros, V., Stocker, T.F. et al., Eds., *A Special Report of Working Groups I and II of the Intergovernmental Panel on Climate Change*, Cambridge University Press, Cambridge, U.K., 2012.)

# 124  Climate Change and Water Resources

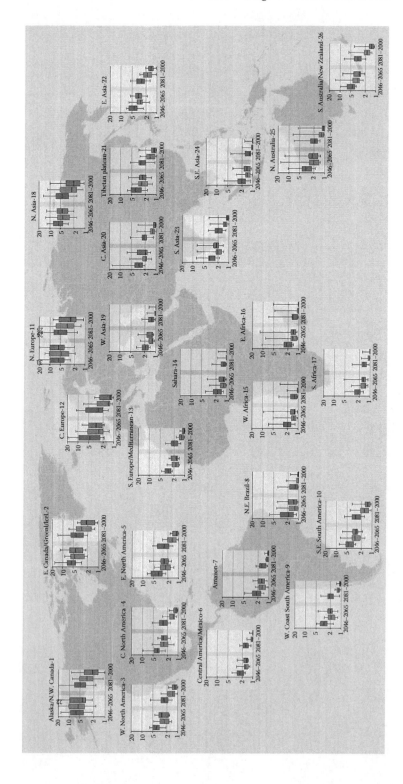

# Climate Change Impacts

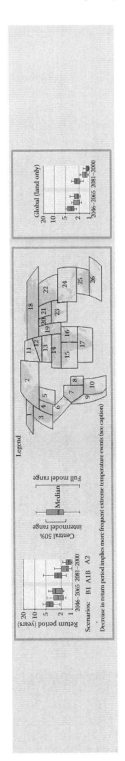

**FIGURE 5.5** Projected return periods for the maximum daily temperature that was exceeded on average once during a 20-year period in the late twentieth century (1981–2000). A decrease in return period implies more frequent extreme temperature events (i.e., less time between events on average). The box plots show results for regionally averaged projections for two time horizons, 2046–2065 and 2081–2100, as compared to the late twentieth century, and for three different SRES emissions scenarios (B1, A1B, and A2). Results are based on 12 global climate models (GCMs) contributing to the CMIP3. The level of agreement among the models is indicated by the size of the colored boxes (in which 50% of the model projections are contained) and the length of the whiskers (indicating the maximum and minimum projections from all models). Values are computed for land points only. The "globe" inset box displays the values computed using all land grid points. (From IPCC, *Managing the risks of extreme events and disasters to advance climate change adaptation*, in Field, C.B., Barros, V., Stocker, T.F. et al., Eds., *A Special Report of Working Groups I and II of the Intergovernmental Panel on Climate Change*, Cambridge University Press, Cambridge, U.K., 2012.)

already being experienced in many parts of the world today and are expected to further increase the frequency and magnitude of flood events in the future. Among the flooding events, there are wide range of flooding events that can be influenced by climate change, which include flash floods, inland river floods, extreme precipitation-causing landslides, and coastal river flooding, combined with the effects of sea-level rise and storm surge-induced coastal flooding (Bates et al. 2008).

In addition to floods and landslides, the Himalayan regions of Nepal, Bhutan, and Tibet are projected to be exposed to an increasing risk of glacial lake outbursts (Lal 2011; Mirza 2010). To evaluate the global risk of flooding for the end of this century, Hirabayashi et al. (2013) developed a global flood risk map (Figure 5.6) based on the outputs of 11 climate models. The risk was estimated by calculating the return period of a 100-year flood of the twentieth century, for the twenty-first century, and the time series of simulated annual maximum daily river discharge were fitted, respectively, to an extreme distribution function.

First, the magnitude of river discharge having a 100-year return period in the twentieth century was calculated for each location. Then, the return period of this magnitude of river discharge was computed for the time series of the twenty-first-century river discharge at each location. Because the global river routing model used in the study did not consider human interventions to regulate flood water, these projections provide potential risks of flooding irrespective of nonclimatic factors such as land-use changes, river improvement, or flood mitigation efforts. Under the RCP8.5 scenario (high greenhouse gas emissions) employed, for this century, small return periods for the twentieth-century 100-year flood were projected in Southeast Asia, Indian

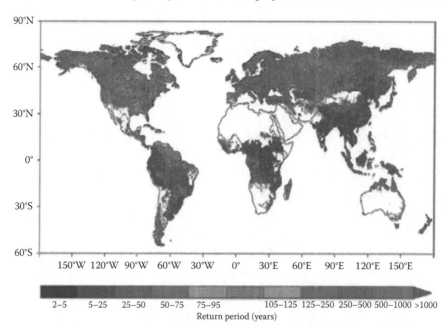

**FIGURE 5.6** Projected change in global flood frequency for the case for the RCP8.5 scenario. (From Hirabayashi, Y. et al., *Nat. Climate Change*, 3, 816, 2013, doi:10.1038/nclimate1911.)

subcontinent, central and eastern Africa, and the upper parts of South America, suggesting that the flood risks in these areas are particularly high. In certain areas of the world (northern Europe, Scandinavia, patches in North America, and the southern part of Latin America), however, flood frequency is projected to decrease.

### 5.3.3.2 Drought

Drought is multifaceted and is broadly categorized into three major types. A meteorological drought is defined by a prolonged period of low or insufficient precipitation, an agricultural drought is defined by soil moisture deficit, and a hydrological drought is characterized by flow reductions in rivers, and from reservoirs, with reduced groundwater levels. A fourth type, socioeconomic drought (Wilhite and Glantz 1985), is also sometimes considered especially in policy development and associates the supply and demand of economic goods with elements of meteorological, hydrological, and agricultural drought. Socioeconomic drought occurs when the demand for any economic good is not met because of shortage of water caused by elements of weather. Like floods, droughts are also driven by nonclimatic variables, which along with climate change intensify the vulnerability of human systems. For example, while the frequency of occurrence of droughts is very strongly influenced by natural climatic variability, it may also be influenced by changes in land cover. Therefore, the effects of climate change on drought and its impacts are likely to be extremely complex.

Hegerl et al. (2007) point out to the strong possibility that anthropogenic activities have contributed to the increase in the droughts observed towards the end of the twentieth century. Global trends of drought correspond well with trends of precipitation and temperature, which are consistent with expected responses to anthropogenic forcing. Global and regional projections of hydrological drought (e.g., Hirabayashi et al. 2008) indicate a higher likelihood of hydrological drought by the end of this century, especially in the duration of drought days (when the streamflow is below a specific threshold) in North and South America, central and southern Africa, the Middle East, southern Asia from Indochina to southern China, and central and western Australia. Some regions, including eastern Europe to central Eurasia, inland China, and northern North America, project increases in drought. In contrast, wide areas over eastern Russia project a decrease in drought days.

Despite progress in forecasting and modeling techniques, there is still uncertainty affecting the projections of trends in meteorological drought for the future because of insufficient knowledge of the physical causes of meteorological droughts and of the links to the large-scale atmospheric and ocean circulation.

### 5.3.4 Water Quality

#### 5.3.4.1 Surface Water Quality

Changes in surface water quality have implications on human and ecological health. While groundwater is relatively free of organic and other contamination (although it has its own unique problems as discussed later in this section), surface water is more prone to pollution. From a drinking water point of view, changes in surface water quality will dictate the type and level of treatment that will be required. Water quality is a dynamic condition of a system and is defined by measurement of multiple parameters.

Hydrological models (with a water quality component) that use GCM data are good tools in order to comprehend and predict the potential effects of climate change on water quality in rivers and streams. The IPCC (2007) suggests that two main drivers of climate change—higher water temperature and variations in runoff—are likely to produce adverse changes in water quality affecting human health, ecosystems, and water use. Higher surface water temperatures will promote algal blooms and increase microbial content, while more intense rainfall will lead to an increase in suspended solids (turbidity) in lakes and reservoirs due to increased soil erosion and contaminant transport (e.g., pesticides, heavy metals, and organics). These effects will especially be a source of major concern in water bodies where water levels are expected to reduce. Delpha et al. (2009) performed an exhaustive literature review to project the potential impacts of climate change on water quality parameters, as shown in Table 5.4.

### TABLE 5.4
### Impacts of Climate Change on Water Quality Parameters

| Water Quality Parameters | | | Climate Change Factors Affecting WQ | Water Body |
|---|---|---|---|---|
| Physiochemical | Basic parameters | pH | Droughts, temperature increase, rainfall | Rivers, lakes |
| | | DO | Droughts, temperature increase, rainfall | Rivers, lakes |
| | | Temperature | Droughts, temperature increase | Rivers |
| | DOC | | Temperature and rainfall increase | Streams and lakes |
| | Nutrients | | Temperature and rainfall increase, droughts, heavy rainfall | River, lakes, streams, groundwater |
| Micropollutants | Inorganic | Metals | Temperature and rainfall increase, droughts, heavy rainfall | River, high alpine lakes, streams |
| | Organics | Pesticides | Temperature and rainfall increase, drying and rewetting cycles | Surface water and groundwater |
| | | Pharmaceuticals | Temperature increase, rainfall | Streams, groundwater |
| Biological | Pathogens | | Temperature and rainfall increase | Surface waters |
| | Cyanobacteria | | Temperature and rainfall increase | Lakes |
| | Cyanotoxins | | Temperature increase | Lakes |
| | Green algae, diatoms, fish | | Temperature increase | Freshwaters |
| | Others | | Temperature increase | Soils |

*Source:* Adapted from Delpha, I. et al., *Environ. Int.*, 35, 1225, 2009.

# Climate Change Impacts

### 5.3.4.2 Groundwater Quality

Climate change, coupled with anthropogenic influence, will impact groundwater quality through the influences of recharge, discharge, and land use on groundwater systems. The coastal regions, in particular, are vulnerable to degraded groundwater quality due to climate change impacts, which affect recharge (sea-level rise, changes in precipitation patterns and timings, and evapotranspiration), and increased groundwater pumping, which will result in aggravated salinity intrusion in many coastal regions (Green et al. 2011). Decreased groundwater levels caused due to reduced recharge of groundwater may lead to an increased rate of pumping to meet demands. This is most likely to further degrade groundwater quality by disturbing the balance of the freshwater/saline water boundary, resulting in saline water intrusion in not only coastal basins but inland aquifers as well. Nutrient transport rates, particularly nitrogen (N) and phosphorus (P), beneath agricultural lands may also be sensitive to climate change (Green et al. 2011). Table 5.5 presents a list of potential impacts on climate change due to various scenarios.

## 5.4 CLIMATE CHANGE IMPACTS ON WATER USE SECTORS

The World Water Assessment Report (2012) indicates that about 70% of the world's freshwater is used for irrigation, 22% for industry, and 8% for domestic use. However, there is a sizeable difference in this water use distribution at regional scales. For example, while in some high-income countries the agricultural water use accounts for only 30% of the total water use, this figure is as high as 90% in low- and middle-income countries. Similarly the proportion of industrial water use in some low- and middle-income countries is a mere 10% compared to about 60% in high-income countries. Given that the economic returns of water in the agriculture sector are quite low when compared to the industrial sector and that low-income countries have typically low irrigation efficiencies, the added stress of climate change will have a lasting impact on the economy of these countries. Described hereafter are the effects of climate change on certain water use sectors.

### 5.4.1 AGRICULTURE

Climate change impacts on agriculture can be broadly classified into two groups. The impacts in the first group have a direct relation with water, which primarily include changes in the irrigation demand due to changes in evapotranspiration and effective rainfall. In the second group, the impacts are less related to water, e.g., weed and pest proliferation, wilting, and loss in soil fertility. This section focuses on the former. Climate change impacts on irrigation are mainly because of

1. Change in the trends of average precipitation and temperature over longer durations
2. Change in seasonal variability and the occurrence of extreme events like floods, droughts, and frosts

## TABLE 5.5
## Potential Scenarios and Impacts on Groundwater Quality due to Climate Change

| Scenario | Foreseen Impact on Groundwater | Potential Impact on Aquifers | Potential Impacts on Ecosystems | Uncertainty Related to Impact |
|---|---|---|---|---|
| Increased leaching due to more intense rainfall | Increased leaching of water-soluble contaminants such as nitrates | Increased concentration of pollutants | Potential impact on ecosystem—eutrophication and pollution | Changes in precipitation intensity vary regionally (this change is mainly foreseen for dry and warm climate) |
| Sea-level rise | Salt water intrusion in coastal aquifers | Increased groundwater salinity | More seawater exchange to coastal lagoons. Changes in groundwater flow pattern in coastal ecosystems | The amount of intrusion will depend on coastal aquifer system water level and amount of water extraction |
| Changed agricultural practices | Increased leaching of water-soluble nutrients due to longer growing season and/or intensified irrigation. Increased need for pesticides in cold climate | Increase in agriculture can lead to increased pollution. Lower groundwater levels due to higher irrigation may add to the problem | Eutrophication, salinization, reduced discharge to ecosystems | Increased $CO_2$ can lead to less transpiration counteracting the irrigation needs and risk of increased leaching |
| Changed snow accumulation and melt | Increased winter time groundwater recharge in temperate climate with seasonal snow cover. Changes to the timing of snowmelt and corresponding recharge | Increase risk of salt intrusion from road runoff as more salt is used and recharge occurs in winter | No direct impacts known on that change water quality in ecosystems | |

*Source:* Klove, B. et al., *J. Hydrol.*, in press.

Climate change impacts in the midlatitudes, where agriculture is already precarious and often heavily dependent on irrigation, include higher temperatures and more variable rainfall, with likely substantial reductions in precipitation. Water resource availability will be altered by changed rainfall patterns and increased rates of evaporation. Rainfed farming that accounts for more than 80% of global crop area and 60% of global food output is especially susceptible to the impacts of climate change, more so in the arid and semiarid regions in the mid and low latitudes, while productivity may rise for a time in the higher latitudes (notably North America and northern Europe) (Bates et al. 2008).

Fischer et al. (2007) estimated the global irrigation water requirement, with and without climate change (Figure 5.7), and found that under the socioeconomic development pathways of the A2r reference scenario, without climate change, agricultural water requirements are projected to increase by about 45% in 2080.

Impacts of climate change on irrigation water requirements by 2080 are an additional +20% in global irrigation water needs in 2080. Two thirds of the increase (75%–80% in developing countries, but only 50%–60% in developed countries) results from an increase in daily water requirements, and one-third occurs because of extended crop calendars in temperate and subtropical zones.

From a food production point of view, crop systems will be under increasing pressure to meet growing global demand in the future. The World Bank (2012) reports

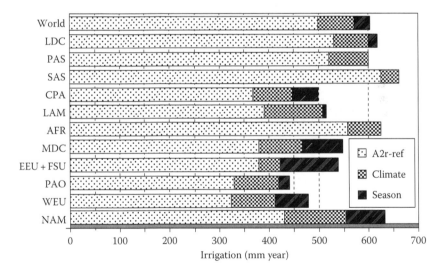

**FIGURE 5.7** Impacts climate change on average regional net irrigation water requirements (mm per year) in 2080, for A2r scenario and Hadley GCM. Diagram indicates values under reference climate (A2r-ref), increases because of warming and changed precipitation patterns (climate), and increases caused by expanded crop calendars (season). *Notes*: (1) Reference scenario indicated the irrigation water requirement in 2080, without considering the impacts of climate change. (2) MDC, developed countries; LDC, developing countries; NAM, North America; WEU, other developed countries (mainly Europe, including Turkey); PAO, developed Pacific Asia; EEU + FSU, eastern Europe and former USSR; AFR, sub-Saharan Africa; LAM, Latin America; MEA, Middle East and north Africa; CPA, east Asia; SAS, south Asia; and PAS, developing countries in Southeast Asia. (From Fischer, G. et al., *Technol. Forecast. Social Change*, 74, 1083, 2007.)

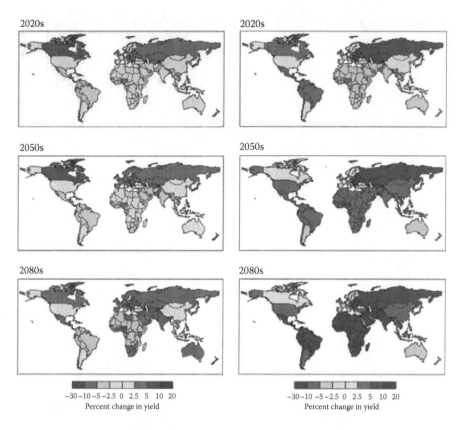

**FIGURE 5.8** Potential changes (%) in national cereal yields for the 2020s, 2050s, and 2080s (compared with 1990) under the HadCM3 SRES B1a scenario with and without $CO_2$ effects. (From Parry, M.L. et al., *Global Environ. Change*, 14(1), 53, 2004.)

that significant crop yield impacts are already being felt at 0.8°C warming and if the average temperatures increase by 1.5°C–2°C, the most heavily affected regions in the world would be sub-Saharan Africa, Southeast Asia, and south Asia, where crop yields will suffer severely resulting in high production. These impacts would have strong repercussions on food security and are likely to negatively influence economic growth and poverty reduction in the impacted regions. In a global assessment, Parry et al. (2004) made projections of the potential changes in national cereal yield (wheat, rice, maize, and soybean) with and without climate change for different future timelines, as shown in Figure 5.8. A reduction in yield is forecasted for most regions of the world with more severity in Asia and Africa.

### 5.4.2 Industry (Hydropower)

While global assessment reports suggest that industrial water accounts for around 22% of the water use, a majority of this (60%–70%) is used in different forms of energy or power generation, notably thermoelectric and hydropower (WWAP 2012). Hydropower is the largest renewable energy source, and it produces

Climate Change Impacts

around 16% of the world's electricity and over four-fifths of the world's renewable electricity. While most water sectors have a consumptive use of water, the hydropower sector operates on nonconsumptive use of water. Currently, more than 25 countries in the world depend on hydropower for 90% of their electricity supply (e.g., 99.3% in Norway), and 12 countries are 100% reliant on hydro (IRENA 2012). The World Bank (2009) recognizes the need and role of hydropower in a world of growing demand for clean, reliable, and affordable energy and has been encouraging the growth of hydropower globally through its investment in water development and infrastructure projects. Runoff, which is dependent on rainfall, is the key resource for hydropower generation. Because of the uncertainty of the future global climate and its impacts on river flows, the hydropower generation sector faces considerable risk.

Hamududu and Killingtveit (2012) carried out an assessment (see Table 5.6) using 12 GCMs and estimated the runoff at national scales to determine the hydropower

### TABLE 5.6
### Regional Changes in Hydropower Generation by 2050 from 2005 Conditions

| Continent | Region | Generation TWh | Change TWh | % Change of Total |
|---|---|---|---|---|
| Africa | Eastern | 10.97 | 0.11 | 0.59 |
|  | Central | 12.45 | 0.04 | 0.22 |
|  | Northern | 15.84 | −0.08 | −0.48 |
|  | Southern | 34.32 | −0.07 | 0.83 |
|  | Western | 16.03 | 0.00 | 0.03 |
|  |  | 89.60 | 0.00 | 0.05 |
| Asia | Central | 217.34 | 2.29 | 2.58 |
|  | Eastern | 482.32 | 0.71 | 0.08 |
|  | Southeastern | 57.22 | 0.63 | 1.08 |
|  | Southern | 141.54 | 0.70 | 0.41 |
|  | Western | 70.99 | −1.66 | −1.43 |
|  |  | 996.12 | 2.66 | 0.27 |
| Australasia/Oceania |  | 39.80 | −0.03 | 0.00 |
| Europe | Eastern | 50.50 | −0.60 | −1.00 |
|  | Northern | 227.72 | 3.32 | 1.46 |
|  | Southern | 96.60 | −1.79 | −1.82 |
|  | Western | 142.39 | −1.73 | −1.28 |
|  |  | 517.21 | −0.80 | −0.16 |
| America | Northern Central/Caribbean | 654.70 | 0.33 | 0.05 |
|  | Southern | 660.81 | 0.30 | 0.03 |
|  |  | 1315.50 | 0.63 | 0.05 |
| Global |  | 2931 | 2.46 | 0.08 |

*Source:* Hamududu, B. and Killingtveit, A. *Energies* 5(2): 305–322, 2012
*Note:* Russia and Turkey have been included in Asia.

potential in various regions of the world by 2050 and the corresponding change from 2005 conditions, considering climate change impacts on hydropower production.

For global projections, there is a very little change in hydropower production but the production in individual countries and regions may be significantly impacted. Apart from Europe, all regions are likely to experience an increase in hydropower production.

Asia has the maximum hydropower potential, but because many Asian countries do not have the financial resources and infrastructure to effectively respond to climate change, it will certainly have implications on hydropower availability in the region.

In North America, the runoff is expected to decrease as a result of higher temperatures in the summer–autumn months until 2030 lowering hydropower potential in the short term. However, long-term increase of annual and seasonal precipitation, especially in parts of Canada, has the capacity to increase hydroelectric output. Among all the regions in the world, Latin America relies most on hydropower for energy production. Almost all climate change models predict increased temperatures and rainfall across all of Latin America. A significant amount of the region's installed hydropower resources are located along the Paraná River, meaning most of these are not likely to be affected negatively (reduced power production). In fact, it is expected that some of the hydroelectric facilities in eastern South America may be able to upgrade and increase the amount of electricity they produce.

Generally water availability, and subsequently hydropower production, is likely to increase across northern Europe and decrease in the remaining parts over the next several decades, where there is likely to be a decline in hydropower production potential. Another study by Lehner (2001) also reported that overall across Europe, developed hydropower potential is predicted to decrease 7%–12% by the year 2070.

Climate is already a major factor in African hydroelectric production. Recurring droughts have plagued hydroelectric dams and led to power rationing across the continent. Climate change impacts vary across the continent. For example, a gradual overall reduction in generation capacity in the Zambezi River basin is projected over the next 60 years (Yamba et al. 2011), whereas the hydropower production in the Nile River basin is expected to increase because of increased streamflow (Beyene et al. 2010).

### 5.4.3 WATER SUPPLY AND SANITATION

Like all other water use sectors, domestic water use is also a function of water availability, which is affected by changes in the amount and patterns of precipitation and changes in temperatures and other meteorological variables. Issues pertaining to water availability, both surface water and groundwater, have already been discussed previously in Section 5.3.1. This section seeks to address the effects of climate change on the water supply and sanitation sector as a whole. The WHO (2009) suggests that out of the various climate change effects, floods and droughts will impact this sector the most, as listed in the following:

- Floods affect the basin water supply infrastructure that can take years to repair, especially in low-income countries.
- Flooding of sanitation facilities will not only cause a breakdown in service but also is a source of water-borne and water-related diseases via the spread of human excrement in the surrounding areas.
- In groundwater-dependent countries, increasing droughts will lead to drying up of well, and greater distances will have to be travelled to access drinking water.

Further degradation of raw water quality may require a change in the type and nature of treatment to meet potable standards, which may not be technically and economically feasible in some developing countries.

Water is intrinsic to all human basic and psychological needs, and access to safe drinking water and sanitation is key to meeting these needs. The United Nations Human Rights Council in 2010 affirmed a resolution indicating that access to water and sanitation is a human right, which is also a component of the Millennium Development Goals (MDGs). However, as reported by the WHO (2013) despite of good progress over the years, much remains to be done, especially in the sanitation sector. Figure 5.9a and b shows the current coverage of access to improved water supply and sanitation facilities.

According to the WHO/UNICEF joint monitoring program, "An improved drinking-water source is defined as one that, by nature of its construction or through active intervention, is protected from outside contamination, in particular from contamination with faecal matter." Similarly, "An improved sanitation facility is defined as one that hygienically separates human excreta from human contact." As seen in Figure 5.9, the situation for drinking water coverage is much better than that of sanitation, with the main hotspots in the African continent where in some countries the coverage is as less than 50%. The situation in Africa is also the poorest with respect to improved sanitation coverage, with less than 50% coverage in most of the countries. Much improvement is also desired in Asia, especially parts of south Asia. While the WHO (2013) reports that the drinking water and sanitation coverage currently stand at 89% and 64%, respectively, climate change effects and coupled nonclimatic drivers like population growth and urbanization are likely to effect the progress of providing this basic human right throughout the world.

## 5.5 EXAMPLES TO EVALUATE CLIMATE CHANGE IMPACTS

### 5.5.1 WATER RESOURCES: CASE OF THE BAGMATI RIVER BASIN, NEPAL

#### 5.5.1.1 Introduction

The Bagmati River basin (BRB) is one of the major basins of Nepal and sustains much of the socioeconomic activities of the country. It is located within 26°45′N–27°49′N and 85°02′E–85°57′E, with a catchment area of about 3750 km$^2$

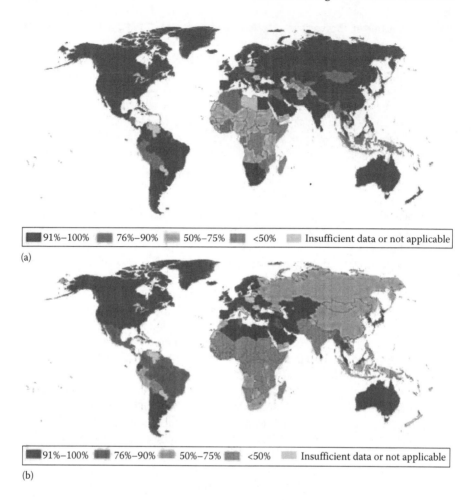

**FIGURE 5.9** Proportion of population using (a) improved sources of drinking water and (b) improved sanitation in 2011. (From WHO, *Progress on Sanitation and Drinking Water*, World Health Organization Press, Geneva, Switzerland, 2013.)

in Nepal. The climate of the BRB varies from cold temperate in higher mountains and warm temperate at midelevation levels to subtropical in the southern lowlands (<1000 m msl) with a mean annual temperature of 20°C–30°C. The mean relative humidity varies from 70% to 86%. The average annual rainfall in the basin is about 1800 mm with 80% of the total annual precipitation occurring during the summer (June–September). Snowfall is negligible in the basin. The BRB in Nepal is divided into three parts as upper (Kathmandu valley), middle (mountains/hills), and lower (Terai) considering the physiographic variation. This study, however, considers the upper (Kathmandu valley) and middle (mountains/hills) parts of the Bagmati Basin up to the Pandheradobhan gauging station, as shown in Figure 5.10, with a total catchment area of 2789 km$^2$.

Climate Change Impacts 137

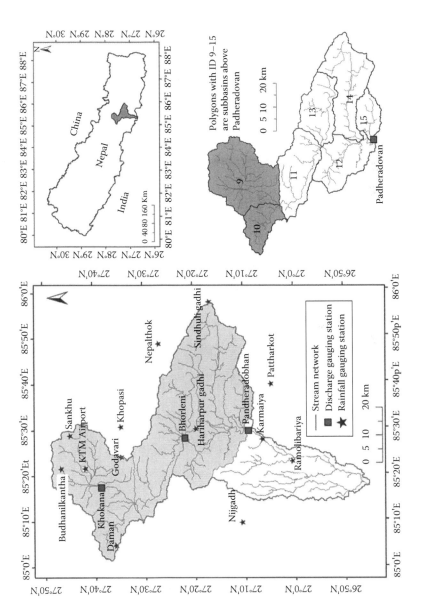

**FIGURE 5.10** The BRB. (From Babel, M.S. et al., *Theor. Appl. Climatol.*, 113, 585, 2013.)

The basin's water is widely used for drinking, irrigation, industrial, and other purposes in the Kathmandu valley. In recent time, increasing population density, unplanned rapid urbanization, land conversion to agriculture, and unregulated and illegal quarries have been responsible for degradation of the river water's quality and quantity (Babel et al. 2011a). The uncertainty of future water availability on a basin scale has seriously impaired water resource planning in the BRB, thereby increasing the risk of failure of water-related programs and projects. To address the issue, this case study presents the analysis of future changes in local climate and their impact on the hydrology of the BRB to help in managing water more efficiently and making necessary plans of adaptation in changing climatic conditions.

### 5.5.1.2 Methodology

The statistical downscaling model (SDSM) version 4.2 (Wilby and Dawson 2004) was used for this study. This model uses the principle of developing multiple linear regression transfer function between large-scale predictors and local climatic variables (predictand), and these transfer functions are used for downscaling future climate data as predicted by GCMs. Downscaling of low-resolution climate data for future periods obtained from GCM HadCM3 under two IPCC emission scenarios, A2 and B2, was done using the SDSM.

The rainfall–runoff process was simulated using the semidistributed hydrological model Hydrologic Engineering Center's Hydrologic Modeling System (HEC-HMS), version 3.3, developed by the US Army Corps of Engineers. HEC-HMS is chosen since it provides the user a choice among a number of loss, direct runoff, base flow, and channel routing methods (McColl and Aggett 2007). Four scenario runs, each of 30-year periods, were developed for Special Report of Emission Scenarios (SRES) A2 and B2. The changes relative to the baseline period (1970–1999) were calculated for three future periods 2020s, 2050s, and 2080s. To analyze the climate change impact on streamflow, monthly, seasonal, and annual variations on water availability were computed for each of the future time periods. To assess spatial variations of climate change impact within the basin, the changes in future water availability were estimated for the upper and middle parts of the basin.

### 5.5.1.3 Results and Discussion

*5.5.1.3.1 Temperature and Precipitation*

The basin average annual mean of $T_{max}$ is predicted to increase by 2.1°C under A2 scenario and by 1.5°C under B2 scenario in 2080s. Table 5.7 shows the interseasonal variations in increase of $T_{max}$ for the basin. Scenario A2 shows higher increase of $T_{max}$ in spring, while scenario B2 shows higher increase in summer during all three future periods (Table 5.7). Projected precipitation does not show any significant trend for both A2 and B2 scenarios. There are wide temporal and spatial variations throughout the basin. Scenario A2 shows a decrease of basin average precipitation during winter and spring and increase during the summer (Figure 5.11a). On the other hand, scenario B2 shows an increase in precipitation during all the seasons (as seen in Figure 5.11b).

## TABLE 5.7
### Basin Average Change in Seasonal T_max for Three Future Time Periods Relative to the Baseline Period (1980s)

| Scenario →<br>Period ↓ | Winter A2 | Winter B2 | Spring A2 | Spring B2 | Summer A2 | Summer B2 | Autumn A2 | Autumn B2 | Annual A2 | Annual B2 |
|---|---|---|---|---|---|---|---|---|---|---|
| 2020s | 0.4 | 0.4 | 0.7 | 0.5 | 0.5 | 0.5 | 0.5 | 0.5 | 0.5 | 0.5 |
| 2050s | 1.0 | 0.9 | 1.2 | 0.9 | 1.2 | 1.1 | 1.0 | 0.8 | 1.1 | 0.9 |
| 2080s | 1.8 | 1.4 | 2.4 | 1.6 | 2.1 | 1.7 | 1.9 | 1.3 | 2.1 | 1.5 |

*Source:* Babel, M.S. et al., *Theor. Appl. Climatol.*, 113, 585, 2013.

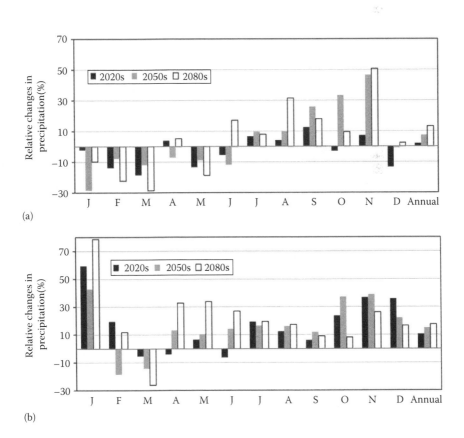

**FIGURE 5.11** Changes in basin average monthly precipitation for three future periods relative to the baseline period (1980s). (a) Scenario A2 and (b) scenario B2. (From Babel, M.S. et al., *Theor. Appl. Climatol.*, 113, 585, 2013.)

### 5.5.1.3.2 Streamflow

The impact on streamflow and water resources was analyzed for wet season and dry season. The wet season (June–September) is the same as the summer season as considered in this study, and dry season (October–May) comprises winter, spring, and autumn. To further represent the results, dry season is divided into two parts, premonsoon (January–May) and postmonsoon (October–December). The plot of monthly average flow for the baseline period and three future periods shows that the hydrograph ordinates for future time periods diminish during premonsoon months (January–May), except April, while they significantly increase during the monsoon (June–September) and postmonsoon (October–December) periods. The monthly average peak shifts to August in the 2080s, while it is obtained in July for the baseline period, 2020s, and 2050s (Figure 5.12a). Hydrograph ordinates decrease during

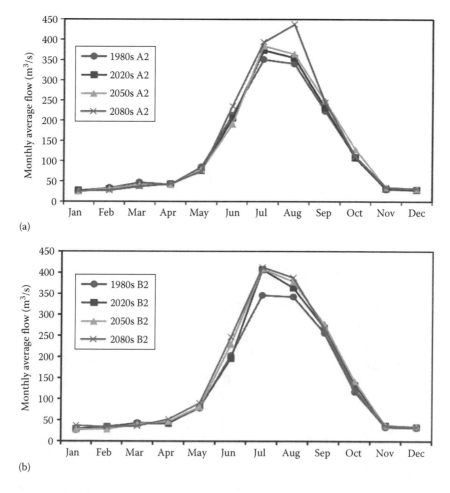

**FIGURE 5.12** Monthly average flow hydrographs for the baseline period and three future periods at the Pandheradobhan gauging station. (a) Scenario A2 and (b) scenario B2. (From Babel, M.S. et al., *Theor. Appl. Climatol.*, 113, 585, 2013.)

# Climate Change Impacts

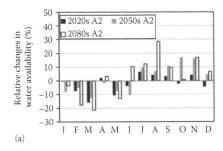
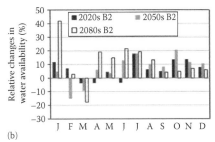

**FIGURE 5.13** Percentage changes in monthly water availability during three future periods relative to the baseline period (1980s) for the whole BRB (a) under scenario A2 and (b) under scenario B2. (From Babel, M.S. et al., *Theor. Appl. Climatol.*, 113, 585, 2013.)

premonsoon and increase during postmonsoon; the shifting of the peak from July to August clearly indicates slight changes in seasonal water availability. Scenario A2 predicts a very significant increase in the average monthly flow during the monsoon period, and the month of August has the highest increase with 28.6% increase in 2080. As seen in Figure 5.12b, scenario B2 indicates an increase in hydrograph ordinates during most of the time of the year, except March, and the peak of the hydrograph occurs in July for all future time periods. This scenario also predicts the slight widening of the monsoon hydrograph with an increase in pre- and postmonsoon ordinates.

### 5.5.1.3.3 Water Availability

When analyzing seasonal water variation for the entire BRB, scenario A2 shows a decrease in premonsoon (January–May) water availability and an increase in monsoon (June–September) and postmonsoon (October–December) water availability in the future (see Figure 5.13a). This implies that according to A2 scenario, the premonsoon season is expected to become drier and the monsoon and postmonsoon seasons may become wetter. Monthly results represent that all premonsoon months (January–May), except April, are expected to experience decrease in water availability, thereby worsening the water stress situation in the basin. The monsoon season's months from July to September and postmonsoon months from October to December may experience an increase in water availability. The change in monthly water availability is predicted to vary from 21.54% decrease in May to 28.64% increase in August during 2080s. Scenario B2 shows an increase in water availability during both dry and wet seasons (Figure 5.13b). On a monthly scale, scenario B2 shows an increase in water availability in all months, except February and March during all three future periods. However, there is a wide variation in magnitude.

### 5.5.1.4 Conclusions

This study quantifies the changes in future climate and its impact on the hydrology of the BRB in Nepal. The Hadley Centre Coupled Model, version 3 (HadCM3) GCM, was used to capture future temperature and precipitation. The resolution of the HadCM3 is very coarse for hydrological analysis at the basin level. Therefore, an SDSM was employed to simulate future climate at the station level using large-scale

GCM predictors. Downscaling results indicate that the SDSM model was able to estimate both mean and extreme values of temperature and mean values of precipitation with considerable reliability. Results show a higher rise in temperature during summer as compared to winter. Future average annual basin precipitation is predicted to increase under both A2 and B2 scenarios. However, dry season is expected to become drier and wet season is expected to become wetter under A2 scenario. It is anticipated that the annual water availability during all three future periods may increase under both A2 and B2 scenarios, indicating that the basin as a whole becomes wetter when water accounting is done annually. There may be a wide variation in seasonal and monthly water availability. Under A2 scenario, the premonsoon water availability may decrease, indicating a worsening situation of water stress during the dry season. However, an increase in the postmonsoon water availability may relieve the water stress situation to some extent. In contrast, under B2 scenario, water availability is expected to increase during both wet and dry seasons. Higher water availability during the wet season under both A2 and B2 scenarios may worsen the flood situation in the future.

### 5.5.2 Water Use Sector: Case of Rice Cultivation in Northeast Thailand

#### 5.5.2.1 Introduction

The objective of the study was to assess the impacts of future climate change on rice yield in northeast Thailand using the CERES-rice model (DSSAT version 4.0). Northeast Thailand, comprising 19 provinces, lies between latitude 14.50°N–17.50°N and longitude 102.12°E–104.90°E. The region has about 9.3 million ha of agricultural land, of which about 7.9 million ha is under rainfed farming. Up to 75% of this land is devoted to rice, and the planting area varies considerably from year to year. Rice production in the region relies mainly on rainfall; irrigation is limited to about 20% of the total rice producing area. The region has a tropical climate, with average temperature ranging from 19.6°C to 30.2°C. October–February is the cool season, while March–May is the hot season, with highest temperatures observed in the month of April. Rainfall in the region is highly unpredictable, mainly concentrated in the rainy season, i.e., May–October. The average annual rainfall varies from 1270 to 2000 mm within the region. Soils of the study area are highly acidic, saline, and low in fertility. The present study used the climatic and soil data, as well as other information, from crop experiments conducted at the Rice Research Centers (RRCs) in three provinces, namely, Khon Kaen, Roi Et, and Ubon Ratchathani, representing northeast Thailand (Figure 5.14).

The average rice yield in the region is 1.9 t ha$^{-1}$, which is the lowest in the country, with the countrywide average being 2.5 t ha$^{-1}$. Rainfed rice is grown under poor conditions, i.e., poor crop management with low inputs, and is highly subjected to climatic variability. The major production constraints are high rainfall variability, drought, submergence, and inherent low soil fertility. The main varieties of Jasmine rice, namely, Khao Dok Mali 105 (KDML105) and Rice Department 6 (RD6), are medium-maturing varieties and cover almost 80% of the rice fields in northeast Thailand.

Climate Change Impacts 143

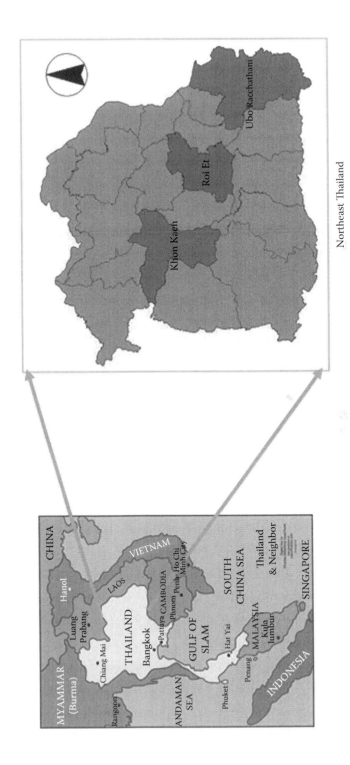

**FIGURE 5.14** Northeast Thailand and three study provinces. (From Babel, M.S. et al., *Climate Res.*, 46, 137–146, 2011b.)

### 5.5.2.2 Methodology

The MS Windows-based CERES-rice model (Singh et al. 1993) released with DSSAT version 4.0 (Hoogenboom et al. 2003) in 2004 by the International Consortium for Agricultural Systems Application, University of Hawaii, United States, was used in this study. The model is based on the understanding of plants, soil, weather, and management interaction to predict growth and yield. Yield-limiting factors like water and nutrient stresses (N and P) are considered by the model. The future climate data were collected from the Southeast Asia START Regional Center at Chulalongkorn University, Thailand. The future climate data were predicted using the global climate model (GCM) ECHAM4 (ECMWF atmospheric GCM coupled with the University of Hamburg's ocean circulation model) developed for the global resolution of 280° × 280° km by the Max Planck Institute, Germany. These data were developed considering world growth forced by a level of atmospheric $CO_2$ according to the IPCC SRES A2 scenario, one of the most pessimistic projections. These data were further downscaled at the regional level using the RCM providing regional climates for impact studies (PRECIS) for the study area at 25° × 25° km. The downscaled data for the periods of 2020–2029, 2050–2059, and 2080–2089 for the grid that falls nearest to the study locations in the three provinces were used.

The predicted future climate scenario was applied to the calibrated CERES-rice model for the study sites to determine the impacts on rice yield during the three future periods. The impacts were then determined by computing the changes in the yield averaged for each of the three future decades (2020–2029, 2050–2059, and 2080–2089), with respect to the yield as obtained for the actual daily weather data collected for 10 consecutive years from 1997 to 2006 for the study sites.

### 5.5.2.3 Results and Discussion

#### 5.5.2.3.1 Future Climate Projections

The $CO_2$ concentrations and the changes in average maximum and minimum temperatures and rainfall at Khon Kaen, Roi Et, and Ubon Ratchathani for the future periods 2020–2029, 2050–2059, and 2080–2089 relative to the baseline period of 1980–1989 are provided in Table 5.8. The average $CO_2$ concentration during the 1980–1989 period was 330 ppm. Results indicated that there may be an increase in maximum and minimum temperatures, as well as in rainfall, at all three locations. At Ubon Ratchathani, the increase in maximum temperature could be as high as 3.51°C for 2080–2089, and the increase in rainfall could be 45.20% for 2050–2059 relative to the base period (1980–1989).

#### 5.5.2.3.2 Yield and Its Components for Observed and Future Periods

The simulated yields and yield components for the cultivar KDML105 for observed and future climate periods under the ECHAM4 A2 SRES scenario at Ubon Ratchathani are given in Table 5.9. Although the number of panicles per unit area remained almost the same (varying from 27.30 to 36.20 no. m$^{-2}$, i.e., a reduction of about 18% for 2050–2059 and an increase of about 9% for 2080–2089 compared to 1997–2006), the total number of grains per unit area was considerably reduced, with a maximum reduction of 35% for 2080–2089 compared to 1997–2006.

## TABLE 5.8
### Average Changes of $CO_2$, $T_{max}$, $T_{min}$, and Percent Changes of Annual Rainfall in Three Different Future Periods Relative to the Base Period of 1980–1989

|  |  | Khon Kaen |  |  | Roi Et |  |  | Ubon Ratchathani |  |  |
|---|---|---|---|---|---|---|---|---|---|---|
|  | $CO_2$ | Increase in |  |  | Increase in |  |  | Increase in |  |  |
| Time Period | Conc. (ppm) | $T_{max}$ (°C) | $T_{min}$ (°C) | Rainfall (%) | $T_{max}$ (°C) | $T_{min}$ (°C) | Rainfall (%) | $T_{max}$ (°C) | $T_{min}$ (°C) | Rainfall (%) |
| 2020–2029 | 437 | 0.32 | 2.14 | 5.90 | 0.35 | 2.21 | 19.2 | 1.47 | 0.93 | 10.1 |
| 2050–2059 | 555 | 3.30 | 3.15 | 2.60 | 1.59 | 3.20 | 40.5 | 1.72 | 2.14 | 45.2 |
| 2080–2089 | 735 | 3.25 | 5.19 | 5.20 | 3.20 | 5.10 | 20.1 | 3.51 | 3.06 | 2.98 |

*Source:* Babel, M.S. et al., *Climate Res.*, 46, 137, 2011b.

## TABLE 5.9
### Simulated Yield and Yield Components of KDML105 at Ubon Ratchathani

| Period | Yield (kg/ha) | Panicle (no./m²) | Grains (no./m²) | Total Biomass (kg/ha) | Anthesis Duration (Days) | Maturity Duration (Days) | Harvest Index |
|---|---|---|---|---|---|---|---|
| 1997–1906 | 2732 | 33.4 | 10613 | 6353 | 81 | 110 | 0.43 |
| 2020–2029 | 2427 | 31.7 | 8990 | 6742 | 87 | 113 | 0.36 |
| 2050–2059 | 2200 | 27.3 | 8149 | 6463 | 96 | 120 | 0.30 |
| 2080–2089 | 1855 | 36.2 | 6869 | 6625 | 85 | 107 | 0.28 |

*Source:* Babel, M.S. et al., *Climate Res.*, 46, 137, 2011b.

This reduction was caused by the rise in temperature, which decreased the grain-filling duration from 32 to 28 days. The duration between anthesis and maturity was reduced for future periods, which would affect spikelet sterility and, hence, reduce the final grain yield. The harvest index was also reduced for future periods (from 0.43 in 1997–2006 to 0.28 in 2080–2089), indicating that, although the total biomass yield remained almost the same (varying from 6353 to 6742 kg ha⁻¹), the grain yield was reduced significantly from 2732 kg ha⁻¹ in 1997–2006 to 1855 kg ha⁻¹ in 2080–2089, a reduction of almost 35%. Similar trends in yield and yield components were obtained at Khon Kaen and Roi Et.

*5.5.2.3.3 Effect of Predicted GCM Scenario on Rice Yield*

The simulated KDML105 rice yields for the observed weather (1997–2006) and predicted weather for the 2020s, 2050s, and 2080s are given in Table 5.10. A significant decline in the yield is projected for the future periods. By taking the average of the three locations to represent northeast Thailand, a decline in rice yield of 17.81%, 27.59%, and 24.34% in the 2020s, 2050s, and 2080s, respectively, was expected in the region.

### TABLE 5.10
### Simulated Rice Yield and Changes (%) for Three Future Periods

|  | 1997–2006 | 2020–2029 |  | 2050–2059 |  | 2080–2089 |  |
|---|---|---|---|---|---|---|---|
| Location | Yield (kg/ha) | Yield (kg/ha) | Change (%) | Yield (kg/ha) | Change (%) | Yield (kg/ha) | Change (%) |
| Ubon | 2732 | 2427 | −11.16 | 2200 | −19.47 | 1855 | −32.10 |
| Khon Kaen | 2807 | 2101 | −25.15 | 1883 | −32.91 | 1901 | −32.27 |
| Roi Et | 2128 | 1764 | −17.11 | 1481 | −32.11 | 1944 | −8.64 |
| Average | 2556 | 2097 | −17.81 | 1855 | −27.59 | 1900 | −24.34 |

*Source:* Babel, M.S. et al., *Climate Res.*, 46, 137, 2011b.

#### 5.5.2.3.4 Effect of Temperature and $CO_2$ Levels on Rice Yield

Simulations were conducted using the calibrated CERES-rice model to determine the effect of increase in $CO_2$ and temperature on rice yield. The other weather parameters, i.e., rainfall and solar radiation, were considered the same as for the base period (1997–2006) weather. For the $CO_2$ levels tested (330, 400, 500, 600, 700 ppm), CERES-rice predicted an increase in yield for both cultivars KDML105 and RD6 with increasing $CO_2$ concentration (Table 5.11). In contrast, at all $CO_2$ concentrations, the model predicted a decline in yield with increase in temperature. At the base $CO_2$ concentration (330 ppm), the model predicted a decline in yield of 33.89% for KDML105 with a 5°C increase in temperature. This loss in yield may be caused by heat-induced spikelet sterility or increased crop respiration loss during grain filling, which reduces the grain-filling capacity and thus reduces the grain yield. At ambient temperature, for change in $CO_2$ concentration from 330 to 700 ppm, the model predicted an increase in average yield of 16.91% for KDML105. The advantage of elevated $CO_2$ on rice yield is nullified by the rise in temperature effects. There were

### TABLE 5.11
### Mean Predicted Change (%) in the Potential Yield of KDML105 for Different Temperatures and $CO_2$ Scenarios

|  | Temperature Increment (°C) |  |  |  |  |  |  |
|---|---|---|---|---|---|---|---|
| $CO_2$ (ppm) | 0 | 1 | 2 | 3 | 4 | 5 | Average |
| 330 | 0.00 | −5.05 | −11.53 | −15.48 | −24.45 | −33.89 | −15.07 |
| 400 | 4.06 | −1.24 | −6.66 | −13.14 | −20.02 | −29.10 | −11.02 |
| 500 | 9.48 | 3.07 | −0.44 | −7.80 | −16.33 | −23.65 | −5.94 |
| 600 | 13.03 | 7.83 | 2.93 | −3.29 | −11.93 | −20.68 | −2.02 |
| 700 | 16.91 | 12.01 | 6.26 | 0.77 | −8.27 | −17.75 | 1.65 |
| Average | 8.70 | 3.32 | −1.89 | −7.79 | −16.20 | −25.01 | −6.48 |

*Source:* Babel, M.S. et al., *Climate Res.*, 46, 137, 2011b.
*Note:* Changes are averaged across all sites for the period 1997–2006.

declines in yield to varying degrees for every 1°C rise in temperature; in contrast, with an increase in $CO_2$ concentration, yields increased to varying degrees.

#### 5.5.2.4 Conclusions

The study investigated the effects of climate change on rice production in northeast Thailand using CERES-rice crop growth model. The simulated weather data downscaled using RCM (PRECIS) were in good agreement with the observed weather in terms of seasonal pattern, indicating that PRECIS provided acceptable weather data for future periods. The $CO_2$ concentration, temperature, and rainfall are found to increase in the future at the study area. The combined effect of these changes may adversely affect the future rice yield. The vulnerability of rainfed rice production to climate variability and changes may lead to further large yearly fluctuations in the yield in the study area.

## 5.6 SUMMARY

In recent years, there has been increasing scientific evidence that climate change is impacting water resources and related economic sectors worldwide. Global assessments of climate change on hydrology suggest that there has been a discernible and contrasting change in the pattern of runoff: the regions lying in the higher latitudes have been experiencing an increase, while parts of west Africa, southern Europe, and southern Latin America have had a decrease. Climate change is likely to reduce groundwater availability (recharge) in Africa and Latin America and the Caribbean, whereas there will be an increase in east Asia and the Pacific and Europe and central Asia. A general rise in the mean sea level is expected because of glacier melting, thermal expansion of water, and other related reasons. Water-related disasters are expected to be aggravated by climate change. Southeast Asia, Indian subcontinent, central and eastern Africa, and the upper parts of South America are particularly prone to flooding, while North and South America, central and southern Africa, the Middle East, southern Asia from Indochina to southern China, and central and western Australia face imminent drought risk. Higher surface water temperatures will promote algal blooms and increase microbial content, while more intense rainfall will lead to an increase in suspended solids (turbidity) in lakes and reservoirs due to soil erosion and contaminant transport (e.g., pesticides, heavy metals, and organics). Salinity intrusion in coastal aquifers will become even more pronounced in the face of climate change. Of all the water-related sectors, agriculture is likely to be the most affected because of variability in the frequency and magnitude of precipitation patterns, coupled with increased air temperatures. Hydropower production is likely to be largely unaffected globally, but individual countries and regions may face significant changes. It must be noted that while climate change affects the hydrology of a region, there are certain other nonclimatic drivers (e.g., land-use change and pollution), which influence climate indirectly and have the potential to exacerbate the impacts that climate change is likely to bring about.

With progress in technology and an improved level of understanding about climate science, projections of the future climate are now possible in much smaller resolutions. Despite this, there is still a certain degree of uncertainty associated with these projections. Nevertheless, the existing knowledge of future climatic conditions provides sufficient information to policy and decision makers to develop adaptation plans to address climate change impacts.

# REFERENCES

Alavian, V., Qaddumi, H.M., Dickson, E. et al. 2009. *Water and Climate Change: Understanding the Risks and Making Climate-Smart Investment Decisions.* Washington, DC: International Bank for Reconstruction and Development/ The World Bank.

Alcamo, J., Flörke, M., and Marker, M. 2007. Future long-term changes in global water resources driven by socio-economic and climatic change. *Hydrological Sciences Journal* 52: 247–275.

Allison, I., Bindoff, N.L., Bindschadler, R.A. et al. 2009. *The Copenhagen Diagnosis: Updating the World on the Latest Climate Science.* Sydney, New South Wales, Australia: The University of New South Wales Climate Change Research Centre (CCRC).

Anandhi, A., Srinivas, V.V., Nanjundiah, R.S., and Kumar, D.N. 2008. Downscaling precipitation to river basin in India for IPCC SRES scenarios using support vector machine. *International Journal of Climatology* 28: 401–420.

Arnell, N.W. 2004. Climate change and global water resources: SRES emissions and socio-economic scenarios. *Global Environmental Change* 14: 31–52.

Arnell, N.W. and Delaney, E.K. 2006. Adapting to climate change: Public water supply in England and Wales. *Climatic Change* 78: 227–255.

Arnell, N.W. and Gosling, S.N. 2013. The impacts of climate change on river flow regimes at the global scale. *Journal of Hydrology* 486: 351–364.

Babel, M.S., Agarwal, A., Swain, D.K., and Herath, S. 2011b. Evaluation of climate change impacts and adaptation measures for rice cultivation in northeast Thailand. *Climate Research* 46: 137–146.

Babel, M.S., Bhusal, S.P., Wahid, S.M., and Agarwal, A. 2013. Climate change and water resources in the Bagmati River Basin, Nepal. *Theoretical and Applied Climatology* 113: 585–600. doi:10.1007/s00704-013-0910-4.

Babel, M.S., Pandey, V.P., Rivas, A.A., and Wahid, S.M. 2011a. Indicator-based approach for assessing the vulnerability of freshwater resources in the Bagmati River basin, Nepal. *Environmental Management* 48(5): 1044–1059.

Bates, B.C., Kundzewicz, Z.W., Wu, S., and Palutikof, J.P. (eds.). 2008. Climate change and water. Technical paper of the Intergovernmental Panel on Climate Change. Geneva, Switzerland: IPCC Secretariat.

Beare, S. and Heaney, A. 2002. Climate change and water resources in the Murray Darling Basin, Australia; impacts and adaptation. Conference Paper 02.11. Canberra, Australian Capital Territory, Australia: Australian Bureau of Agricultural and Resource Economics. Retrieved from http://www.abarepublications.com/product.asp?prodid=12389 (accessed January 2014): 1–33.

Beyene, T., Lettenmaier, D.P., and Kabat, P. 2010. Hydrologic impacts of climate change on the Nile River Basin: Implications of the 2007 IPCC scenarios. *Climatic Change* 100(3–4): 433–461.

Boe, J., Terray, L., Habets, F., and Martin, E. 2007. Statistical and dynamical downscaling of the Seine basin climate for hydro-meteorological studies. *International Journal of Climatology* 27: 1643–1655.

Boyer, C., Chaumont, D., Chartier, I., and Roy, A.G. 2010. Impact of climate change on the hydrology of St. Lawrence tributaries. *Journal of Hydrology* 384: 65–83.

Brecht, H., Dasgupta, S., Laplante, B., Murray, S., and Wheeler, D. 2012. Sea-level rise and storm surges: High stakes for a small number of developing countries. *Journal of Environment and Development* 21(1): 120–138.

Christensen, J.H. and Christensen, O.B. 2007. A summary of the PRUDENCE model projections of changes in European climate by the end of this century. *Climate Change* 81: 7–30.

Cromwell, J.E., Smith, J.B., and Raucher, R.S. 2009. No doubt about climate change and its implications for water suppliers. In Smith, J., Howe, C., and Henderson, J. (Eds.) *Climate Change and Water: International Perspectives on Mitigation and Adaptation.* London, U.K.: IWA Publishing, pp. 3–13.

De Silva, C.S., Weatherhead, E.K., Knox, J.W., and Rodriguez-Diaz, J.A. 2007. Predicting the impacts of climate change—A case study of paddy irrigation water requirements in Sri Lanka. *Agricultural Water Management* 93: 19–29.

Delpha, I., Jung, A.-V., Baures, M., Clement, M., and Thomas, O. 2009. Impacts of climate change on surface water quality in relation to drinking water production. *Environment International* 35: 1225–1233.

EPRI. 2009. Review of general circulation models and downscaling techniques. *Climate Science Newsletter.* Palo Alto, CA: Electric Power Research Institute, October.

Eriksson, M., Jianchu, X., Shrestha, A.B., Vaidya, R.A., Nepal, S., and Sandström, K. 2009. *The changing Himalayas: impact of climate change on water resources and livelihoods in the greater Himalayas.* International centre for integrated mountain development (ICIMOD). Retrieved from http://www.worldwatercouncil.org/fileadmin/world_water_council/documents_old/Library/Publications_and_reports/Climate_Change/PersPap_01._The_Changing_Himalayas.pdf (accessed January 2014).

FAO. 2011. *Climate Change, Water and Food Security.* Rome, Italy: FAO, pp. 55–60.

Fischer, G., Tubiello, F.N., Velthuizen, H., and Wiberg, D.A. 2007. Climate change impacts on irrigation water requirements: Effects of mitigation, 1990–2080. *Technological Forecasting and Social Change* 74: 1083–1107.

Fowler, H.J., Blenkinsop, S., and Tebaldi, C. 2007. Linking climate change modeling to impact studies: Recent advances in downscaling techniques of hydrological modeling. *International Journal of Climatology* 27: 1547–1578.

Green, T.R., Taniguchi, M., Kooi, H. et al. 2011. Beneath the surface of global change: Impacts of climate change on groundwater. *Journal of Hydrology* 405: 532–560.

Hamududu, B. and Killingtveit, A. 2012. Assessing climate change impacts on global hydropower. *Energies* 5(2): 305–322.

Hegerl, G.C., Zwiers, F.W., Braconnot, P. et al. 2007. Understanding and attributing climate change. In *Climate Change 2007: The Physical Science Basis.* Contribution of Working Group I to the Fourth Assessment Report of the Intergovernmental Panel on Climate Change. Cambridge, U.K.: Cambridge University Press, pp. 663–745.

Hessami, M., Gachon, P., Ouarda, T.B.M.J., and St-Hilaire, A. 2008. Automated regression-based statistical downscaling tool. *Environmental Modelling and Software* 23: 813–834.

Hirabayashi, Y., Kanae, S., Emori, S., Oki, T., and Kimoto, M. 2008. Global projections of changing risks of floods and droughts in a changing climate. *Hydrological Sciences Journal* 53(4): 754–772.

Hirabayashi, Y., Mahendran, R., Koirala, S. et al. 2013. Global flood risk under climate change. *Nature Climate Change* 3: 816–821. doi:10.1038/nclimate1911.

Hiscock, K., Sparkes, R., and Hodgens, A. 2012. Evaluation of future climate change impacts on European groundwater resources. In Treidel, H., Martin-Bordes, J.L. and Gurdak, J.J. (Eds.) *Climate Change Effects on Groundwater Resources: A Global Synthesis of Findings and Recommendations.* London, U.K.: Taylor & Francis Group, pp. 351–366.

Hoogenboom, G., Jones, J.W., Porter, C.H., and Wilkens, P.W. (eds.). 2003. *Decision Support System for Agrotechnology Transfer,* Version 4.0. Vol. 1: Overview. Honolulu, HI: University of Hawaii.

Ines, A.V.M. and Hansen, J.W. 2006. Bias correction of daily GCM rainfall for crop simulation studies. *Agriculture and Forest Meteorology* 138: 44–53.

IPCC. 2007. Climate Change 2007: Impacts, Adaptation and Vulnerability. In: Parry, M.L., Canziani, O.F., Palutikof, J.P., van der Linden, P.J., and Hanson C.E. (Eds.) *Contribution of Working Group II to the Fourth Assessment Report of the Intergovernmental Panel on Climate Change (IPCC)*. Cambridge: Cambridge University Press, UK, 2007a.

IPCC. 2012. Managing the risks of extreme events and disasters to advance climate change adaptation. In Field, C.B., Barros, V., Stocker, T.F. et al. (Eds.) *A Special Report of Working Groups I and II of the Intergovernmental Panel on Climate Change*. Cambridge, U.K.: Cambridge University Press.

IRENA. 2012. *Renewable Energy Technologies, Cost Analysis Series: Hydropower*. Vol. 1: Power Sector, Issue 3/5. IRENA Secretariat. Retrieved from http://www.irena.org/DocumentDownloads/Publications/RE_Technologies_Cost_Analysis-HYDROPOWER.pdf (accessed Janury 2014).

Jiang, T., Chen, Y.D., Xu, C., Chen, X., and Singh, V.P. 2007. Comparison of hydrological impacts of climate change simulated by six hydrological models in the Dongjiang Basin, South China. *Journal of Hydrology* 336: 316–333.

Jones, R.N., Chiew, F.H.S., Boughton, W.C., and Zhang, L. 2006. Estimating the sensitivity of mean annual runoff to climate change using selected hydrological models. *Advances in Water Resources* 29: 1419–1429.

Klove, B., Ala-Aho, P., Bertrand, G. et al. in press. Climate change impacts on groundwater and dependent ecosystems. *Journal of Hydrology*. In Press.

Lal, M. 2011. Implications of climate change in sustained agricultural productivity in South Asia. *Regional Environmental Change* 11(S1): 79–94.

Lehner, B., Henrichs, T., Döll, P., and Alcamo, J. 2001. *EuroWasser—Model-Based Assessment of European Water Resources and Hydrology in the Face of Global Change*. Kassel World Water Series 5. Kassel, Germany: Center for Environmental Systems Research, University of Kassel.

Madani, K., Lund, J.R., and Jenkins, M.W. 2010. Estimated impacts of climate warming on California's high elevation hydropower. *Climatic Change* 102: 521–538.

Maraun, D., Wetterhall, F., Ireson, A.M. et al. 2010. Precipitation downscaling under climate change. Recent developments to bridge the gap between dynamical models and the end user. *Reviews of Geophysics* 48: RG3003.

McColl, C. and Aggett, G. 2007. Land-use forecasting and hydrologic model integration for improved land-use decision support. *Journal of Environmental Management* 84: 494–512.

Milly, P.C.D., Dunne, K.A., and Vecchia, A.V. 2005. Global patterns of trends in streamflow and water availability in a changing climate. *Nature* 438: 347–350.

Mirza, M.M.Q. 2010. Climate change, flooding in South Asia and implications. *Regional Environmental Change* 11(S1): 95–107.

Moriondo, M. and Bindi, M. 2006. Comparison of temperatures simulated by GCMs, RCMs and statistical downscaling: Potential application in studies of future crop development. *Climate Research* 30: 149–160.

Parry, M.L., Rosenzweig, C., Iglesias, A., Livermore, M., and Fischer, G. 2004. Effects of climate change on global food production under SRES emissions and socio-economic scenarios. *Global Environmental Change* 14(1): 53–67.

Ramirez, J. and Jarvis, A. 2010. Downscaling global circulation model outputs: The delta method. Decision and policy analysis working paper no. 1, May. Cali, Colombia: International Center for Tropical Agriculture, CIAT.

Ranjan, P., Kazama, S., and Sawamoto, M. 2006. Effects of climate change on coastal fresh groundwater resources. *Global Environmental Change* 16: 388–399.

Savé, R., Herralde, F.D., Aranda, X., Pla, E., Pascual, D., Funes, I., and Biel, C. 2012. Potential changes in irrigation requirements and phenology of maize, apple trees and alfalfa under global change conditions in Fluvià watershed during XXIst century: Results from a modeling approximation to watershed-level water balance. *Agricultural Water Management* 114: 78–87.

Schaefli, B., Hingray, B., and Musy, A. 2007. Climate change and hydropower production in Swiss Alps: Quantification of potential impacts and related modelling uncertainties. *Hydrology and Earth System Science* 11(3): 1191–1205.

Singh, U., Ritchie, J.T., and Godwin, D.C. 1993. *A Users' Guide to CERES-Rice Simulation Manual, V 2.10. IFDC-SM-4*. Muscle Shoals, AL: IFDC.

Stehlik, J. and Bardossy, A. 2002. Multivariate stochastic downscaling model for generating daily precipitation series based on atmospheric circulation. *Journal of Hydrology* 256: 120–141.

Thodsen, H. 2007. The influence of climate change on stream flow in Danish rivers. *Journal of Hydrology* 333: 226–238.

Tisseuil, C., Vrac, M., Lek, S., and Wade, A.J. 2010. Statistical downscaling of river flows. *Journal of Hydrology* 385: 279–291.

Treidel, H., Martin-Bordes, J.J., and Gurdak, J.J. (eds.). 2012. *Climate Change Effects on Groundwater Resources: A Global Synthesis of Findings and Recommendations*. London, U.K.: Taylor & Francis Group.

Vicuna, S., Leonardson, R., Dale, L., Hanemann, M., and Dracup, J. 2008. Climate change impacts on high elevation hydropower generation in California's Sierra Nevada: A case study in the Upper American River. *Climatic Change* 87: 123–137.

Wang, S., McGrath, R., Sweeney, C., and Nolan, P. 2006. The impact of the climate change on discharge of Suir River Catchment (Ireland) under different climate scenarios. *Natural Hazards Earth System Science* 6: 387–395.

WHO. 2009. *Vision 2030: The Resilience of Water Supply and Sanitation in the Face of Climate Change. Summary and Policy Implications*. Geneva, Switzerland: World Health Organization Press.

WHO. 2013. *Progress on Sanitation and Drinking Water*. Geneva, Switzerland: World Health Organization Press, Geneva, Switzerland, 2013.)

Wilby, R.L., Charles, S.P., Zorita, E., Timbal, B., Whetton, P., and Mearns, L.O. 2004. Guidelines for use of climate scenarios developed from statistical downscaling methods. Supporting material of the IPCC. Available from the DDC of IPCC TGCIA, 27.

Wilby, R.L. and Dawson, C.W. 2004. *Using SDSM Version 3.1—A Decision Support Tool for the Assessment of Regional Climate Change Impacts*. Nottingham, U.K. Retrieved from http://unfccc.int/resource/cd_roms/na1/v_and_a/Resoursce_materials/Climate/SDSM/SDSM.Manual.pdf (accessed January 2014).

Wilhite, D.A. and Glantz, M.H. 1985. Understanding the drought phenomenon: The role of definitions. *Water International* 10(3): 111–120.

World Bank. 2009. *Directions in Hydropower*. Washington, DC: World Bank.

World Bank. 2012. *Turn Down the Heat: Why a 4°C Warmer World Must Be Avoided*. Washington, DC: World Bank.

World Bank. 2013. *Turn Down the Heat: Climate Extremes, Regional Impacts, and the Case for Resilience*. Washington, DC: World Bank.

WWAP (World Water Assessment Program). 2012. *The United Nations World Water Development Report 4: Managing Water Under Uncertainty and Risk* (Vol. 1), *Knowledge Base* (Vol. 2), and *Facing the Challenges* (Vol. 3). Washington, DC: UNESCO Publishing.

Xu, J., Grumbine, R.E., Shreshta, A. et al. 2007. The melting Himalayas: Cascading effects of climate change on water, biodiversity, and livelihoods. *Conversation Biology* 23(3): 520–530.

Yamba, F.D., Walimwipi, H., Jain, S., Zhou, P., Cuamba, B., and Mzezewa, C. 2011. Climate change/variability implications on hydroelectricity generation in the Zambezi River Basin. *Mitigation and Adaptation Strategies to Global Climate Change* 16: 617–628.

# 6 Economics of Climate Change

*Sujata Manandhar, Vishnu Prasad Pandey, Futaba Kazama, and So Kazama*

## CONTENTS

6.1 Introduction .................................................................................................. 153
6.2 Development of Climate Change Economics .............................................. 154
6.3 Economics of Climate Change Impacts ....................................................... 155
    6.3.1 About Climate Change Impacts ....................................................... 155
    6.3.2 Methodological Approaches ............................................................. 159
6.4 Economics of Climate Change Adaptation .................................................. 166
    6.4.1 About Climate Change Adaptation ................................................... 166
    6.4.2 Methodological Approaches ............................................................. 170
6.5 Case Study .................................................................................................... 173
    6.5.1 Context .............................................................................................. 174
    6.5.2 Approach ........................................................................................... 174
    6.5.3 Results ............................................................................................... 176
    6.5.4 Lessons Learned ................................................................................ 178
6.6 Summary ...................................................................................................... 178
References ............................................................................................................ 179

## 6.1 INTRODUCTION

Climate change is one of the biggest environmental issues of the century. The climate plays such a major part in our planet's environmental system that even minor changes have impacts that are large and complex. The impacts of climate change are diverse and could be damaging to billions of people across the world. Climate change impacts particularly in water resources will have cascading effects on human health and many parts of the economy and society, as various sectors such as agriculture, energy and hydropower, navigation, health, and tourism directly depend on water—as does the environment (UN 2009). It necessitates strong actions to reduce the risk of very damaging and potentially irreversible impacts of climate change in water resources, societies, and economies.

    Neither science in general nor economics in particular can resolve the fundamentally moral issues posed by climate change (Toman 2006). Analyses limited to physical impacts of climate change on water resources may miss critical economic feedbacks, leading to erroneous conclusions and interpretations (Jeuland 2010).

However, including economic understanding can help inform the impacts and risks of climate change in water resources in a quantitative way and understand the potential for adaptation to anticipated or realized climate change impacts. Kolstad and Toman (2001) also opine that impact/damage estimates are fundamental to understanding the climate problem. Once the aggregated cost of climate change is expressed in monetary (economic) terms, it is possible to compare this cost with the anticipated cost of adapting to climate change. The adaptation route to be chosen should be the one that yields the highest net benefit, having taken account of the risks and uncertainties surrounding climate change (Stern 2006). Although not very prominent in policy advice, monetary impact estimates have also been good predictors of real policy (Tol 1999). It provides powerful means not only for galvanizing the discussion about climate change policy but also for investment decision making. Furthermore, estimates of the costs of adapting environmental and infrastructure goods and services to climate change can provide insight into the very real costs of inaction or, conversely, the benefits of maintaining and protecting societal goods and services through effective policies that avoid the most severe climate impacts (CIER 1992). For this reason, studies on economic implications of climate change in water resources are intensifying (Stern 2006). It helps develop better understanding of important economic uncertainties associated with climate change and evaluate, design, and manage water resources systems in an efficient manner.

This chapter aims to highlight the importance of economics of climate change and quantify socioeconomic impacts and adaptation costs/benefits referring to climate change effects in water resources. It also reviews existing methods/frameworks and demonstrates the suitability of the methods to quantify, analyze, and interpret socioeconomic impacts and adaptation costs/benefits of climate change with a specific case study.

## 6.2  DEVELOPMENT OF CLIMATE CHANGE ECONOMICS

Climate change is fraught with basic uncertainties; nonetheless, economics can contribute useful information to the debate on how to address the issue. Economics has figured prominently in the assessment of impacts on human society of climate change and in the assessment of the pros and cons of various response strategies, with respect to both adaptation and mitigation. Therefore, economics is believed to play an important role in assessing climate change impacts and the effects of various individual and policy response strategies (Toman 2006).

Earlier climate change studies by economists evaluated the transition between two climate equilibriums or, at the most, two climate paths that smoothly changed over time from today's climate to one characterized by a doubling of atmospheric concentrations of $CO_2$ and other warming gases since the pre-Industrial Revolution (Gaskins and Weyant 1993; Manne and Richels 1991; Mendelsohn et al. 1994). Cline (1992) was the first to perform the most detailed economic analysis of the potential impacts of climate change extending the analysis beyond a doubling, with $CO_2$ emissions that were derived from simple models of economic growth. The study also presented a sensitivity analysis of the benefits and costs of avoiding climate change with respect to the discount rate that converts future damages and costs into present values. Furthermore, Nordhaus (1994) developed a first dynamic economic growth model (called dynamic integrated model of climate and economy [DICE]) useful for estimating the costs and benefits of different paths for slowing climate change and for analyzing the impact of control strategies over time. The study also acknowledged that its equilibrium climate submodel is not applicable to a greater than doubling of $CO_2$ equivalent gases (Hall and Behl 2006). Nevertheless, the model has been the basis for economic analysis beyond a doubling and is extended to analyze abrupt climate change. The science of climate change advanced considerably along with the development of economic models, and newer studies increasingly emphasized adaptation, variability, extreme events, other (nonclimate change) stress factors, and the need for integrated assessment of damages (Fankhauser 1996). Considering the lack of generally accepted framework for characterizing the regional economic impacts of, and responses to, climate change, Abler et al. (2000) took a step in developing such a framework. It further considered two modeling (static and dynamic) frameworks for responses to climate change. Tol et al. (2004) noted that despite huge improvements in studying economic impacts of climate change, quantitative assessments of the uncertainty were still rare. Later, Stern (2006) published a review on the economics of climate change, which emphasized on using a consistent approach towards uncertainty. Moreover, Anda et al. (2009) expanded on economics of climate change under uncertainty and proposed an application of the real option analysis in order to formulate rules for the selection of a climate policy and estimate the economic value of the future flexibility created by interim climate policy. Thereafter, studies on climate change economics have been continuously expanding in various spatial (country, city, basin/catchment) scales and sectors including the water sector (Hallegatte et al. 2011; Hughes et al. 2010; Jiang and Grafton 2012; Mideksa 2010; Zhou et al. 2012).

## 6.3 ECONOMICS OF CLIMATE CHANGE IMPACTS

### 6.3.1 About Climate Change Impacts

Climate change affects people and nature in countless ways, such as through social disruption, economic decline, and displacement of populations, among others. Besides this, health impacts associated with such socioeconomic dislocation and population displacement are also substantial (IPCC 2007). The risks of climate change are sector specific and depend not only on changes in climate system but also on the physical and socioeconomic implications of a changing climate. Chapter 5 sheds light on the climate change impacts and risks in the water sector. This section provides a brief overview of the connection between climate change impacts and the water sector from the perspective of quantifying in economic terms.

The water sector is the one that often mediates the climate change impacts. Some examples are given in Box 6.1. People will feel the impact of climate change most strongly through changes in the distribution of water around the world and its seasonal and annual variability. Climate change will alter patterns of water availability by intensifying the water cycle. As the water cycle intensifies, severe floods, droughts, and storms occur more often. Billions of people will lose or gain water.

## BOX 6.1  SELECTED EXAMPLES OF SOCIOECONOMIC IMPACTS OF CLIMATE CHANGE

1. *Drought impacts in Ethiopia*: Disaster losses, mostly weather and water related, have grown much more rapidly than population or economic growth, suggesting a negative impact of climate change. Given that the use of livestock for plowing dominates Ethiopia's agriculture, droughts frequently kill cattle, effectively destroying productive assets in an important sector of the economy. During the 1984–1985 and 2002–2003 droughts, Ethiopia's GDP declined by around 10% and over 3%, respectively (World Bank 2008). A further study by Mideksa (2010) shows that climate change (mainly drought) will decrease agricultural production and output in the sectors linked to agriculture, which is likely to reduce Ethiopia's GDP by about 10% from its benchmark level. It also boosts the degree of income inequality, which retards economic growth and fuels poverty.
2. *Flood impacts in Japan*: The Japanese government concerns the increase of heavy downpour in future and wants to count the damage costs. Kazama et al. (2009) evaluated the cost of flood damage using numerical simulations based on digital map data and the flood control economy investigation manual submitted by the Ministry of Land, Infrastructure, Transportation, and Tourism in Japan. The economic predictions, which estimate flood damage caused by extreme rainfall for the return periods of 5, 10, 30, 50, and 100 years, are as follows: (1) The cost of flood damage increases nearly linearly with increases in extreme precipitation; (2) assuming that flood protection is completed for a 50-year return period of extreme rainfall, the benefit of flood protection for a 100-year return period of rainfall is estimated to be US$210 billion; (3) the average annual expected damage cost for flooding is predicted to be approximately US$10 billion per year, based on the probability of precipitation for a return period of 100 years and assuming that flood control infrastructures will be completed within the 50-year return period and will be able to protect from flooding with a 50-year return period; and (4) urban and rural areas are predicted to suffer high and low costs of damage, respectively. These findings will help to derive measures to enhance flood protection resulting from climate change.
3. *Impacts of sea-level rise in the coastal zones*: Socioeconomic impacts in the coastal zone are generally a product of the physical changes in climate drivers (such as $CO_2$ concentration, sea surface temperature, sea level, storm intensity, storm frequency, storm track, wave climate, and runoff) (IPCC 2007). The impacts are influenced

by the magnitude and frequency of existing processes and extreme events, for example, the densely populated coasts of East, South, and Southeast Asia are already exposed to frequent cyclones. In Thailand, loss of land due to a sea-level rise of 50 and 100 cm could decrease national GDP by 0.36% and 0.69% (US$300–$600 million) per year, respectively; due to location and other factors, the manufacturing sector in Bangkok could suffer the greatest damage, amounting to about 61% and 38% of the total damage, respectively (Ohno 2000). The annual cost of protecting Singapore's coast is estimated to be between US$0.3 and $5.7 million by 2050 and between US$0.9 and $16.8 million by 2100 (Ng and Mendelsohn 2005).

4. *Climate change impacts on water resources in the United States*: There is tremendous variation in water resources and water supply systems not only across the United States but also within regions and particular watersheds. Hurd et al. (1999, 2004) have approached this issue from a region-specific perspective using hydroeconomic models of four major water resources regions (i.e., Colorado River, Missouri River, Delaware River, and the Apalachicola–Flint–Chattahoochee Rivers). They have developed national-level estimates of economic damages for 15 scenarios of incremental climate change based on the regional model results and a model to extrapolate to unmodeled regions. They have estimated total annual damages to consumptive and nonconsumptive water users by as much as $43.1 billion (1994$) under an incremental level of climate change where temperatures rose by 5°C and 0% change in precipitation. Later, Backus et al. (2010) estimate there is a 50–50 chance that cumulative direct and indirect macroeconomic losses in GDP through 2050 will exceed nearly $1.1 trillion (2008$), not including flood risks, that is, approximately 0.2% of the cumulative GDP projected between 2010 and 2050. They estimate a 50–50 chance of nondiscounted annual losses of $60 billion (2008$) by 2050.

5. *Coastal storm impacts in Dade County, Florida*: Florida's coastline can expect a dramatic increase in major storm surge events and associated property damage along with sea-level rise. Damage costs associated with storm surge events (assuming no increase in storm intensity) will increase from 10% to 40%, depending on the extent of sea-level rise and other factors. Property losses in Dade County in Florida alone will exceed $12 billion if sea-level rises by 2 ft. (exclusive of future increases in coastal population or property values) (BPC 2009). Besides the market impacts, Heinz Center (2000) showed that family roles and responsibilities after a disastrous coastal storm undergo profound changes associated with household and employment disruption, economic hardship, poor living conditions, and the disruption

> of public services such as education and preventive health care. Damaged homes and utilities, extreme temperatures, contaminated food, polluted water, debris- and mud-borne bacteria, and mildew and mold often bring health problems. In some cases, relationships after a disastrous climate-related event can become so stressful within the family that family desertion and divorce may increase. Accounting for the full range of costs for nonmarket impacts is difficult; however, recognizing them in disaster cost accounting is very important.

That means impacts can be negative or positive, depending on the time, region, and sector one is looking at. Warming (increased temperature) may result in some negative impacts by inducing sudden shifts in regional weather patterns like the monsoons or the El Niño, which will have severe consequences for water availability and flooding in tropical regions, eventually threatening the livelihoods and causing economic damage in billions, and by melting glaciers, which will initially increase flood risk and then strongly reduce water supplies, eventually threatening one-sixth of the world's population, predominantly in the Indian subcontinent, parts of China, the Andes in South America, etc. (Stern 2006). Warming may also lead to some positive effects. For example, less sea ice will improve the accessibility of Arctic harbors, will reduce the costs of exploitation of oil and minerals in the Arctic, and may even open up new transport routes between Europe and East Asia (Wilson et al. 2004). The impacts of sea-level rise are overwhelmingly adverse, but benefits have also been identified, including opportunities for increased use of fishing vessels and coastal shipping facilities, expansion of areas suitable for aquaculture, and reduced hull strengthening and icebreaking costs (IPCC 2007). Besides climate change impacts on water availability and hydrological risks, there are also consequences on water quality. For example, floods and droughts will also modify water quality by direct effects of dilution or concentration of dissolved substances (Delpla et al. 2009), and changes in precipitation, wind speed, incoming solar radiation, and air temperature directly influence river water quality by altering changes in stream flow and river water temperatures (Rehana and Mujumdar 2012). Climate change is likely to impact more severely on the poorer people of the world, because they are more exposed to the weather, closer to the biophysical and experience limits of climate, and their adaptive capacity is lower (Tol et al. 2004). But it is equally evident that developed countries are not insulated from water-related disastrous consequences of climate change.

Climate change can produce economic harm in different ways. It can originate either directly from a change in climate itself (through changes in temperature, precipitations, or storms and other extreme events) or indirectly by inducing changes in ecosystems or social systems. For example, higher flooding from more severe storms might directly damage property, disrupt commerce, and take lives; in other cases, warmer temperatures have been associated to rises in sea level that erode ocean-front property and increase the cost of maintaining coastal homes and highways.

# Economics of Climate Change

Whatever be the ways of harming, the social and economic consequences of climate change impacts can be classified broadly into two types: "market/tangible" impacts and "nonmarket/intangible" impacts.

- *Market/tangible impacts*: They directly affect the economy, where prices exist and a valuation can be made relatively easily (e.g., asset losses due to sea-level rise, flooding). They allow for an uncontroversial assessment of monetary values, so that assessment problems are mainly in the technical domain, not in the ethical domain (IPCC 1996; Stern 2006; Tol 2009).
- *Nonmarket/intangible impacts:* They affect humans and the environment in a broad way, where market prices tend not to exist (e.g., human health, biodiversity, and ecosystem impacts) (IPCC 1996; Stern 2006; Tol 2009). Intangibles consist of nonmonetary inputs to the human welfare; implicitly, economic valuation methods translate the intangibles' utility into their monetary utility equivalent and further into money (Tol 1994). However, monetary estimates for nonmarket goods are still debatable.

There can be different actions such as mitigation and adaptation to "manage" the climate change impacts. In this chapter, we will elaborate only on adaptation.

### 6.3.2 Methodological Approaches

The economic impact of climate change is usually measured as the extent to which the climate of a given period affects social welfare in that period (Fankhauser and Tol 2005). Efforts to quantify the economic impacts of climate-related changes in water resources are hampered by a lack of data and by the fact that the estimates are highly sensitive to different estimation methods and to different assumptions regarding how changes in water availability will be allocated across various types of water uses (Changnon 2005; Schlenker et al. 2005; Young 2005). However, increasing attention to the quantification of economic impacts of climate change by scientists and economists has led to the development of a range of methods or approaches. This section sheds light on those methods or approaches.

*Integrated assessment models (IAMs)*: IAMs simulate the process of human-induced climate change, from emissions of GHGs to the socioeconomic impacts of climate change. It is shown in Figure 6.1 as a simple unidirectional chain, a simplification as in the real climate–human system; there will be feedbacks between many links in the chain. The focus of initial IAMs is on economic sectors for which prices exist but later improvements are made to capture "nonmarket" impacts as well. Some of the IAMs are detailed as follows:

- *The Mendelsohn model/global impact model (GIM)*: Assessment models have taken two approaches to calculating climate impacts: "top-down" and "bottom-up." "Top-down" models rely on aggregate damage functions, the simplest of which calculate global damages as a function of only global-mean temperature change, while "bottom-up" IAMs sought to capture

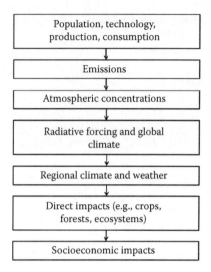

**FIGURE 6.1** Modeling climate change from emissions to impacts. (From Hope, C., Integrated assessment models, in: Helm, D., ed., *Climate-Change Policy*, Oxford University Press, Oxford, U.K., 2005, pp. 77–98.)

the individual direct effects of climate change across the landscape. The Mendelsohn model or GIM uses the strengths of both the "top-down" and "bottom-up" approaches. Following the spirit of the "top-down" approach, this model on the basis of economic theory attempts to calculate the economic welfare associated with climate change. On the other hand, keeping to the spirit of "bottom-up" models, it adds important spatial detail to the model in order to capture climate forecasts more accurately and differentiate impacts by country. In a nutshell, GIM combines (1) future world scenarios, (2) geographically detailed climate simulations from a general circulation model (GCM), (3) sectoral data for different countries, and (4) climate response functions by the market sector (Mendelsohn et al. 2000). It estimates impacts only for five "market" sectors: agriculture, forestry, energy, water, and coastal zones. It is based on scientific evidence up to the mid-to-late 1990s and considers adaptation to climate change but omits other potentially important factors such as social and political stability and cross sectoral impacts (Stern 2006).

- *The "Tol" model*: Most impact studies before the development of the "Tol" model took a static approach (models just investigated the effect of a single, changed climate [usually, 2× $CO_2$, i.e., climate if the atmospheric concentration of carbon dioxide would be doubled] on the current system). This is a useful starting point, but climate will change gradually and is not likely to stop at 2× $CO_2$. Moreover, populations and economics will grow, and technologies and institutions will evolve. Considering these points, Tol (2002) developed a model of climate change impacts that takes account of the dynamics of climate change and the systems affected by it. Costs are

Economics of Climate Change 161

weighted either by output or equity-weighted output. It estimates impacts for a wider range of market and nonmarket sectors, agriculture, forestry, water, energy, coastal zones, and ecosystems, as well as mortality from vector-borne diseases, heat stress, and cold stress. It is based on scientific evidence up to the mid-to-late 1990s and considers adaptation to climate change but omits other potentially important factors such as social and political stability and cross sectoral impacts (Stern 2006).

- *Policy Analysis of the Greenhouse Effect 2002 (PAGE2002) IAM*: The PAGE2002 IAM functions based on "Monte Carlo" simulation approach. The model generates a probability distribution of results rather than just a single-point estimate. Specifically it yields a probability distribution of future income under climate change, where climate-driven damage and the cost of adapting to climate change are subtracted from a baseline gross domestic product (GDP) growth projection. It can take account of the range of risks by allowing outcomes to vary probabilistically across many model runs, with the probabilities calibrated to the latest scientific quantitative evidence on particular risks. PAGE2002 is flexible enough to include market impacts and nonmarket impacts, as well as the possibility of catastrophic climate impacts. Unfortunately, PAGE2002 do share many of the limitations of other formal models. It must rely on sparse or nonexistent data and faces difficulties in valuing direct impacts on health and the environment (Stern 2006).

- *A hydroeconomic model (IIAWM)*: Jiang and Grafton (2012) introduced a hydroeconomic model (integrated irrigated agriculture water model [IIAWM]) to examine the effects of severe climate change and to investigate whether water trading can significantly offset these effects. Hydroeconomic models represent spatially distributed water resources systems, infrastructure, management options, and economic values in an integrated manner. In these tools, water allocations and management are either driven by the economic value of water or economically evaluated to provide policy insights and reveal opportunities for better management. A central concept is that water demands are not fixed requirements but rather functions where quantities of water use at different times have varying total and marginal economic values (Harou et al. 2009). IIAWM simulates climate change in two steps: (1) calibrates using historical climate to provide a baseline for the other simulations, simulates the climate change scenarios, and compares their differences in percentage terms relative to the baseline results and (2) repeats the process by allowing for water trading between regions to calculate the effects of water trade. The model can be used for assessment of economic impacts at a regional scale.

- *A macroeconomic general equilibrium model (GRACE)*: The Global Responses to Anthropogenic Changes in the Environment (GRACE) model integrates impacts of climate change on different activities of the economies. It is a standard multiregional computable general equilibrium model based on a set of constant elasticity of substitution (CES) production and preference trees, using the GTAP version 7 social accounting matrices. The version of

GRACE applied by Aaheim et al. (2012) was developed to address impacts of climate change and economic aspects of adaptation. Economic effects of climate change are integrated by their impacts on deliveries in an extended input–output matrix based on the national accounts. The basic idea is that economic behavior described in general equilibrium models captures important parts of the responses to climate change among economic agents. However, adaptation is a process with many barriers, which are partly related to the need for time to adapt. General equilibrium models suppose in principle that adaptation happens instantly when resource constraints shift, technologies change, or new information arrives. To study adaptation to climate change, one therefore needs to consider how possible macroeconomic consequences of these barriers can be represented. Finally, it is important to note that economic estimates of climate change impacts depend heavily on the underlying climate scenarios and downscaling of the output from global circulation models.

*Engineering rules of thumb*: Larsen et al. (2008) coupled future climate change projections with engineering rules of thumb to estimate how thawing permafrost, increased flooding, and increased coastal erosion affect annualized replacement costs for Alaska public infrastructure. They introduced the ISER Comprehensive Infrastructure Climate Life-Cycle Estimator (ICICLE) model that follows several steps: (1) acquiring climate projections, (2) creating a database of public infrastructure, and (3) estimating the replacement costs and life spans for existing infrastructure, with and without the effects of climate change (assuming planners will adapt structures strategically). The basis for the model is the calculation of the net present value of infrastructure replacement over time, under different conditions. This model not just estimates future costs for public infrastructures under climate change but also shows that the potential risks for man-made systems are considerable. The given model and its estimation is just a preliminary work; developer highlights the need to make further improvements in both modeling techniques and cost estimates in the future.

*Hydroeconomic modeling framework*: There are a very large number of potential linkages and feedbacks between climate change and water resources systems (Figure 6.2a), and tractability precludes inclusion of all of them at one time in a given analysis. Thus, Jeuland (2010) developed a hydroeconomic modeling framework for integrating climate change into the problem of water resources infrastructure decision making. The framework allows for consideration of a reduced set of physical and economic impacts of climate change (see Figure 6.2b). It first includes explicit functional linkages between climatic factors (such as temperature and precipitation), specified based on GCM or other future projections, and many hydrological model components that these factors influence. It further allows analysis of economic uncertainties and relative changes in the real value, and productivity of the goods and services generates by hydrological systems. It is applicable to explore the simultaneous influence of a number of physical changes in runoff, net evaporation from reservoirs, and crop water requirements in irrigation, as well as economic changes in the value of water and energy and the value of carbon offsets. However, for the developer, still it is an empirical question where different linkages will be most important and highlights the research needs to better understand them.

# Economics of Climate Change

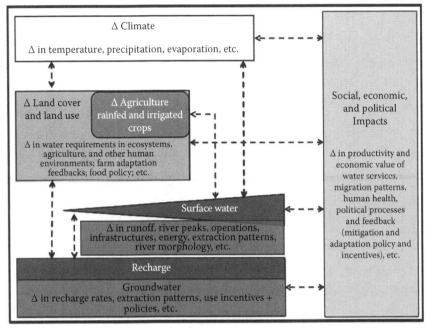

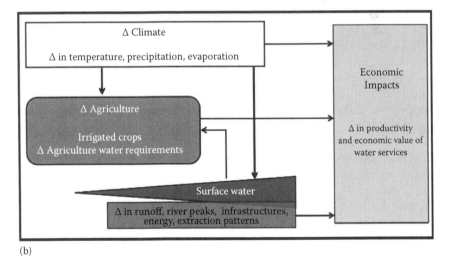

**FIGURE 6.2** A conceptual diagram showing (a) key linkages and feedbacks between climate change, water resources, and human systems and (b) the relationships specifically considered in the hydroeconomic modeling framework developed by Jeuland (2010).

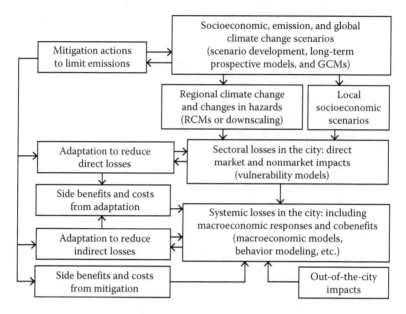

**FIGURE 6.3** The different components necessary to assess climate change impacts. (From Hallegatte, S. et al., *Clim. Change*, 104, 51, 2011.)

*Framework to assess local economic impacts of climate change*: Hallegatte et al. (2011) presented a conceptual and methodological framework (Figure 6.3), which is interdisciplinary, working across natural science, technology, engineering, and economics to assess local economic impacts of climate change. The proposed methodological framework focuses on model-based analysis of future scenarios, including a framing of uncertainty for these projections, as a valuable input into the decision-making process. Furthermore, it includes assessment of local impacts in three cases: (1) a case with no adaptation; (2) a case with an imperfect adaptation, that is, a guess based on observation of current adaptation to climate natural variability; and (3) a case with perfect adaptation, assuming that responses are optimally planned and implemented by perfectly rational agents. A number of challenges are unique to climate change impact assessment and others are unique to the problem of working at local scales. Thus, Hallegatte et al. (2011) identified the need for additional research, to develop more integrated and systemic approaches to address climate change as a part of the urban development challenge as well as to assess the economic impacts of climate change and response policy at a local scale.

In this fashion, economists have developed a range of techniques for calculating market impacts and nonmarket impacts, but the resulting estimates of nonmarket impacts are still problematic in terms of concept, ethical framework, and practicalities. Many would argue that it is better to present costs in human lives and environmental quality side-by-side with income and consumption, rather than trying to summarize them in monetary terms. Nevertheless, modelers have tried to do their best to assess the full costs of climate change and the costs of avoiding it on a comparable basis and thus make their best efforts to include "nonmarket" impacts. Box 6.2 includes some methodological issues on valuation of human life.

> **BOX 6.2   VALUATION OF HUMAN LIFE: SOME METHODOLOGICAL ISSUES**
>
> To encompass nonmarket health impacts, many studies use the value of a statistical life (VSL). A VSL can be estimated from evidence on market choices that involve implicit trade-offs between risk and money, such as smoking a cigarette or driving a car (Viscusi and Aldy 2003). They can also be estimated based on stated preferences (e.g., from consumer surveys of the WTP to avoid risks to human life). Meta-analyses of studies suggest that estimates of VSL may depend on the age, income, gender, education, health, etc., of the respondents and on the risk change context, as well as the estimation method used (Viscusi and Aldy 2003). Viscusi and Aldy (2003) note that even though values depend on the context, for example, the type of risk and the probability of occurrence of the considered event, most estimates lie between US$1 million and $10 million. The VSL meets serious ethical challenges, including the difference in VSL between rich and poor individuals and the possible difference between individual choices (used to assess VSL) and collective choices (for which VSLs are used). The World Bank and the World Health Organization sometimes choose to use physical indicators of risks to human life such as the "disability-adjusted life years" (DALYs), to quantify health effects (see Murray and Lopez 1996). These indicators can be somewhat less controversial and a complement to more formal economic impact assessment, which otherwise requires valuation of all direct impacts. On the other hand, they do not allow a direct comparison of costs and benefits of a policy in common unit and can hence, for example, not indicate if a policy measure ought to be implemented or not. The DALY concept uses life years lost due to premature death and fraction of years of healthy life lost as a result of illness or disability to measure the burden of disease. Contrary to the VSL, age is taken into account in the DALY, through weights that are incorporated to discount year of life lost at different ages.
>
> *Source*: Hallegatte et al., *Clim. Change*, 104, 51, 2011.

*Approach based on willingness to pay (WTP) and willingness to accept or avoid change (WTA)*: Changes in economic welfare due to climate change are measurable in terms of reductions in economic benefits or increases in opportunity costs (Hurd and Rouhi-Rad 2013). Adding up the estimated values across all affected individuals could approximate the total economic value or welfare change. In practice, economists have developed various approaches for estimating the WTP and WTA, respectively (refer to Young 2005). Economic benefits and costs based on estimates of individual WTP or WTA and aggregated across affected individuals are used to develop economic demand and supply schedules. Demand and supply schedules describe the marginal benefits and marginal costs, respectively, for varying quantities of the particular good or service. Subtracting the schedule of marginal costs from the schedule of marginal benefits (i.e., the water supply curve is subtracted from the water demand curve), the result is a schedule for the marginal net benefits.

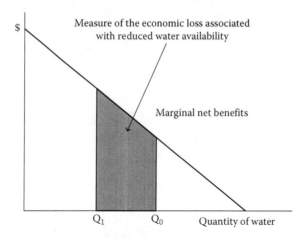

**FIGURE 6.4** Valuing changes in water supply: conceptual basis. (From Hurd, B. and Rouhi-Rad, M., *Clim. Change*, 117, 575, 2013.)

As shown in Figure 6.4 for a given water use or collection of water users, the line labeled marginal net benefits corresponds to the marginal value of water. When water availability falls, for example, from $Q_0$ to $Q_1$, the shaded area is the loss in economic welfare arising from the reduction in water availability. This change in economic value is the damage or loss in economic welfare or equivalently the loss in producer and consumer surplus. Summing the value changes and water supply costs for each water-using industry or sector, in each region, gives an estimate of the total aggregate WTP to avoid the loss of water.

## 6.4 ECONOMICS OF CLIMATE CHANGE ADAPTATION

### 6.4.1 About Climate Change Adaptation

Adaptation is the only means to reduce the now-unavoidable costs of climate change over the next few decades and additionally offers an opportunity to adjust economic activity in vulnerable sectors and support sustainable development. A broad definition of adaptation, following the IPCC, is any adjustment in natural or human systems in response to actual or expected climatic stimuli or their effects, which moderates harm or exploits beneficial opportunities. Significant impacts of climate change are already observed around the world. It is no longer possible to prevent the climate change that will take place over the next two to three decades, but it is still possible to protect our societies and economies from its impacts to some extent. Therefore, adaptation to climate change—that is, taking steps to build resilience and minimize costs—is essential (Adger et al. 2005). The objective of adaptation is to reduce vulnerability to climatic change and variability, thereby reducing their negative impacts (Figure 6.5). Unlike mitigation, adaptation will reduce the damage costs of climate change and in most cases provide local benefits, realized without long lead times. Stern (2006) argues persuasively that the risks of inaction are quite high (and largely uncertain or unknown), when compared to the costs of action. Without adaptation policies and initiatives in place,

Economics of Climate Change

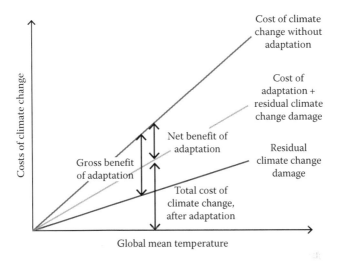

**FIGURE 6.5** The role of adaptation in reducing climate change damages. (From Stern, N., *The Economics of Climate Change*, HM Treasury, London, U.K., 2006.)

the impacts of climate change are likely to be significant and pervasive (Bettencourt et al. 2006). Selected examples of climate change adaptation in the water sector are provided in Box 6.3; details are available in Chapter 10.

Adaptation can operate at two broad levels:

1. *Building adaptive capacity*—creating the information and conditions (regulatory, institutional, and managerial) that are needed to support adaptation
2. *Delivering adaptation actions*—taking steps that will help to reduce vulnerability to climate risks or to exploit opportunities

Adaptations can be short, medium, and long term. Costs vary substantially among different types of adaptations; and the adaptations need to be staged and integrated with the capital replacement and rehabilitation cycles. Estimates of the costs of adaptation are required, at the national level for designing effective adaptation strategies and at the international level for identifying the financial flows needed for an effective international response to climate change. While the previous statement is true for all sectors affected by climate change, it is of particular relevance in the case of water resources, on which humans, livelihoods, and species depend critically (Ortega 2011). From an economic perspective, an efficient adaptation strategy should be based on the most cost-effective combination of measures to achieve a determined goal. Changing to meet altered conditions and new ways of managing water are autonomous adaptations, which are not deliberately designed to adjust with climate change. On the other hand, planned adaptations (including a wide range of actions addressing securing water supply, river flooding, etc.) take climate change specifically into account (IPCC 2007). In doing so, planners need to do economic analysis (compare costs and benefits) of various adaptation options to choose the best/profitable options. In some cases,

## BOX 6.3 SELECTED EXAMPLES OF ADAPTATION TO CLIMATE CHANGE

1. *Water supply in the Berg River Basin in South Africa* (Callaway et al. 2006): The Berg River is a major source of water for Cape Town and the surrounding agricultural land in South Africa. In the last 30 years, water consumption in Cape Town has increased threefold and is expected to continue to grow in the future, as a result of population growth (migration of households to the city from rural communities) and economic development. At the same time, climate models predicted likely decrease in the average annual runoff in the catchment by as much as 25% during the period 2010–2040 due to climate change. To cope with the increasing development pressures and probable future water problems due to climate change, two climate change adaptation strategies are identified and their net benefits are compared. Two strategies are Strategy A, constructing a storage reservoir to cope with development pressures and then adding capacity to cope with climate change, and Strategy B, implementing water markets to cope with development pressures and then building a dam to cope with climate change. The following table shows present value estimates for costs and benefits of adjustments for increasing development pressures and climate change in the 2080s.

| Estimated Benefit or Cost Measure | Strategy A | Strategy B |
|---|---|---|
| Development action (no climate change) | Construct dam, no water markets | Water markets, no dam |
| Net benefits of development action | 15 billion | 17 billion |
| Additional adaptation action (development + climate change) | Increase dam capacity, no water markets | Construct dam + water markets |
| Net benefits of adaptation (reduction in damages from adaptation minus costs of adapting) | 0.2 billion | 7 billion |
| Cost of not planning for climate change that does occur | −0.2 billion | −7 billion |
| Cost of planning for climate change that does not occur | −0.2 billion | −1 billion |

*Note:* All monetary estimates are expressed in present values for constant R and for the year 2000, discounting over 30 years at a real discount rate of 6%.

Both the dam and water market options individually have similarly large projected net present values, but adding the possibility of adaptation to climate change shows the benefits of adopting both simultaneously. Increasing the water storage capacity of the Berg Dam could have a significant benefit for welfare. The effect is particularly strong if efficient water markets are introduced (net benefit of 7 billion, discounted over 30 years). Under this flexible and economically efficient approach, the costs of not adapting to climate change that does occur are much greater than the costs of adapting to climate change that does not occur (−7 billion vs. −1 billion in the case of efficient markets).

2. *Adapting to tropical cyclones and storm surges in Bangladesh* (World Bank 2010a): Bangladesh has been identified as the most vulnerable country in the world to tropical cyclones, with a severe cyclone striking Bangladesh every three years on average and causing damages and losses in billions. Faced with a significant chronic risk from storm surges induced by cyclones, the government of Bangladesh has already put in place some adaptation measures (creating polders, foreshore afforestation, early warning and evacuation system, etc.), and others are in planning process. Primarily focusing on the upgrading and expanding of existing adaptation measures to prevent the inundation risks identified under the baseline and climate change scenarios, World Bank determined the cost of adaptation in 2050 by comparing the damages and costs under climate change against the counterfactual baseline scenario that excludes climate change. The following table shows the cost of adapting to climate change by 2050 ($ millions).

| Adaptation Option | Baseline Scenario Investment Cost | Climate Change Scenario Investment Cost | Climate Change Scenario Annual Recurrent Cost | Additional Cost due to Climate Change Investment Cost | Additional Cost due to Climate Change Annual Recurrent Cost |
|---|---|---|---|---|---|
| Polders | 2462 | 3354 | 18 | 893 | 18 |
| Foreshore afforestation | | 75 | | 75 | |
| Cyclone shelters | | 1219 | 24 | 1219 | 24 |
| Cyclone-resistant housing | | 200 | | 200 | |
| Early warning and evacuation system | | 39 | 8+ | 39 | 8+ |
| Total | 2462 | 4888 | 50+ | 2426 | 50+ |

Under the baseline scenario, all of the total investment costs of $2462 million are for upgrading polders. The costs for additional cyclone shelters are not included since an adequate number of these are assumed to be already under construction or have been planned. These investments prevent damages from the average cyclone with a 10-year return period, as has been experienced during the past 50 years. The potential damages from a single such storm currently are $1802 million and are expected to rise to $4607 million by 2050. On an average, four such storms can be expected over the next 40 years. The potential benefits exceed the investment costs by several times, even when the future benefits and costs are discounted.

the adaptation benefits may also spread across different sectors. The costs of adaptation in the health sector will be significantly lower if measures for water quality depletion are implemented, or double counting can otherwise take place if measures are costed in the two sectors (water and health) (Ortega 2011).

### 6.4.2 Methodological Approaches

Adaptation is not an easy or cost-free option; it also entails costs. The key question is, how much will it cost, because the costs of adaptation themselves are part of the cost of climate change, and what will it achieve? Decisions about the timing and amount of adaptation require comparison of costs and benefits, which demands more quantitative information. The adaptation route that is chosen should be the one that yields the highest net benefit than their costs, having taken account of the risks and uncertainties surrounding climate change (Adger et al. 2005; Loe et al. 2001). Hence, adaptation economics can be a useful tool to decision makers and water resources planners. A simple framework for thinking about the costs and benefits of adaptation is shown in Table 6.1. The columns in Table 6.1 reflect two climate scenarios: one with climate change and one without ($C_0$ and $C_1$, respectively). The two rows

---

**TABLE 6.1**
**A Framework for Thinking about the Costs and Benefits of Adaptation**

| Adaptation Type | Existing Climate ($C_0$) | Altered Climate ($C_1$) |
|---|---|---|
| Adaptation to existing climate ($A_0$) | Box A: Existing climate. Society is adapted to existing climate: $C_0$, $A_0$ or base case. | Box C: Altered climate. Society is adapted to existing climate: $C_1$, $A_0$. |
| Adaptation to altered climate ($A_1$) | Box B: Existing climate. Society is adapted to altered climate: $C_0$, $A_1$. | Box D: Altered climate. Society is adapted to altered climate: $C_1$, $A_1$. |

*Sources:* Drawing on a framework originally presented by Fankhauser (1997) and modified by Callaway, J.M., The benefits and costs of adapting to climate variability and climate change, in: Morlot, J.C. and Agrawala, S. (eds.), *The Benefits of Climate Change Policies*, OECD, Paris, France, 2004, pp. 113–157.

*Note:* There will be various costs and benefits of adapting to climate change and can be thought of along the following lines:

- Climate change damage is the welfare loss associated with moving from the base climate (Box A) to a changed climate without adaptation (Box C): W ($C_1$, $A_0$) − W ($C_0$, $A_0$).
- Net benefits of adaptation are the reduction in damage achieved by adapting to the changed climate (net of the costs of doing so), subtracting Box C from the bottom Box D: W ($C_1$, $A_1$) − W ($C_1$, $A_0$).
- Climate change damage after adaptation is the difference between social welfare in Box D and in Box A: W ($C_1$, $A_1$) − W ($C_0$, $A_0$).

### TABLE 6.2
### Framework for Adaptation Planning

| Cost of Planning | Risk of Climate Change | |
| for Climate Change | Low | High |
| --- | --- | --- |
| Low | Low risk | Plan for climate change |
| High | Don't plan for climate change | High risk |

*Source:* Stern, N., *The Economics of Climate Change*, HM Treasury, London, U.K., 2006.

represent two adaptation options: one is the best to pursue without climate change and the other with climate change ($A_0$ and $A_1$, respectively). *Box A* represents the initial situation, where society is adapted to the current climate ($C_0$, $A_0$). *Box D* represents a situation where society adapts ($A_0$–$A_1$) to a change in climate from $C_0$ to $C_1$. *Box C* represents a situation where society fails to adapt to the change in climate. Finally, *Box B* represents a counterfactual situation where society undertakes adaptation ($A_0$–$A_1$), but the climate does not in the end change. This is an example of the type of situation that could arise if climate does not change in the anticipated way.

Uncertainty over the nature of future climate change is implicit in the given framework and is one of the principal challenges facing climate policy. Table 6.2 therefore modifies the framework to illustrate the trade-offs facing those planning adaptation under uncertainty. The decision to implement an adaptation strategy should take account of the balance of risks and costs of planning for climate change that does not occur and vice versa. Where the cost of planning for climate change is low, but the risks posed by climate change are high, there is a comparatively unambiguous case for adaptation. In contrast, where the costs of adaptation are high but the risks posed by climate change are low, the proposed adaptation responses may be disproportionate to the risks faced. Where the costs of planning for climate change and the risks of climate change are both low, there is little risk to the situation and the downsides are small, regardless of the choice made. In contrast, where the costs of both "mistakes" are high, the stakes and risks are very high for the planner.

In conjunction with the development of climate change adaptation planning framework, various methods/approaches are also introduced for quantification of adaptation costs. This section briefly highlights the methods/approaches:

*Top-down approach to estimate climate change adaptation costs*: Hughes et al. (2010) introduced a top-down approach to estimate the costs of adapting to climate change on a consistent basis for different climate scenarios. The work represents the most extensive and careful effort that has been made to estimate the costs of adapting to climate change for infrastructure and the water sector in particular. It consists of the following steps: (1) constructing baseline projections of infrastructure investment, (2) adding alternative climate scenarios, (3) projecting infrastructure quantities under the alternative climate scenarios, (4) applying the dose–response relationship to estimate changes in unit costs for alternative climate scenarios, (5) estimating the change in total investment and maintenance costs for the baseline projections, (6) estimating the change in investment and maintenance costs due to the difference between the baseline infrastructure quantities and the alternative climate scenario quantities, and (7) special adjustments. Detailed methodology is available in World Bank (2010b). The study relied upon a very detailed inventory of infrastructure assets and took no account of changes in the amounts of infrastructure over the time horizon. Poor inventory of infrastructure assets may lead to error in estimation.

*Dynamic and Interactive Vulnerability Assessment (DIVA) model*: DIVA is a dynamic, interactive, and flexible model that enables its users to produce quantitative information on a range of coastal vulnerability indicators, for user-selected climatic and socioeconomic scenarios and adaptation strategies, on national, regional, and global scales covering all coastal nations. DIVA first downscales to relative sea-level rise (RSLR) by combining the sea-level rise scenarios due to global warming with the vertical land movement. The latter is a combination of glacial-isostatic adjustment according to the geophysical model and an assumed uniform 2 mm/year subsidence in deltas. Human-induced subsidence (due to ground fluid abstraction or drainage) is not considered due to the lack of consistent data or scenarios. Based on the RSLR, three types of biophysical impacts are assessed: (1) dry-land loss due to coastal erosion, (2) flooding, and (3) salinity intrusion in deltas and estuaries. Besides biophysical impact assessment, DIVA assesses the social and economic consequences of the physical impacts. Economic consequences are expressed in terms of damage costs and adaptation costs. DIVA implements the adaptation options according to various complementary adaptation strategies. The "cost–benefit adaptation" strategy balances costs and benefits of adaptation. Detailed methodology is given in Hinkel et al. (2010).

*Conceptual framework for assessing costs of adaptation to climate change*: To fulfill the necessity of overarching conceptual framework for the estimation of costs of adaptation to the impacts of climate change in freshwater systems, Ortega (2011) introduced a conceptual framework. Key elements of the framework include (1) prerequisites for the estimation, (2) adaptation target and adaptation options, and (3) the cost-effectiveness logic. The model starts with the definition of the climate change impact and the unit of measurement (see Figure 6.6). It puts forward a precondition of identification of the "adaptation offer." It then defines the adaptation target. It focuses on considering the adaptation deficit as well as the residual damage. In addition to these points, the framework relies on the concept that an efficient adaptation strategy should be based on the most cost-effective combination of measures to achieve a determined goal from an economic perspective.

*Framework for economic pluvial flood risk assessment*: Zhou et al. (2012) developed a pluvial flood risk assessment framework to identify and assess adaptation options in the urban context. An integrated approach is adopted by incorporating climate change impact assessment, flood inundation modeling, economic tool, and risk assessment, thereby developing a step-by-step process for cost–benefit analysis (CBA) of climate change adaptation measures. CBA is a well-known tool for analyzing economic scenarios. CBA compares the benefits of a project or a policy with its corresponding costs of implementation on welfare-economics ground. This provides

# Economics of Climate Change

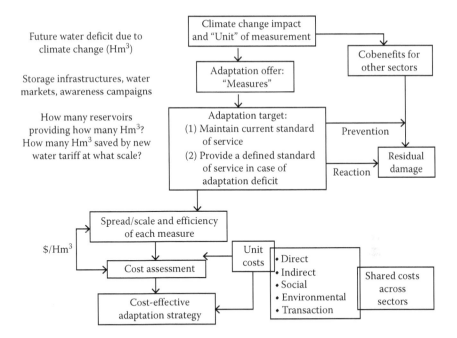

**FIGURE 6.6** A conceptual framework for the assessment of the costs of adaptation to climate change: the case of freshwater systems. (*Note*: $Hm^3$ = cubic hectometer.) (From Ortega, J.M., *Economía Agraria y Recursos Naturales*, 11(1), 5, 2011.)

valuable information on the net benefits of a proposed project/policy and can help decision makers to prioritize various adaptation investments.

The economic costs associated with impacts and adaptation to climate change are a topic of growing concern for national, state, and local governments throughout the world. Reviews in Sections 6.3.2 and 6.4.2 show that research efforts have been made to assess the aggregate costs of climate change impacts and adaptation at various spatial scales. However, methodologies for the estimation of total adaptation costs to the impacts of climate change are highly variable among aforementioned studies. Those studies nonetheless provide general guidance on estimates of climate change impacts as well as adaptation costs. Among countless methods/approaches applied in analyzing climate change impact and adaptation costs, CBA approach has been widely used (Leichenko et al. 2011; Lind 1995; Lunduka et al. 2012; World Bank 2010b; Zhou et al. 2012). Eventually, the chapter drags to a general framework of CBA to provide an overview on assessment of the potential costs of key climate change impact and adaptation options by means of a case study referring to Leichenko et al. (2011).

## 6.5 CASE STUDY

This section presents a case study of an economic analysis of climate change impacts and adaptations in water resources sector in New York State carried out by Leichenko et al. (2011).

### 6.5.1 Context

Water resources sector in New York State is an essential part of the economy and culture of the state. The principal impacts expected from climate change will be on various types of infrastructure that will be subject to increased risks from flooding as sea-level rises as well as significant impacts from droughts and inland flooding. These impacts, without adaptation, are likely to be at least in the tens of billions of dollars. In order to respond to the possible adverse impacts of climate change, New York State Energy Research and Development Authority (NYSERDA) has carried out an integrated assessment for effective climate change adaptation in New York State. The purpose of this study is to provide information on the economic impacts of climate change and adaptation for use by public officials, policy makers, and members of the general public. The study is also intended to provide information that will assist the New York State Climate Action Council with identification and prioritization of adaptation areas for the state.

### 6.5.2 Approach

The following six interrelated steps were carried out (Figure 6.7), a process involved in estimation of costs and benefits of climate change impacts and adaptation:

*Step 1*: Description of the major economic components of the water sector that are potentially vulnerable to the impacts of climate change (e.g., the built environment in the ocean coastal zones), review of economic data, and compilation of data on economic values of the key components in the water sector.
*Step 2*: Identification of the facets of climate change (e.g., sea-level rise) that are likely to have significant impacts on the water sector, developing a

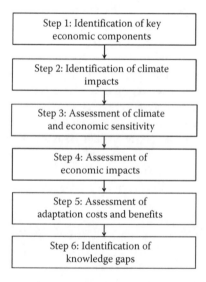

**FIGURE 6.7** Process involved in estimation of costs and benefits of climate change impacts and adaptation.

climate sensitivity matrix for the water sector based on review of existing sectoral literature, and interviews with government officials and some related stakeholders.

*Step 3*: Further refinement of the climate sensitivity matrix in order to specify that climate-related changes identified in step 2 will have the most significant potential costs for the key economic components; draw information from other relevant studies to identify economic components at risk and the costs of adaptation to climate change.

*Step 4*: Assess quality and comprehensiveness of available data, supplement them where possible, and extend on an estimated basis for future time periods; estimate range and value of possible economic impacts (as direct costs) based on the definition of the most important economic components and potential climate-related changes (steps 1 through 3) assuming "business as usual" frame with no steps to adapt to climate change; refer to Table 6.3 for an example of various impact costs to be considered in economic impact assessment of sea-level rise and storm surge on the water supply and wastewater treatment systems.

## TABLE 6.3
## Impact and Adaptation Costs to Be Considered in Case of Sea-Level Rise and Storm Surge Impacts on the Water Supply and Wastewater Treatment Systems

| Impact Costs | Adaptation Costs |
|---|---|
| • Costs related to damage of WWTPs due to flooding (asset values and repair cost for plants)<br>• Replacement costs of the water, sanitary, and storm pipes; lift stations; sewage treatment plant and related infrastructures<br>• Costs related to repositioning of the intakes for the pump station and the water supply system in case of saltwater intrusion due to sea-level rise<br>• Substantial institutional and operating costs relating to the integrated operation of the river with water supply system, which releases specified flows to the river from reservoirs<br>• Costs involved in undertaking turbidity control measures in case of intense precipitation and increased turbidity problems in watersheds<br>• Costs related to losses to water system consumers and for emergency measures (in case of drought) | • Costs of raising key equipment at the wastewater treatment plant to adapt to both current variability and future sea-level rise<br>• Costs to build levees/floodwalls, bridge/road modifications, channel modifications, closure structures, dry detention basins, flood proofing, and pump stations<br>• Costs related to other possible adaptation measures such as increased use of groundwater supplies, increased storage at existing reservoirs, withdrawals and treatment from other surface waters, and hydraulic improvement to existing aqueducts and additional tunnels<br>• Costs of a drought emergency measure i.e., costs to bring water from alternative sources by laying pipelines |

*Source:* Leichenko, R. et.al. An economic analysis of climate change impacts and adaptations in New York State. In: Rosenzweig, C. et al. (eds.), *Responding to Climate Change in New York State*, pp. 1–439. Albany: New York State Energy Research and Development Authority, 2011.

*Step 5*: Estimate the costs and benefits of a range of adaptations recognized; use standard concept of CBA to frame the work at this step; combine extrapolated information and results from interviews with experts to identify and assess relevance of other adaptation cost and benefits studies; refer to Table 6.3 for various costs related to impacts and adaptation; refer to the following literatures for more details: Leichenko et al. (2011), Lunduka et al. (2012), World Bank (2010a), and Zhou et al. (2012).

*Step 6*: Identify knowledge gaps; recommend further economic analyses, based on assessments in steps 1 through 5.

### 6.5.3 Results

The main relationships of climate and economic sensitivity in the water sector in New York State are shown in Table 6.4. The costs of climate change are expected to be substantial in the water sector, both for upland systems and for those parts of the system such as drainage and Wastewater treatment plants (WWTPs) located near the coastal area. An estimate for climate change impacts resulting from increased flooding of coastal WWTPs and annual costs and benefits of adaptation are given in Table 6.5. While these costs are expected to be significant, they are just a part of the total impact costs for the water sector, which will be quite high. These costs will include the cost of infrastructure for improving system resilience and intersystem linkages, the costs of drought (to both consumers and water agencies), and the increased costs of maintaining water quality standards with changing temperature and precipitation patterns. Adaptation costs for the sector will also be higher than what is presented in the table and will include costs for adaptation of urban drainage and sewer systems, the costs of managing droughts, and the costs of preventing inland flooding. However, it is important to note that much of the drainage, wastewater, and water supply infrastructure in New York is antiquated and inadequately maintained, with an estimated cost for upgrades of tens of billions of dollars. An important policy opportunity will be using the need for infrastructure improvement as a simultaneous chance to adapt to anticipate climate change impacts, thereby reducing future risk and saving water currently lost through leaks or inefficient operations.

As other examples in the water sector where climate change impacts are expected to be substantial, upstate WWTPs will be subject to flooding, and water supply systems will be subject to increased droughts as climate change progresses. To improve the efficiency of economic analysis of impacts and adaptations costs in the water sector, there are many knowledge gaps to which resources can be directed. They include undertaking of a series of comprehensive CBA of potential adaptations to aid in long-term planning, building upon current studies of the New York City system and other systems, integrating population projections into climate change planning, and more advanced planning for power outages and their impacts on wastewater treatment plants and other facilities. For detailed information on given method and results, readers may follow Leichenko et al. (2011).

Economics of Climate Change 177

### TABLE 6.4
### Climate and economic sensitivity matrix: water resources sector

| Element | Temperature | Precipitation | Extreme Events: Heat | Sea Level Rise and Storm Surge | Economic Risks and Opportunities "−" Is Risk "+" Is Opportunity | Annual Incremental Impact Costs of Climate Change at Mid-Century without Adaptation | Annual Incremental Adaptation Cost and Benefits of Climate Change at Mid-Century |
|---|---|---|---|---|---|---|---|
| Coastal flooding | X | X | | X | • Damage to wastewater treatment plants<br>• Blockage from sea level rise of system outfalls<br>• Salt water intrusion into aquifers | Coastal flooding of WWTPs $116–203M (million) | Costs: $47M Benefits: $186M |
| Inland flooding | | X | | | • Increased runoff leading to water quality problems<br>• Damage in inland infrastructure | High direct costs statewide estimated $237M in 2010 | Restore natural flood area, decrease permeable surfaces, possible use of levees, and control turbidity |
| Urban flooding | | X | | | • Drainage system capacity exceeded; combined sewer overflows<br>• Damage to infrastructure | Violation of standards | Very high costs of restructuring drainage systems |
| Droughts | X | X | | | • Reduction in available supplies to consumers<br>• Loss of hydroelectric generation<br>• Impacts on agricultural productivity | 1960s drought in New York City system reduced surface safe yield from 1800 to 1290 million gallons per day | Increased redundancy and interconnectedness costs for irrigation equipment |
| Power outages | X | X | X | | • Loss of functionality of wastewater treatment plants and other facilities | Violation of standards | Flood walls |
| Total estimate costs of key elements (values in $2010 US.) | | | | | | $353–440M | Costs: $47M Benefits: $186M Unknown |

▨ Analyzed example  ▨ From literature  ☐ Qualitative information

*Source:* (From Leichenko, R. et al., An economic analysis of climate change impacts and adaptations in New York State, in: Rosenzweig, C. et al., eds., *Responding to Climate Change in New York State*, New York State Energy Research and Development Authority (NYSERDA), Albany, NY, 2011, pp. 1–439.)

**TABLE 6.5**
**Illustrative Key Impacts and Adaptations: Water Resources Sector**

| Element | Time Slice | Annual Costs of Current Climate Hazards without Climate Change ($ Million) | Annual Incremental Costs of Climate Change Impacts, without Adaptation ($ Million) | Annual Costs of Adaptation ($ Million) | Annual Benefits of Adaptation ($ Million) |
|---|---|---|---|---|---|
| All New York State WWTP damages from 100-year coastal event | Baseline | 100 | — | — | — |
| | 2020s | 143 | 14–43 | 23 | 91 |
| | 2050s | 291 | 116–203 | 47 | 186 |
| | 2080s | 592 | 415–533 | 95 | 379 |

*Source:* Leichenko, R. et al., An economic analysis of climate change impacts and adaptations in New York State, in: Rosenzweig, C. et al. (eds.), *Responding to Climate Change in New York State*, pp. 1–439. Albany, New York State Energy Research and Development Authority (NYSERDA), NY, 2011.

*Note:* Values are in US$ 2010.

### 6.5.4 Lessons Learned

In estimating the costs of climate change in the water sector (in New York State), relatively standard and sophisticated methods can be applied; however, data are often inadequate and the uncertainties in the future climate are large, compounded by uncertainties in other drivers such as population and real income growth. Nevertheless, attempt is or can be made to quantify the costs of key impacts and adaptations in water resources sector using simple CBA. Undertaking a series of comprehensive CBA of potential adaptations provides a good basis for estimates of adaptation to climate change and aids in the long-term planning of the water resources sector. As the example of Table 6.5 indicates, costs of climate change impacts may be large; however, adaptations are available and their benefits may be substantial. However, estimated costs and benefits depend on the input assumptions. One should try to set assumptions in such a way that it keeps magnitude of costs and benefits within the same range for a fairly wide set of assumptions. Skillfully chosen and scheduled adaptation can markedly reduce the impacts of climate change in excess of their costs.

## 6.6 SUMMARY

Economics of climate change can provide an important forum to consider fundamental economic issues that will enhance understanding and improve climate policy deliberations.

It offers the monetary dimension of costs and benefits of climate change impacts and adaptation. Estimates of the costs of adaptation can provide insight into the very real costs of inaction or, conversely, the benefits of maintaining and protecting societal goods and services through effective policies that avoid the most severe climate impacts. Therefore, interest in estimating the costs of climate change impacts and adaptation is expanding as the need for action has become clear. Various economic methods/approaches have been developed to estimate and compare the costs and address decision making in the face of uncertainty. However, inconsistencies in methods and scale to be considered (i.e., spatial and temporal) are still the issues of active debate. In addition, there are still unanswered questions on application of these methods and improving the quality of information on the possible impacts and benefits. Identifying, assessing, and communicating the implications of economic uncertainty and knowledge gaps remain a major challenge—for example—in the characterization of long-term technology change and valuation of nonmarket impacts. The quantification of nonmarket impacts is less assured than market impacts, but this is still helpful in extending the economic assessment. Regardless of some of these limitations of existing methods, they are still useful tools in making climate change–related policy decisions and implementing adaptation strategies.

## REFERENCES

Aaheim, A., Amundsen, H., Dokken, T., and Wei, T. 2012. Impacts and adaptation to climate change in European economies. *Global Environmental Change* 22: 959–968.

Abler, D., Shortle, J., Rose, A., and Oladosu, G. 2000. Characterizing regional economic impacts and responses to climate change. *Global and Planetary Change* 25: 67–81.

Adger, W.N., Arnell, N.W., and Tompkins, E.L. 2005. Successful adaptation to climate change across scales. *Global Environmental Change* 15: 77–86.

Anda, J., Golub, A., and Strukova, E. 2009. Economics of climate change under uncertainty: Benefits of flexibility. *Energy Policy* 37: 1345–1355.

Backus, G., Lowry, T., Warren, D. et al. 2010. *Assessing the Near-Term Risk of Climate Uncertainty: Interdependencies Among the U.S. States*. SAND2010-2052. Albuquerque, NM: Sandia National Laboratories.

Bettencourt, S., Croad, R., Freeman, P. et al. 2006. *Not If but When: Adapting to Natural Hazards in the Pacific Islands Region*. Washington, DC: World Bank.

BPC. 2009. Climate change and the economy, expected impacts and their implications. BPC. Available at: http://masgc.org/climate/cop/Documents/Climate%20Change%20 and%20The%20Economy%20-%20Expected%20Impacts%20and%20Their%20 Implications.pdf (accessed 30 April 2013).

Callaway, J.M. 2004. The benefits and costs of adapting to climate variability and climate change. In: Morlot, J.C. and Agrawala, S. (eds.), *The Benefits of Climate Change Policies*, pp. 113–157. Paris, France: OECD.

Callaway, J.M., Louw, D.B., Nkomo, J.C., Hellmuth, M.E., and Sparks, D.A. 2006. The Berg River dynamic spatial equilibrium model: A new tool for assessing the benefits and costs of alternatives for coping with water demand growth, climate variability, and climate change in the Western Cape. AIACC working paper 31. Washington, DC: International START Secretariat, AICC.

Changnon, S.A. 2005. Economic impacts of climate conditions in the United States: Past, present, and future—An editorial essay. *Climatic Change* 68: 1–9.

CIER. 1992. *Economic Impacts of Climate Change on Georgia*. College Park, MD: The Center for Integrative Environmental Research, University of Maryland, 2008. Retrieved on April 30, 2013. Available at: http://www.cier.umd.edu/climateadaptation/.

CIER. 2008. Economic Impacts of Climate Change on Georgia. College Park, MD: The Center for Integrative Environmental Research, University of Maryland, 2008. Retrieved on April 30, 2013. Available at: http://www.cier.umd.edu/climateadaptation/Georgia%20 Economic%20Impacts%20of%20Climate%20Change.pdf.

Cline, W. 1992. *Global Warming: The Economic Stakes*. Washington, DC: Institute for International Economics.

Delpla, I., Jung, A.V., Baures, E., Clement, M., and Thomas, O. 2009. Impacts of climate change on surface water quality in relation to drinking water production. *Environment International* 35: 1225–1233.

Fankhauser, S. 1996. Climate change costs. Recent advancements in the economic assessment. *Energy Policy* 24 (7): 665–673.

Fankhauser, S. 1997. The costs of adapting to climate change. Working Paper 13, Washington, DC: Global Environment Facility.

Fankhauser, S. and Tol, R.S.J. 2005. On climate change and economic growth. *Resource and Energy Economics* 27: 1–17.

Gaskins, D. and Weyant, J. 1993. Model comparisons of the costs of reducing $CO_2$ emissions. *American Economic Review* 83 (2): 318–323.

Hall, D.C. and Behl, R.J. 2006. Integrating economic analysis and the science of climate instability. *Ecological Economics* 57: 442–465.

Hallegatte, S., Hanriet, F., and Morlot, J.C. 2011. The economics of climate change impacts and policy benefits at city scale: A conceptual framework. *Climatic Change* 104: 51–87.

Harou, J.J., Velazquez, M.P., Rosenberg, D.E., Azuara, J.M., Lund, J.R., and Howitt, R.E. 2009. Hydro-economic models: Concepts, design, applications, and future prospects. *Journal of Hydrology* 375: 627–643.

Heinz Center. 2000. *The Hidden Costs of Coastal Hazards: Implications for Risk Assessment and Mitigation*, 220pp. Washington, DC: Island Press.

Hinkel, J., Nicholls, R.J., Vafeidis, A.T., Tol, R.S.J., and Avagianou, T. 2010. Assessing risk of and adaptation to sea-level rise in the European Union: An application of DIVA. *Mitigation and Adaptation Strategies for Global Change* 15: 703–719.

Hope, C. 2005. Integrated assessment models. In: Helm, D. (ed.), *Climate-Change Policy*, pp. 77–98. Oxford, U.K.: Oxford University Press.

Hughes, G., Chinowsky, P., and Strzepek, K. 2010. The costs of adaptation to climate change for water infrastructure in OECD countries. *Utilities Policy* 18: 142–153.

Hurd, B. and Rouhi-Rad, M. 2013. Estimating economic effects of changes in climate and water availability. *Climatic Change* 117: 575–584.

Hurd, B.H., Callaway, J.M., Smith, J., and Kirshen, P. 1999. Economic effects of climate change on U.S. water resources. In: Mendelsohn, R. and Neumann, J. (eds.), *The Impact of Climate Change on the United States Economy*. Cambridge, U.K.: Cambridge University Press.

Hurd, B.H., Callaway, M., Smith, J., and Kirshen, P. 2004. Climatic change and US water resources: From modeled watershed impacts to national estimates. *Journal of the American Water Resources Association* 40(1): 129–148.

IPCC. 1996. Climate change 1996: Economic and social dimensions of climate change. In: Bruce, J.P., Lee, H., and Haites, E.F. (eds.), *Contribution to the Working Group III to the Second Assessment Report of the Intergovernmental Panel on Climate Change (IPCC)*. Cambridge, U.K.: Cambridge University Press.

IPCC. 2007. Climate change 2007: Impacts, adaptation and vulnerability. In: Parry, M.L., Canziani, O.F., Palutikof, J.P., van der Linden, P.J., and Hanson, C.E. (eds.), *Contribution of Working Group II to the Fourth Assessment Report of the Intergovernmental Panel on Climate Change (IPCC)*. Cambridge, U.K.: Cambridge University Press.

Jeuland, M. 2010. Economic implications of climate change for infrastructure planning in transboundary water systems: An example from the Blue Nile. *Water Resources Research* 46: W11556, doi:10.1029/2010WR009428.

Jiang, Q. and Grafton, R.Q. 2012. Economic effects of climate change in the Murray-Darling Basin, Australia. *Agricultural Systems* 110: 10–16.

Kazama, S., Sato, A., and Kawagoe, S. 2009. Evaluating the cost of flood damage based on changes in extreme rainfall in Japan. *Sustainability Science* 4(1): 61–69.

Kolstad, C.D. and Toman, M. 2001. The economics of climate policy. Discussion Paper 00-40REV. Washington, DC: Resource for the Future.

Larsen, P.H., Goldsmith, S., Smith, O. et al. 2008. Estimating future costs for Alaska public infrastructure at risk from climate change. *Global Environmental Change* 18: 442–457.

Leichenko, R., Major, D.C., Johnson, K., Patrick, L., and O'Grady, M. 2011. An economic analysis of climate change impacts and adaptations in New York State. In: Rosenzweig, C., Solecki, W., DeGaetano, A., O'Grady, M., Hassol, S., and Grabhorn, P. (eds.), *Responding to Climate Change in New York State*, pp. 1–439. Albany, NY: New York State Energy Research and Development Authority (NYSERDA).

Lind, R.C. 1995. Intergenerational equity, discounting, and the role of cost–benefit analysis in evaluating global climate policy. *Energy Policy* 23(4–5): 379–389.

Loe, R., Kreutzwiser, R., and Moraru, L. 2001. Adaptation options for the near term: Climate change and the Canadian water sector. *Global Environmental Change* 11: 231–245.

Lunduka, R.W., Bezabih, M.E., and Chaudhury, A. 2012. *Stakeholder-Focused Cost Benefit Analysis in the Water Sector: A Synthesis Report*. London, U.K.: International Institute for Environment and Development (IIED).

Manne, A. and Richels, R. 1991. Towards a comprehensive approach to global climate change mitigation. *American Economic Review* 81(2): 140–145.

Mendelsohn, R., Nordhaus, W., and Shaw, D. 1994. The impact of global warming on agriculture: A Ricardian analysis. *American Economic Review* 84(4): 753–771.

Mendelsohn, R.O., Morrison, W.N., Schlesinger, M.E., and Andronova, N.G. 2000. Country-specific market impacts of climate change. *Climatic Change* 45(3–4): 553–569.

Mideksa, T.K. 2010. Economic and distributional impacts of climate change: The case of Ethiopia. *Global Environmental Change* 20: 278–286.

Murray, C., Lopez, A. 1996. The global burden of disease. Cambridge: Harvard University Press.

Ng, W.S. and Mendelsohn, R. 2005. The impact of sea-level rise on Singapore. *Environment and Development Economics* 10: 201–215.

Nordhaus, W.D. 1994. *Managing the Global Commons: The Economics of Climate Change*. Cambridge, MA: MIT Press.

Ohno, E. 2000. Economic evaluation of impact of land loss due to sea level rise in Thailand. *Environmental Systems Research* 28: 445–452.

Ortega, J.M. 2011. Costs of adaptation to climate change impacts on fresh-water systems: Existing estimates and research gaps. *Economía Agraria y Recursos Naturales* 11(1): 5–28.

Rehana, S. and Mujumdar, P.P. 2012. Climate change induced risk in water quality control problems. *Journal of Hydrology* 444–445: 63–77

Schlenker, W., Hanemann, W.M., and Fisher, A.C. 2005. Will U.S. agriculture really benefit from global warming? Accounting for irrigation in the hedonic approach. *The American Economic Review* 95: 395–406.

Stern, N. 2006. *The Economics of Climate Change*. London, U.K.: HM Treasury.

Tol, R.S.J. 1994. The damage costs of climate change: A note on tangibles and intangibles, applied to DICE. *Energy Policy* 22(5): 436–438.

Tol, R.S.J. 1999. The marginal costs of greenhouse gas emissions. *The Energy Journal* 20(1): 61–81.

Tol, R.S.J. 2002. Estimates of the damage costs of climate change—Part II: Dynamic estimates. *Environmental and Resource Economics* 21: 135–160.

Tol, R.S.J. 2009. The economic effects of climate change. *Journal of Economic Perspectives* 23: 29–51.

Tol, R.S.J., Downing, T.E., Kuik, O.J., and Smith, J.B. 2004. Distributional aspects of climate change impacts. *Global Environmental Change* 14: 259–272.

Toman, M. 2006. Values in the economics of climate change. *Environmental Values* 15: 365–379.

UN. 2009. *Guidance on Water and Adaptation to Climate Change*. Geneva, Switzerland: United Nations, ISBN: 978-92-1-117010.

Viscusi, W. and Aldy, J.E. 2003. The value of a statistical life: a critical review of market estimates throughout the world. *Journal of Risk and Uncertainty* 27(1): 5–76.

Wilson, K.J., Falkingham, J. Melling, H., and de Abreu, R. 2004. Shipping in the Canadian Arctic: Other possible climate change scenarios. *Proceedings of the IEEE International* 3: 1853–1856.

World Bank. 2008. Ethiopia—A country study on the economic impacts of climate change. Report No. 46946-ET. Washington, DC: The World Bank Group.

World Bank. 2010a. *Economics of Climate Change: Bangladesh*. Washington, DC: World Bank Group.

World Bank. 2010b. *Countries of Adapting to Climate Change: New Methods and Estimates*. Washington, DC: World Bank Group.

Young, R.A. 2005. *Determining the Economic Value of Water: Concepts and Methods*. Washington, DC: Resources for the Future Press.

Zhou, Q., Mikkelsen, P.S., Halsnaes, K., and Nielsen, K.A. 2012. Framework for economic pluvial flood risk assessment considering climate change effects and adaptation benefits. *Journal of Hydrology* 414–415: 539–549.

# 7 Climate Change Vulnerability Assessment

*Vishnu Prasad Pandey, Sujata Manandhar, and Futaba Kazama*

## CONTENTS

7.1 Introduction ................................................................................................. 183
7.2 Evolution of Climate Change Vulnerability Assessment ............................ 184
    7.2.1 Vulnerability Concepts and Definitions ........................................... 184
    7.2.2 Quantifying Vulnerability ................................................................. 186
    7.2.3 Nomenclature of Vulnerability ......................................................... 187
    7.2.4 Vulnerability-Related Terminologies ............................................... 188
7.3 Vulnerability Assessment and Adaptation Planning ................................... 188
7.4 Steps in Vulnerability Assessment .............................................................. 190
    7.4.1 Structuring Vulnerability Assessment .............................................. 190
    7.4.2 Gathering Relevant Data and Expertise ........................................... 191
    7.4.3 Assessing Vulnerability Components ............................................... 191
7.5 Vulnerability Components ........................................................................... 192
    7.5.1 Exposure ............................................................................................ 192
    7.5.2 Sensitivity .......................................................................................... 193
    7.5.3 Adaptive Capacity ............................................................................. 195
7.6 Vulnerability Index ...................................................................................... 196
    7.6.1 Computation of Vulnerability Index ................................................. 196
    7.6.2 Interpretation and Communication of Vulnerability Results ........... 197
7.7 Uncertainties in Vulnerability Assessment ................................................. 198
    7.7.1 Defining Uncertainty ........................................................................ 199
    7.7.2 Uncertainty Levels ............................................................................ 199
    7.7.3 Uncertainty Assessment Methodologies ........................................... 199
    7.7.4 Combining Uncertainty from Multiple Sources ............................... 201
7.8 Summary ...................................................................................................... 201
Appendix: Illustration of Vulnerability Assessment ............................................ 202
References ............................................................................................................. 206

## 7.1 INTRODUCTION

Current trends and future projections in the earth's climate indicate that the planet's resources would be exposed to vastly different environment in the future. The change in exposures would have more dramatic impacts on many sectors, including water, and make them more vulnerable than the past century, during which our conservation traditions evolved. Increasing advocacies for climate change adaptation has created contexts for vulnerability assessment of the changes. As a result, literatures on climate change vulnerability assessment in different sectors are growing. Vulnerability assessment could be considered as an umbrella under which a broad spectrum of issues related to water resources availability and management can be analyzed. The vulnerability knowledgebase further assists to formulate Integrated Water Resources Management (IWRM) policies and protection measures and provides a basis for decision makers to prioritize the appropriate actions. It also draws attention to the main sources of vulnerability and scans the opportunities to seek adaptive management of water resources. Three key motivations for carrying out vulnerability assessments are the following (U.S. Fish and Wildlife Service 2011): (1) help in setting management and planning priorities, (2) assist in informing and crafting adaptation strategies, and (3) enable more efficient allocation of resources.

This chapter aims to highlight the concepts and techniques for assessing and interpreting vulnerability of climate change in water sector. It also provides general steps for carrying out vulnerability assessment and illustrates the methodology with a hypothetical case. This chapter is expected to help understand vulnerability assessment in the broader context of adaptation planning.

## 7.2  EVOLUTION OF CLIMATE CHANGE VULNERABILITY ASSESSMENT

### 7.2.1  Vulnerability Concepts and Definitions

The concept of vulnerability has a long history in risk-hazard and social science literature; however, it was introduced only in 1992 in water sector, after the International Conference on Water and Environment in 1992 (called as *Dublin Conference*) asserted freshwater as a vulnerable resource. Even though vulnerabilities of various sectors to a wide range of risks are assessed over decades, the word *vulnerability* has no universal definition. It is conceptualized in very different ways by scholars from different knowledge domains, and even within the same domain. There is a bewildering array of terms (Brooks 2003) that either express similar ideas (e.g., risk, sensitivity, and fragility) or inversely similar ideas (e.g., resilience, adaptability, adaptive capacity, and stability). Many publications (e.g., Adger 1999; Brooks 2003; Downing et al. 2001; Downing and Patwardhan 2004; Eakin and Luers 2006; Füssel 2007; Füssel and Klein 2006; Kelly and Adger 2000; Moss et al. 2001; O'Brien et al. 2004, 2007; Olmos 2001; etc.) feature the concept of *vulnerability* in climate change research. Existence of the competing conceptualizations and terminologies has become problematic particularly in climate change research because this field is characterized by intense collaboration between scholars from different disciplines, including climate science, risk assessment, development, economics, and policy analysis.

The scientific use of *vulnerability* has its roots in geography and natural hazard literatures but this is now a central concept in a variety of other research contexts such as ecology, public health, poverty and development, secure livelihoods and famine,

# Climate Change Vulnerability Assessment

sustainability science, land use changes, and climate change impacts and adaptation (Füssel 2007). More than three decades back, Timmermann (1981) considered vulnerability as a term of such a broad use that careful description is unnecessary except as a rhetorical indicator of areas of greatest concern. Later, Liverman (1990) noted that vulnerability can be equated to concepts such as resilience, marginality, susceptibility, adaptability, fragility, and risk. We could easily add exposure, sensitivity, coping capacity, and criticality to that list. As depicted in Figure 7.1, the concept of vulnerability is widening from intrinsic risk factors to a much broader multidimensional concept. Because of multiple dimensions, there is no single *correct* or *best* conceptualization of vulnerability that would fit all assessment contexts (Kasperson et al. 2005). Therefore, it should be defined in the context of the study being undertaken. Important conceptual and semantic ambiguities regarding the vulnerability include the following questions (Füssel and Klein 2006):

- Is it the starting point or an intermediate element or the outcome of an assessment?
- Should it be defined in relation to an external stressor such as climate change or in relation to an undesirable outcome such as famine?

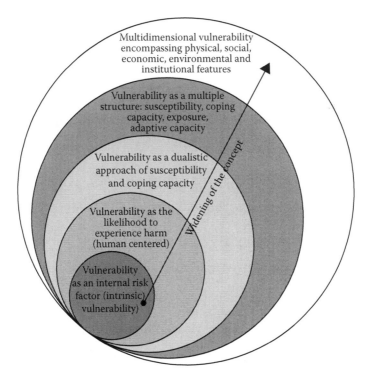

**FIGURE 7.1** Widening of the vulnerability concepts. (From Birkmann, J., Measuring vulnerability to promote disaster-resilient societies: Conceptual frameworks and definitions, in: Birkmann, J., ed., *Measuring Vulnerability to Natural Hazards: Towards Disaster Resilient Societies*, United Nations University Press, Tokyo, Japan, 2006, pp. 9–54. (Accessed 28 March 2013.))

- Is it an inherent property of a system or contingent upon a specific scenario of external stresses and internal responses?
- Is it a static or a dynamic concept?

Despite the aforementioned ambiguities, three general traditions in defining vulnerability can be identified as hazards, poverty, and climate change (Downing and Patwardhan 2004). All of them are rooted on *natural hazard and epidemiology* literatures, the longer tradition, which define vulnerability as "the degree to which an exposure unit is susceptible to harm due to exposure, to a perturbation or stress, in conjunction with its ability (or lack of thereof) to cope, recover, or fundamentally adapt" (Kasperson and Kasperson 2001). This definition clearly distinguishes vulnerability from hazard. The second tradition—*poverty and development literatures*—relates vulnerability to social units (people) and integrates vulnerability across a range of stresses and across the range of capacities. It defines vulnerability as "an aggregated measure of human welfare that integrates environmental, social, economic and political exposure to a range of harmful perturbation" (Bohle et al. 1994). The third tradition, that is, *climate change*, integrates hazard, exposure, impacts, and adaptive capacity and defines vulnerability as promoted by Intergovernmental Panel on Climate Change (IPCC): "The degree to which a system is susceptible to, or unable to cope with, adverse effects of climate change, including climate variability and extremes. Vulnerability is a function of the character, magnitude, and rate of climate variation to which a system is exposed, its sensitivity, and its adaptive capacity" (IPCC 2007a).

Even the IPCC definition is criticized by Hinkel (2011) as being too vague and for the resulting difficulty in making it operational. However, the definition provided by the IPCC is one of the most generic available and thus adopted in this chapter. Vulnerability of the water sector to climate change can be defined as the extent to which the sector is susceptible to, and unable to cope with, adverse impacts of climate change. The more vulnerable the sector is, the greater impacts are likely to experience from climate change and vice versa.

### 7.2.2 Quantifying Vulnerability

To apply the concept of vulnerability in policy-driven assessments, it is necessary to measure/quantify. Measuring vulnerability is a challenging task because it is often not a directly observable phenomenon (Downing et al. 2001), rather a relative and complex concept with entrenched difficulties in defining criteria for its quantification. The approaches considered so far are diverse, ranging from indicator-based approach to sophisticated hydrological models that calculate exposure under future projections of climate change. They also range from qualitative to quantitative that address a broad range of characteristics of social–ecological systems. With improved scientific understanding of the potentials and observed impacts of climate change, the interest in developing useful definitions and frameworks for conducting climate change vulnerability assessments is growing (Füssel and Klein 2006). Using quantitative or semiquantitative matrices with a set or a composite of proxy indicators, which are widely used in other environmental and social studies (e.g., Kelly and Adger 2000; Moss et al. 2002), could be a suitable approach to assess climate change vulnerability of the water sector. Indicators help to reflect and communicate a complex idea because the simple numbers, descriptive or normative statements, can condense the enormous complexity of real problems into a manageable amount of meaningful information.

Many early to mid-stage studies on vulnerability assessment focused on developing frameworks for assessing the vulnerability of agriculture, public health, and other human systems to climate change, building on approaches used in addressing problems such as poverty, famine, and natural hazards (e.g., Bohle et al. 1994; Downing and Patwardhan 2004; Handmer et al. 1999; Kelly and Adger 2000). Some recent studies are paying attention on assessing the vulnerability of natural systems to climate change (e.g., Nitschke and Innes 2008), multidisciplinary efforts to assess the vulnerability of ecosystem services to humans (Metzger et al. 2005), and the interaction between multiple stressors (Turner et al. 2003). Within each of these areas, however, different definitions and concepts for climate change vulnerability have emerged, which often has led to misunderstandings and challenges in the assessment efforts (Füssel 2007). In this chapter, we followed the general framework adopted by the IPCC (2007a), and subsequently by many others, in which vulnerability assessments are founded on evaluations of exposure, sensitivity, and adaptive capacity to climate change. Vulnerability assessment, therefore, refers to a process for measuring the exposure sensitivity and adaptive capacity of the water sector to climate change and aggregating them in a form of *vulnerability index* (VI).

For a practical application, climate change vulnerability can be assessed using a set of composite or proxy indicators. The indicators/indices are limited in their application by considerable subjectivity in their selection and weighting, availability of data at various scales, and difficulties of testing or validating the different metrics (Luers et al. 2003); however, they are still widely used in vulnerability and related studies.

### 7.2.3 NOMENCLATURE OF VULNERABILITY

Vulnerability is highly dependent on context and scale, and care should be taken to clearly describe its derivation and meaning (Downing and Patwardhan 2004) and to address the uncertainties inherent in vulnerability assessments. The term *vulnerability* can only be used meaningfully with reference to a particular situation such as (1) vulnerability of a specified system to a specified hazard or range of hazards (Brooks 2003); (2) vulnerability of selected variables of concern and to specific sets of stressors (Luers et al. 2003); (3) vulnerability as a function of characteristics of the system, type and number of stressors, their effects on the system, and time horizon of assessment (Füssel 2007); and (4) vulnerability of a system to change with respect to a particular service of the system, a location, a scenario of stressor, and a time slice (Metzger et al. 2005). Aforementioned examples largely agree that the following four dimensions are fundamental to describe a vulnerable system (Füssel 2007): system, attribute of concern, hazard, and temporal reference. The following nomenclature allows to fully describe a vulnerable situation: *vulnerability of a system's attribute(s) of concern to a hazard (in temporal reference)*, whereby the temporal reference can alternatively be stated as the first qualifier. For example, *vulnerability of water supply sector in a specified city to climate change over the next 20 years* can be considered as a fully qualified description of vulnerability.

### 7.2.4 Vulnerability-Related Terminologies

*Adaptation*: It refers to the adjustment in natural or human systems in response to actual or expected climatic stimuli or their effects, which moderates harm or exploits beneficial opportunities. Various types of adaptation can be distinguished including anticipatory and reactive adaptation, private and public adaptation, and autonomous and planned adaptation.

*Adaptive capacity*: It refers to the ability or potential of a system to respond successfully to climate variability and change and includes adjustments in both behavior and in resources and technologies.

*Exposure*: It is a measure of how much of a change in the climate and associated problems a system is likely to experience. It is the degree, duration, and/or extent of climate variables (e.g., temperature, precipitation, extreme weather events) to which a system is in contact with a climate perturbation.

*Mitigation*: It refers to an anthropogenic intervention to reduce the sources or enhance the sinks of greenhouse gases. Mitigation of climate change refers to actions that limit the level and rate of climate change (e.g., fuel switching in the energy sector, sequestration of greenhouse gases, enhancing the sink capacity of biological or other systems for greenhouse gases).

*Resilience*: It is a tendency to maintain integrity when subject to disturbance.

*Risk*: It is the likelihood or probability of the occurrence of harmful events at a locality. Risks will change because of changes in climate and mitigation actions.

*Sensitivity*: It is a measure of whether and how the system is likely to be affected by a given change in climate. It describes the human–environmental conditions that can worsen the hazard, ameliorate the hazard, or trigger an impact.

## 7.3 VULNERABILITY ASSESSMENT AND ADAPTATION PLANNING

Vulnerability to climate change can be assessed in a stand-alone assessment, but in many cases, it will be more effective to include it as a part of broader context of adaptation planning addressing a range of risks. Potential impacts of climate change are shifting the paradigm of natural resources conservation and management toward climate change adaptation, which aims at enhancing coping capacity of the natural systems (or a sector) against impacts of the change (Glick et al. 2009). IPCC (2007b) defines climate change adaptation as "initiatives and measures designed to reduce the vulnerability of natural systems to actual or expected climate change effects." Developing meaningful adaptation strategies requires an understanding of first, the impacts, risks, and uncertainties associated with climate change and second, the vulnerability of the different components of our natural world to

# Climate Change Vulnerability Assessment

those changes (IPCC 2007a). Climate change vulnerability assessment represents a key tool for developing adaptation strategies against the change. Vulnerability assessment can provide two essential types of information needed for adaptation planning: (1) identifying the sectors likely to be most strongly affected by projected changes and (2) understanding why they are likely to be vulnerable. While the first information enables managers to better set priorities for conservation action, the second one provides a basis for developing appropriate management and conservation responses.

Figure 7.2 offers a generalized framework indicating how vulnerability assessment can fit into and support adaptation planning as we move into a future that does not necessarily have past analogs. The framework is based on conservation literatures and starts with identifying conservation targets (e.g., water sector), which are then assessed for their vulnerabilities to climate change in order to determine the sector likely to be most at risk and more likely to persist. Based on an understanding of why the targets are regarded as vulnerable to climate change and other stressors, an array of management options can be identified and evaluated based on technical, financial, and legal considerations. Selected management strategies can then be implemented, with the activities and outcomes subject to monitoring in order to feed into a regular cycle of evaluation, correction, and revision. The elements of the adaptation planning process, including the vulnerability assessment, must take existing stressors into consideration as well as other relevant factors affecting the system. The climate change vulnerability assessment must therefore be viewed as an integral part of a broader adaptation planning and implementation framework.

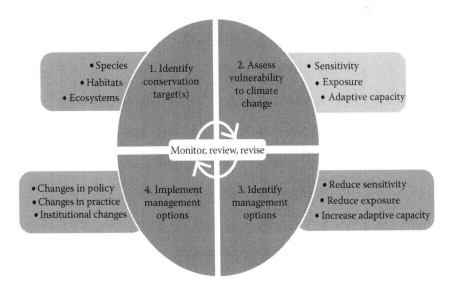

**FIGURE 7.2** Framework for developing climate change adaptation strategies. (From US Fish and Wildlife Service, Scanning the conservation horizon, 2011, available at http://training.fws.gov/EC/Resources/climate_change/vulnerability.html. (Accessed 13 April 2013.))

## 7.4 STEPS IN VULNERABILITY ASSESSMENT

Three general steps are basically recommended to carry out vulnerability assessment before applying the assessment results in adaptation planning.

### 7.4.1 Structuring Vulnerability Assessment

This step consists of defining framework, objectives, and scope of the vulnerability assessment. It may include at least the following substeps:

*Identify end-user (or audience) requirements*: Vulnerability assessments are intended to support decision making. End users of the assessment results could be water resources managers, policy makers, or others in the government or scientific community. Different users (or audiences) will likely warrant different assessment targets, levels of complexity, and approaches to communicate findings. For example, if the primary goal is to raise public awareness of the threats from climate change, it may be sufficient to conduct a review of existing literature on climate change impacts or conduct relatively broad and general assessments and then synthesize that information in understandable and accessible outreach tools. On the other hand, if the intended audience is a river basin authority or policy maker or water resources manager who will be using the data to prioritize investments and/or develop adaptation plans, then much more fine-scale data and assessment results will be necessary. Conducting a more sophisticated and fine-scale analysis and assessment may require additional time and resources to produce a more actionable set of results for water resources managers.

*Engage key stakeholders*: The goal and context of a particular assessment will determine the kind and amount of effort to be made for engaging stakeholders (both internal and external). Basically, engaging and informing stakeholders can help accomplish the following (Vogel et al. 2007): (1) getting less easily available data, (2) refine scope and focus, (3) provide sociopolitical context (e.g., national/regional/local laws/regulations/rules/plans, important subsistence or cultural uses of water resources, and the value systems that determine response to climate change), and (4) build support for adaptation. Engaging the right stakeholders in the right way and at the right times, however, can be the critical factor in determining the success of a vulnerability assessment under some circumstances. In general, categories of stakeholders to be considered include (1) decision makers (e.g., regulators and managers), (2) decision implementers (e.g., managers), (3) end users of the resource (e.g., community, hydropower operators, and farmers), (4) opinion leaders (influential and respected individuals within the region or sector of interest), (5) climate change adaptation planners, and (6) information providers (e.g., scientists, and holders of traditional knowledge; will usually overlap with other groups).

*Establish goals and objectives*: The ultimate goal of a vulnerability assessment is to support adaptation planning in the context of added impacts and complexities from climate change in conjunction with other stressors. Objectives might be a bit different with cases. In some cases, the goals and objectives of vulnerability assessments may depend on factors such as the management jurisdiction or mandate of the

# Climate Change Vulnerability Assessment 191

agency conducting the analysis. To ensure the mandates of the organization looking after the water sector are adequately framed in the vulnerability assessment, establishing goals and objectives of the assessment should be a collaborative endeavor involving the perspective end users as well as scientific and technical staff involved in carrying out the assessment. It will help to match expectations and understanding of mangers and researchers who often speak in different terms. It is always advised to spend adequate time at the beginning of a project to ensure that all the participants have a common understanding of intended outcomes, technical requirements, resource needs, and timelines. It will maximize the likelihood of the assessment helping achieve the conservation and/or management goals.

*Determining right scales (spatial and temporal)*: Climate change vulnerability assessments for the water sector can be done at the scales equivalent to jurisdictions of the management organizations (e.g., local, regional, and national) or river basin. If we consider water resource as the specific target of the vulnerability assessment, it may be relevant to focus on river basin scale. In some cases, even when our management needs are very local, by its nature, climate change will require us to think and plan within the context of larger landscapes. Another key consideration is which climate change scenarios to use and over what time frame. Multiple scenarios are available based on a range of assumptions, including future emission trends, levels of economic activity, and other factors. Identifying the potential impacts of climate change under multiple scenarios and time steps (e.g., 10, 25, 50 years) will be important to inform a range of possible management strategies. Looking at the far future for an assessment tends to have lower degree of certainty than the projections for the near future.

*Select suitable assessment approach*: Assessment approach should be selected based on user needs, available resources, and data. Using more data and/or increasing complexity is not always better unless the need be. Suitable approach may range from qualitative assessments based on expert knowledge to highly detailed quantitative analysis using physically based distributed hydrological and climate models. Selecting a suitable approach may depend on a host of factors, including the availability of already existing information, the level of expertise, time and budget constraints, among others.

## 7.4.2 GATHERING RELEVANT DATA AND EXPERTISE

Relevant data, information, and expertise can be gathered through literature review, reaching out to subject experts, and relevant stakeholders in the study area. Consideration of scale (spatial and temporal) is always important. Projected future state of climate for certain point of time in the future can be obtained from several secondary sources. Climate variables relevant to water sector can be obtained from those sources and be downscaled (if necessary).

## 7.4.3 ASSESSING VULNERABILITY COMPONENTS

This step consists of determining likely exposures, sensitivities, and adaptive capacity of the water sector; computing VI; assessing uncertainties associated with the assessment; and interpreting and disseminating the vulnerability results. Please refer to Sections 7.5 and 7.6 for determining vulnerability components and computing VI and Section 7.7 for uncertainty assessment.

## 7.5 VULNERABILITY COMPONENTS

Vulnerability can be expressed as a function of the following three components (IPCC 2007a): (1) *exposure* of a particular system climate change, (2) its *sensitivity* to those changes, and (3) its *capacity to adapt* to those changes. Figure 7.3 outlines the relationship among the components. Likely consequence (i.e., vulnerability) of climate change on a system depends on degree of change (i.e., exposure) that the system is projected to experience along with its likely responses (i.e., sensitivity) and its ability to reduce/moderate the potential impacts (i.e., its adaptive capacity).

The following steps are generally followed to quantify the vulnerability components using an indicator-based approach: (1) identify determinants of the components, (2) select suitable set of indicators, (3) establish/describe functional relationship of the indicators to the vulnerability, (4) calculate indicator values/scores and evaluate them, and (5) aggregate the indicators in the form of an index (if necessary).

The determinants and indicator of each of the components and their relation with the vulnerability depends on areas of consideration within the water sector (e.g., flood, drought, water management, water supply, hydropower, water resources system). Selection of the indicators depends upon the purpose of study, scale, type of system under consideration, preferences of the researcher, and the availability of data. The indicator should be easy to understand, minimum in number but sufficient to represent the concept, easy to measure and quantify, and replicable to other areas. The following subsections detail the vulnerability components, indicators, and computation of the component index.

### 7.5.1 Exposure

It is a measure of how much of a change in the climate and associated problems a system is likely to experience. It is the degree, duration, and/or extent to which a system is in contact with a climate perturbation. The exposure basically includes variables for climate change/variability, atmospheric concentration of greenhouse gas,

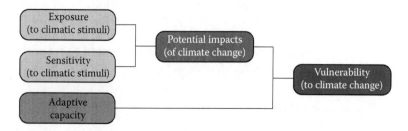

**FIGURE 7.3** Relationship among the vulnerability components.

# Climate Change Vulnerability Assessment

and nonclimatic factors. Depending upon the objectives of the vulnerability assessment, exposure could be depicted from analysis of either historic observation (retrospective assessment) or future modeled projections (prospective assessment) or a combination of both. One or the other may be more appropriate depending upon objectives of the assessment and availability of resources and data. However, a combination of both retrospective and prospective assessments provides the most complete picture in terms of the current and the likely future status. Historic changes are generally analyzed to assess current vulnerability as compared with the past, while future climate projections are analyzed for future vulnerability of the system.

The key exposures related to climate change impacts in the water sector include basic climate and hydrologic variables. The climate variables that directly increase vulnerability are temperature, precipitation, wind, humidity, cloud cover, and solar radiation. The variables also increase vulnerability indirectly, for example, by changing hydrology. The change in mean values of these variables can be used in vulnerability analyses (e.g., changes in average annual temperature or total annual precipitation). However, changes to the extreme values of these variables (e.g., daily minimum or daily maximum temperatures) may be more important for determining vulnerability. Examples of exposure related to hydrologic change include snowmelt, runoff, and stream flow shift. Further to these, soil water storage, groundwater flows, variation in groundwater tables, and evapotranspiration (ET) can also be considered as exposure related to hydrologic change that increases vulnerability of a system to climate change. The basic climate and hydrologic variables can be measured for different time periods—annually, seasonally, or within specific months. Understanding which of these time periods and climate measures are relevant for the particular assessment is a key for determining climate change vulnerability.

Possible indicators of exposure in the water sector and their functional relationships with vulnerability are listed in Table 7.1. The set of indicators for a particular study depends on issues/areas of consideration within the water sector such as water resources management or flood or drought or water supply or hydropower or sea level rise or water quality. The indicators can be combined in a form of index following the approach discussed in Section 7.6.1.

## 7.5.2 Sensitivity

It is a measure of whether and how the system is likely to be affected by a given change in climate. Sensitivity of a system to climate stimuli can be affected by nonclimatic factors/stressors (e.g., a wide range of environmental, economic, social, demographic, technological, and political factors). Many of the critical sensitivity elements for water sector are associated with the earth's physical systems and hydrological processes. All the sectors related to water are sensitive to changes in hydrology such as timing and volume of stream flow, availability of groundwater, frequency and magnitude of extreme events (flood and drought), snow melt, etc. Some processes related to quality of water (e.g., transport of nutrient, sediments) are sensitive to changes in temperature or precipitation. Changes in river flow and water temperatures, for instance, are likely to have an impact on eutrophication and oxygen depletion in wetland systems.

## TABLE 7.1
## Possible Indicators and Their Functional Relationships with Vulnerability

| Component | Indicator | Proxy For | Functional Relationship |
|---|---|---|---|
| Exposure | Change in $P$, $T_{max}$, and $T_{min}$ from base year value (%) | Extent of CVC | ↑↓ |
| | No. of flood/drought events | Extreme climatic events | ↑ |
| | Elevation (masl) | Risk of inundation | ↓ |
| | Population density (persons/km²) | Potential loss in society from climate change impacts | ↑ |
| Sensitivity | $P > 80$ mm (day) | Number of extreme events | ↑ |
| | Maximum $P$ (mm) | Extent of extreme events | ↑ |
| | Total $P$ in rainy season (mm) | Potential to harm from flooding | ↑ |
| | Annual runoff (m³/capita) | Population pressure on WR | ↓ |
| | CV of precipitation | Reliability of available WR | ↑ |
| | Water exploitation (as % of available WR) | Exploitation pressure on WR | ↑ |
| | Areas without irrigation coverage (%) | Expected pressure on WR from future irrigation water use | ↑ |
| | Population without access to water supply and sanitation | Further need of basic services to buffer against CVC | ↑ |
| | Groundwater availability (m³) | Available WR to cope with emergency needs | ↓ |
| | Fertilizer use (kg/ha) | State of health of the WR system | ↑ |
| | Waste water discharge (as% of total WR) | State of health of the WR system | ↑ |
| | Hydroelectric installed capacity (MW) | River flow alterations upstream from man-made structures | ↑ |
| Adaptive capacity | Area under vegetation and wetlands (%) | Degree of natural condition in terms of ecosystem functioning | ↓ |
| | GDP/capita | Access to technology and other resources useful for adaptation | ↓ |
| | Population below poverty line (%) | Capacity to go for adaptation options | ↓ |
| | Nonagricultural employment (%) | Reliability of economic wealth through income diversification | ↓ |
| | Adult literacy rate (%) | Knowledge and awareness to cope with CVC | ↓ |

# Climate Change Vulnerability Assessment

## TABLE 7.1 (continued)
### Possible Indicators and Their Functional Relationships with Vulnerability

| Component | Indicator | Proxy For | Functional Relationship |
|---|---|---|---|
| | Economically active population (%) | Human capital available for adaptation | ↓ |
| | Water governance status | Capacity for the management of various problems of WR | ↓ |

*Sources:* Bae, D.H., *Climate Change Land Manage.*, 3, 32, 2005; Brooks, N. et al., *Global Environ. Change*, 15, 151, 2005; Moss, R.H. et al., *Vulnerability to Climate Change: A Quantitative Approach*, Prepared for the US Department of Energy, Pacific Northwest National Laboratory, Richland, WA, 2001; Pandey, V.P. et al., *Water Sci. Technol./Water Supply*, 9(2), 213, 2009; Pandey, V.P. et al., *Water Sci. Technol.*, 61(6), 1525, 2010; Pandey, V.P. et al., *Ecol. Indic.*, 11(2), 480, 2011; World Economic Forum, Environmental sustainability index, *World Economic Forum*, 2002, available at http://sedac.ciesin.columbia.edu/data/collection/esi.

*Note:* WR is water resources, CV is coefficient of variation, P is precipitation, $T_{min}$ and $T_{max}$ are minimum and maximum temperatures, no. is number, masl is meters above mean sea level, CVC is climate variability and change, MW is megawatt, GDP is gross domestic product, and ↑/↓ is vulnerability increases/decreases with increase in indicator value.

Table 7.1 lists possible indicators of sensitivity in the water sector and their functional relationships with vulnerability. The indicators can be combined in a form of index following the approach discussed in Section 7.6.1.

### 7.5.3 ADAPTIVE CAPACITY

*Adaptive capacity* refers to the ability or potential of a system to respond successfully to climate variability and change and includes adjustments in both behavior and in resources and technologies. Adaptive capacity can be shaped by the availability of economic resources, access to information, social capital, technology, information and skills, infrastructure, institutions, and equity. From the perspective of water resources system, the determinants may have four dimensions: human capacity (HC), economic capacity (EC), natural capacity (NC), and physical capacity (PC) (Pandey et al. 2011). The four dimensions represent the three pillars of a sustainable state: society, economy, and the environment. The HC represents the social aspect, the EC represents the economic aspect, the NC represents the environmental aspect, and the PC represents integration of all the dimensions of sustainability.

Some dimensions of adaptive capacity are generic, while others are specific to particular climate change impacts. Generic indicators include factors such as education, income, and health, whereas indicators specific to a particular impact, such as drought or floods, may relate to institutions, knowledge, and technology (Eriksen and Kelly 2007; Metzger et al. 2005; Tol and Yohe 2007). Although economic development may provide greater access to technology and resources to invest in adaptation, high income per capita is considered neither a necessary nor a sufficient indicator of the capacity to adapt to climate change. There are many examples where social capital, social networks, values, perceptions, customs, traditions, and levels of cognition affect the capability of communities to adapt to risks related to climate change. For natural systems, adaptive capacity is often considered to be an intrinsic trait that may include evolutionary changes as well as *plastic* ecological, behavioral, or physiological responses. In the context of vulnerability of water resources system, a number of factors (both natural and anthropogenic) are likely to influence the ability of the system to adjust to or cope with climate change.

Table 7.1 lists possible indicators of adaptive capacity in the water sector and their functional relationships with vulnerability. The indicators can be combined in a form of index following the approach discussed in Section 7.6.1.

## 7.6 VULNERABILITY INDEX

### 7.6.1 COMPUTATION OF VULNERABILITY INDEX

Quantitative assessment of vulnerability is usually done by constructing a *VI*. VI in an indicator-based approach can be computed by aggregating the selected set of vulnerability indicators. Index in the context of this chapter is a numerical scale calculated from a set of indicators for the analysis units (e.g., district or region or river basin) (hereafter called as *units*) and used to compare them with one another or with some reference point. This numerical value is used in the ordinal sense, that is, on the basis of this index, different units are ranked and grouped to be relatively less or more vulnerable. It is constructed in such a ways that it always lies between 0 and 1 so that it is easy to compare the units under consideration. It can also be expressed as a percentage by multiplying it by 100.

Computation of VI starts with arranging the indicator values in the form of a rectangular matrix ($m \times n$) with $m$ representing the number of units and $n$ representing the number of indicators considered (Table 7.2). The indicator values will obviously

### TABLE 7.2
### Rectangular Matrix of Data Arrangement for Vulnerability Analysis

| Analysis Unit | Indicator | | | | |
|---|---|---|---|---|---|
| | 1 | 2 | . $j$ | . | $n$ |
| 1 | $X_{11}$ | $X_{12}$ | . $X_{1j}$ | . | $X_{1n}$ |
| 2 | . | . | . . | . | . |
| . | | | | | |
| $i$ | $X_{i1}$ | . | . $X_{ij}$ | . | $X_{in}$ |
| . | | | | | |
| $m$ | $X_{m1}$ | . | . $X_{mj}$ | . | $X_{mn}$ |

# Climate Change Vulnerability Assessment

be in different units, scales, and ranges. To make them free of unit and scale in a range of 0–1, they will be normalized based on their functional relationship with vulnerability. Two types of functional relationships are possible: vulnerability increases with increase (or decrease) in the indicator value. Min–max approach can be used to normalize the indicator values: $(X_{ij} - X_{min})/(X_{max} - X_{min})$ if increase in indicator value results increase in vulnerability, otherwise $(X_{max} - X_{ij})/(X_{max} - X_{min})$, where $X_{ij}$, $X_{max}$, and $X_{min}$ are the original values of the $j$th indicator for the $i$th unit and maximum and minimum values of all the units considered, respectively. Normalized scores can now be aggregated together using a composite index approach to compute VI (Equation 7.1), where $w_j$ is the weight to the $j$th indicator and $x_{ij}$ is the normalized value of $X_{ij}$:

$$VI = \sum_{j=1}^{n}(w_j \times x_{ij}) \quad (7.1)$$

Depending upon importance to the vulnerability, equal or differential weights can be assigned to the indicator values in Equation 7.1. Since the vulnerability indices are multivariate in nature, it is also possible to apply multivariate statistical analysis tools (e.g., principal component analysis and cluster analysis) to obtain weights for the indicators. Equal weights are assigned if all the indicators are expected to contribute equally to the vulnerability, while differential weights are assigned if some indicators are considered more important against the others. Differential weights can be calculated by the following two ways: expert judgments (e.g., analytical hierarchy process [Saaty 1980]) and statistics-based weight (e.g., variance-based weight by Iyengar and Sudarshan [1982]). The variance-based method calculates weights using Equation 7.2, where weights are assumed to vary inversely with standard deviations (SD) to ensure that large variation in any one of the indicators would not unduly dominate the contribution of the rest of the indicators and distort interunit comparisons. The Appendix illustrates computation of VI with unequal weights for a hypothetical case:

$$w_j = \frac{1}{\left(SD_i \times \sum_{j=1}^{n}(1/SD_i)\right)} \quad (7.2)$$

## 7.6.2 Interpretation and Communication of Vulnerability Results

VI computed as discussed in Section 7.6.1 lies between 0 and 1, with 1 indicating maximum vulnerability and 0 no vulnerability at all. To interpret the level of vulnerability, a vulnerability scale needs to be prepared like the one shown in Table 7.3. The VI range and interpretation may vary depending upon the scale, type of system or sector under consideration, and set of vulnerability indicators selected.

The policy recommendations are then made based on the VI score, its components, and indicator values. Based on the VI score, $m$ units (under consideration) can be ranked; 1 representing the unit with highest vulnerability and the $m$th as the lowest. Vulnerability component values can be shown in a form of radar diagram to

**TABLE 7.3**
**Interpretation of Vulnerability Index**

| VI | Interpretation |
|---|---|
| Low (0.0–0.2) | Healthy unit in terms of resource richness, development practice, ecological state, and management capacity. No serious policy change is needed. However, it is still possible for the basin with moderate problems in one or two aspects of the assessed components, and policy adjustment should be taken into account after examining the VI structure. |
| Moderate (0.2–0.4) | The unit is generally in a good condition toward realization of sustainable water resources management. However, it may still face high challenges in either technical support or management capacity building. Therefore, policy design of the basin should focus on the main challenges identified after examination of the VI structure, and storing policy interventions should be designed to overcome key constraints of the river basin. |
| High (0.4–0.7) | The unit is under high stress, and great effort should be made to design policy that provides technical support and policy backup to mitigate the high pressure. A long-term strategic development plan should be made accordingly with a focus on rebuilding of management capacity to deal with the main threats. |
| Severe (0.7–1.0) | The unit is highly degraded in water resources system with poor management setting up. Restoration of the basin's water resources management will need high commitment from both government and general public. It will be a long process for the restoration, and an integrated plan should be made at basin level with involvement from international, national and local agencies. |

*Source:* Huang, Y. and Cai, M., *Methodologies Guidelines: Vulnerability Assessment of Freshwater Resources to Environmental Change*, Developed jointly by United Nations Environment Programme and Peking University, in collaboration with Asian Institute of Technology and Mongolia Water Authority, Beijing, China, 2008.

further visualize the sectors/factors that render the water sector more vulnerable. The indicator values may infer extent of intervention needed to improve the situation of vulnerable systems.

## 7.7 UNCERTAINTIES IN VULNERABILITY ASSESSMENT

Each piece of scientific information from a variety of sources (e.g., field studies, modeling experiments, experimental studies, secondary data) and expert knowledge that are integrated together to produce a vulnerability assessment has different levels of certainty and confidence. No one knows exactly how climate may change or how water resources system may respond to the change at a particular location. A useful way to characterize uncertainty in the assessment process is the level of confidence in a given input or outcome. It is important to understand the level of certainty about different vulnerability components, to identify the range of potential vulnerability

# Climate Change Vulnerability Assessment

given the uncertainties, and to determine what we can and cannot say about the vulnerability of the system. Quantifying uncertainty allows for inclusion into a risk assessment or analysis. Risk assessment involves estimating both the probability of occurring an event and the severity of the impacts or consequences of that event. Analyses of risk, therefore, provide an opportunity to address quantifiable uncertainties through probabilistic calculations. While risk assessment may allow for the inclusion of some types of uncertainty, some others may not be handled through exact quantification. Management decisions, however, can proceed in the face of uncertainty even though they are not quantified, much less reduced, or eliminated. Dealing with uncertainty is nothing new in natural resource management. Being transparent about the general magnitude of uncertainty and understanding the range of possibilities given the uncertainty allows managers to articulate the reasoning for making a specific decision.

## 7.7.1 DEFINING UNCERTAINTY

Uncertainties in the context of climate change vulnerability assessments are related to identifying and modeling the sensitivities, levels of exposure, and adaptive capacity of the assessment target (i.e., water sector). IPCC defines uncertainty as "an expression of the degree to which a value (e.g., the future state of the change system) is unknown either due to lack of information or disagreement about what is known." Sources of uncertainty may range from quantifiable errors in the data to ambiguously defined concepts or terminology or uncertain projections of human behavior. Uncertainty can therefore be represented by quantitative measures (e.g., a range of values calculated by various models) and/or by qualitative statements (e.g., reflecting the judgment of a team of experts).

## 7.7.2 UNCERTAINTY LEVELS

Many techniques are available to quantify and communicate uncertainty in vulnerability assessments. Some uncertainties can be quantified using statistics and modeling approach, while others may require more qualitative assessment. IPCC approach represents the longest focused attempt to describe uncertainty in the context of climate change and reports that uncertainty language generally differs with discipline and builds on quantitative analyses for the physical sciences and on qualitative analyses for the socioeconomic discipline. Two different scales of uncertainties and their languages used in IPCC reports are given in Table 7.4 as an example.

## 7.7.3 UNCERTAINTY ASSESSMENT METHODOLOGIES

Some of the uncertainties can be quantified using statistics and modeling approaches, while others may require more qualitative assessment. A combination of these different methods can be used to bound the uncertainty and understand the range of possibility for vulnerability to climate change (Refsgaard et al. 2007). This section

### TABLE 7.4
### Scales and Languages of Uncertainties

| Confidence Scale (Chance of Being Correct) | | Likelihood Scale (Probability of Occurrence) | |
|---|---|---|---|
| Very high confidence | >9 out of 10 | Virtually certain | >99% |
| High confidence | About 8 out of 10 | Very likely | 90%–99% |
| Medium confidence | About 5 out of 10 | Likely | 66%–90% |
| Low confidence | About 2 out of 10 | About as likely as not | 33%–66% |
| Very low confidence | <1 out of 10 | Unlikely | 10%–33% |
| | | Very unlikely | 1%–10% |
| | | Exceptionally unlikely | <10% |

aims to introduce a few methods available to address uncertainty in climate change vulnerability assessments. Readers are expected to refer to Chapter 4 for more details on the methodologies.

*Monte Carlo Method (MCM)*: It is a common quantitative approach for measuring uncertainty in vulnerability assessments. An MCM is a computer-based statistical technique that uses random sampling to convert uncertainties in the input variables of a model (e.g., incomplete knowledge of the climate sensitivity of a water sector) into probability distributions over output variables.

*Expert elicitation*: It is a formal, systematic process to determine subjective judgments about uncertainties from relevant experts (Refsgaard et al. 2007). This approach is often warranted in cases where there are many sources of uncertainty and where critical information may be unavailable. The results of expert elicitation are often characterized quantitatively as probabilities that represent their levels of confidence. However, it is also important to include documentation of the evidence and criteria used by the experts to support their decisions.

*Scenario analysis*: Projecting the future with multiple scenarios from base assessment is a relatively straightforward way to address uncertainties inherent in future predictions (Walker et al. 2003). Scenario uncertainty implies that there is a range of possible outcomes, but the mechanisms leading to these outcomes are not well understood and it is, therefore, not possible to formulate the probability of any one particular outcome occurring. For example, if downscaled climate models are unable to determine whether future conditions in a particular area will be warmer and wetter or warmer and drier, assessing the vulnerability of water sector under both possible scenarios may be warranted. Similarly, given the currently wide range of possible scenarios for sea level rise and the numerous factors that can affect relative sea level at a local or regional level, projecting future impacts and vulnerability based on a number of scenarios and assumptions may offer the most flexibility for determining possible management strategies.

### 7.7.4 COMBINING UNCERTAINTY FROM MULTIPLE SOURCES

Multiple sources of uncertainty may interact to magnify or reduce the overall uncertainty. Therefore, combining multiple sources of uncertainty in each of the components is really the glue that brings vulnerability assessments together into a synthetic product that can be used for decision making and adaptation planning. A key to combining multiple sources of uncertainty is to identify interactions between the different components, such as how temperature and precipitation interact to affect soil moisture and river flows. This can be done qualitatively through conceptual models, diagrams, and narratives, or more quantitatively, through scientific models and computational algorithms. The method used for combining uncertainty should be chosen based on the methods used to assess uncertainty of the components (e.g., qualitative vs. quantitative), the degree of understanding about the interactions between the components, and the resources available for combining the data (e.g., technological capacity and budget).

## 7.8 SUMMARY

Vulnerability assessment can accommodate a broad spectrum of issues related to water resources availability and management. In the context of projected vastly different climate that the Earth would be exposed in the future, and subsequent negative impacts that the water sector will face, vulnerability assessments provide a basis for planning adaptation strategies. It is certain that the concept of vulnerability is widening and becoming more complex to quantify. However, as promoted by IPCC, climate change vulnerability can be expressed as a function of exposure, sensitivity, and adaptive capacity. Indicator-based approach can be applied to quantify it in a form of an index, in which the components of vulnerability (i.e., exposure, sensitivity, and adaptive capacity) are represented as a set of indicators. Depending upon their functional relationship with vulnerability, the indicators can be normalized to make them unidirectional and to standardize into a uniform range. The normalized indicator scores can then be aggregated in a form of VI using a composite index approach (Equation 7.1). Assigning weights to the indicators is a delicate task as indicators are addressing different issues that may not be related and it may yield biased/misleading results if not carefully prioritized. To avoid biasness in vulnerability results because of differential weights, some studies prefer to use equal weights. However, depending upon the objectives and indicators selection, some indicators could be more important than others. In such cases, differential weights can be calculated using the approaches discussed in Section 7.6.1.

For a successful vulnerability assessment, structuring the vulnerability assessment (i.e., defining framework, objectives, and scope of the vulnerability assessment) would be a key step before computing VI. Experience counts at this stage. Another challenge would be identifying appropriate determinants of the vulnerability components, selecting a suitable set of indicators, and establishing their functional relationships with the vulnerability. Possible list of indicators for climate change vulnerability assessments of the water sector is given in Table 7.1. The list, however, is no way a comprehensive one. Selection of the indicators depends upon the purpose of the study, scale, areas within the water sector (e.g., flood or drought or water resources management or water supply or hydropower or sea level rise or water quality), preferences of the researcher, and the availability of data. The indicator should be easy to understand, minimum in number but sufficient to represent the concept, easy to measure and quantify, and replicable to other areas.

Interpretation and communication of VI should carefully be made so that it makes impact on decision making. It varies across spatial and temporal scales and there exists significant cross scale interactions due to the interconnectedness of economic and climate systems. Because of multiple dimensions of vulnerability, it is always recommended to use an appropriate nomenclature while communicating the vulnerability results. For example, *vulnerability of water supply sector in a specified city to climate change over the next 20 years* can be considered as a fully qualified description of vulnerability.

Finally, vulnerability is not the end point but a tool that provides information about the levels and sources of vulnerability to assist in planning and decision making for adaptation to climate change. However, the assessments alone cannot dictate what those adaptation strategies and/or priorities should be. The choice of whether to focus conservation/adaptation efforts on the most vulnerable location or the most viable one or a combination of the two will be based not only on scientific factors but also on social, economic, and legal values. Making decisions in the face of climate change will depend on a combination of sound science and practical experience modulated by societal values.

## APPENDIX: ILLUSTRATION OF VULNERABILITY ASSESSMENT

This section illustrates the method for vulnerability assessment (discussed in Sections 7.4 and 7.5) with a hypothetical case. The case considers a *river basin X* that covers eight administrative units D1–D8 (hereafter *units*) (Figure 7A.1) and exposed to climate variability and change. For reasons like increasing population density, urbanization and associated water-intensive activities, climate change, and other stressors, water resources is already under stress. To formulate adaptation plans and implementation strategies for protecting water resources, a vulnerability assessment is carried out as a basis. The objective therefore is to rank the different units within the basin based on vulnerability score and suggest critical units that need immediate attention for protecting water resources against the climate change in the basin X.

In this case, an approach described in Sections 7.4—expressing vulnerability as a function of exposure, sensitivity, and adaptive capacity—was considered and vulnerability score is computed as discussed in Section 7.5. A set of eight indicators (two for exposure and three each for sensitivity and adaptive capacity) were selected with careful attention to the objective of the vulnerability assessment, review of possible indicators, and availability of data (Table 7A.1). Their description and functional

# Climate Change Vulnerability Assessment

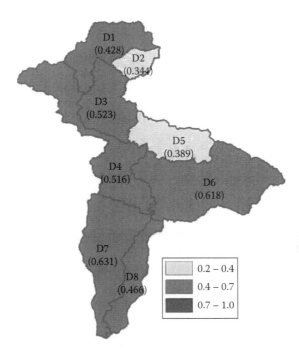

**FIGURE 7A.1** Vulnerability of the eight units (D1–D8) in the hypothetical basin X. Score inside parenthesis indicates VI.

relationship with the vulnerability was established (Table 7A.1). Relevant data for the indicators were collected from several secondary sources (e.g., Central Bureau of Statistics of the Country, Hydrology and Meteorology Department, published papers and reports, gray literatures, and websites of climate-related agencies). The data were processed and arranged in a form of rectangular matrix in Table 7A.2.

To make the indicator values free of units and scale in the range of 0–1, they were normalized following the approach described in Section 7.6.1: $(X_{ij} - X_{min})/(X_{max} - X_{min})$ if increase in indicator value results increase in vulnerability, otherwise $(X_{max} - X_{ij})/(X_{max} - X_{min})$, where $X_{ij}$, $X_{max}$, and $X_{min}$ are the original values of the $j$th indicator for $i$th unit and maximum and minimum values of all the units considered, respectively. The normalized indicator values are tabulated in Table 7A.3.

Weights to the indicators were assigned as per Equation 7.2 (refer Table 7A.3) and VI was calculated as per Equation 7.1 and mapped in Figure 7A.1 using ArcGIS. Based on the VI, the units were ranked from 1 (the most vulnerable) to 8 (the least vulnerable) (Table 7A.3).

The vulnerability of the eight units in the basin X varies from 0.344 to 0.631. As per Table 7.3, the units D2 and D5 are *moderately vulnerable* and the rest are *highly vulnerable*. Based on VI, the units in decreasing order of vulnerability can be listed as D7, D6, D3, D4, D8, D1, D5, and D2. The factors/determinants that make unit D7 the most vulnerable are low adult literacy rate, high rate of change of annual maximum temperature values, and low per capita income compared to other units

## TABLE 7A.1
## Vulnerability Indicators Selected for the Hypothetical Case

| Component | Indicator | Proxy for | Functional Relationship |
|---|---|---|---|
| Exposure | E1: Change in annual rainfall from base year value (%) | Degree of CVC | ↑ |
|  | E2: Change in annual maximum temperature (%) | Degree of CVC | ↑ |
| Sensitivity | S1: Annual precipitation (mm) | Available WR | ↓ |
|  | S2: Water demand/WR (%) | Exploitation pressure on WR | ↑ |
|  | S3: Fertilizer use (kg/ha) | State of health of the WR system | ↑ |
| Adaptive capacity | A1: Area under vegetation and wetlands (%) | Degree of natural condition in terms of ecosystem functioning | ↓ |
|  | A2: GDP index[a] | Access to technology and other resources useful for adaptation | ↓ |
|  | A3: Adult literacy rate (%) | Knowledge and awareness to cope with CVC | ↓ |

*Source:* Pandey, V.P. et al., *Ecol. Indic.*, 11(2), 480, 2011.

*Note:* WR is water resources, GDP is gross domestic product, CVC is climate variability and change, ↑ is vulnerability increases with increase in indicator value, and ↓ is vulnerability decreases with increase in indicator value.

[a] GDP index (%) = $\dfrac{\log(\text{per\_capita\_income}) - \log(\text{min})}{\log(\text{max}) - \log(\text{min})} \times 100$, where the max (or maximum) and min (or minimum) values are set at \$40,000 and \$100, respectively (as used in calculating the Human Development Index by UNDP in 2006)

in the basin X. Moreover, determinants of vulnerability for other units with relatively lower vulnerabilities vary widely (e.g., *higher exploitation of available water resources* for D3 and *high rate of change of annual rainfall* for D4).

For easy visualization and then interpretation, scores of each component (exposure, sensitivity, and adaptive capacity) are also calculated with the approach used for VI and presented them as a radar diagram (Figure 7A.2). The radar diagram clearly shows whether higher vulnerability is related to exposure or sensitivity or adaptive capacity.

This type of analysis, map, and plots, therefore, help understand spatial variation of vulnerability within a river basin and identify critical locations that need urgent attention for water resources protection against the climate change in the basin.

## TABLE 7A.2
### Rectangular Matrix of Vulnerability Indicator Values for the Units in Basin X

|  | Exposure (E) |  | Sensitivity (S) |  |  | Adaptive Capacity (A) |  |  |
|---|---|---|---|---|---|---|---|---|
| Unit ID | E1 | E2 | S1 | S2 | S3 | A1 | A2 | A3 |
| D1 | −3.2 | −0.58 | 2897.7 | 25.1 | 286.6 | 16.2 | 49.0 | 70.3 |
| D2 | 3.8 | 1.65 | 2576.2 | 12.0 | 216.3 | 33.4 | 59.0 | 77.1 |
| D3 | 14.6 | 1.76 | 2381.2 | 21.0 | 213.5 | 52.5 | 46.0 | 63.8 |
| D4 | 29.8 | 1.89 | 2177.5 | 16.0 | 182.7 | 52.8 | 50.0 | 70.8 |
| D5 | 12.6 | 1.64 | 1926.7 | 14.0 | 123.3 | 70.0 | 49.0 | 63.2 |
| D6 | 6.6 | 2.72 | 1723.7 | 10.0 | 117.5 | 29.7 | 36.0 | 32.5 |
| D7 | 17.4 | 1.42 | 1430.1 | 8.9 | 128.2 | 25.5 | 35.0 | 36.2 |
| D8 | 8.4 | 2.38 | 917.0 | 4.8 | 92.8 | 71.9 | 40.0 | 50.1 |
| Max | 29.8 | 2.72 | 2897.7 | 25.1 | 286.6 | 71.9 | 59.0 | 77.1 |
| Min | −3.2 | −0.58 | 917.0 | 4.8 | 92.8 | 16.2 | 35.0 | 32.5 |
| Max–Min | 33.0 | 3.30 | 1980.8 | 20.3 | 193.9 | 55.7 | 24.0 | 44.6 |

## TABLE 7A.3
### Normalized Indicator Scores as per Functional Relationship in Table 7A.1

|  | Exposure (E) |  | Sensitivity (S) |  |  | Adaptive Capacity (A) |  |  |  |  |
|---|---|---|---|---|---|---|---|---|---|---|
| Unit | E1 | E2 | S1 | S2 | S3 | A1 | A2 | A3 | VI | Rank |
| D1 | 0.00 | 0.00 | 0.00 | 1.00 | 1.00 | 1.00 | 0.42 | 0.15 | 0.428 | 6 |
| D2 | 0.21 | 0.68 | 0.16 | 0.35 | 0.64 | 0.69 | 0.00 | 0.00 | 0.344 | 8 |
| D3 | 0.54 | 0.71 | 0.26 | 0.80 | 0.62 | 0.35 | 0.54 | 0.30 | 0.523 | 3 |
| D4 | 1.00 | 0.75 | 0.36 | 0.55 | 0.46 | 0.34 | 0.38 | 0.14 | 0.516 | 4 |
| D5 | 0.48 | 0.67 | 0.49 | 0.45 | 0.16 | 0.03 | 0.42 | 0.31 | 0.389 | 7 |
| D6 | 0.30 | 1.00 | 0.59 | 0.26 | 0.13 | 0.76 | 0.96 | 1.00 | 0.618 | 2 |
| D7 | 0.62 | 0.61 | 0.74 | 0.20 | 0.18 | 0.83 | 1.00 | 0.92 | 0.631 | 1 |
| D8 | 0.35 | 0.90 | 1.00 | 0.00 | 0.00 | 0.00 | 0.79 | 0.60 | 0.466 | 5 |
| SD | 0.30 | 0.30 | 0.32 | 0.33 | 0.34 | 0.37 | 0.34 | 0.37 | — | — |
| 1/SD | 3.33 | 3.36 | 3.08 | 3.07 | 2.94 | 2.67 | 2.97 | 2.69 | — | — |
| Weight | 0.14 | 0.14 | 0.13 | 0.13 | 0.12 | 0.11 | 0.12 | 0.11 | — | — |

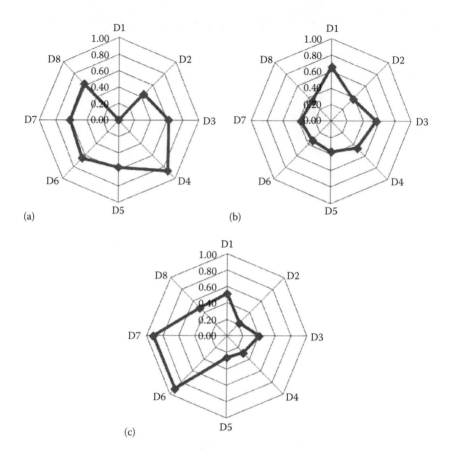

**FIGURE 7A.2** Radar plot of vulnerability component scores for the eight units (D1–D8) in the hypothetical basin X: (a) exposure, (b) sensitivity, and (c) adaptive capacity.

## REFERENCES

Adger, W.N. 1999. Social vulnerability to climate change and extremes in coastal Vietnam. *World Development* 27: 249–269.

Bae, D.H. 2005. The impacts of climate change on water resources and strategies. *Climate Change and Land Management* 3: 32–38.

Birkmann, J. 2006. Measuring vulnerability to promote disaster-resilient societies: Conceptual frameworks and definitions. In: Birkmann, J. (ed.), *Measuring Vulnerability to Natural Hazards: Towards Disaster Resilient Societies*, pp. 9–54. Tokyo, Japan: United Nations University Press.

Bohle, H.G., Downing, T.E., and Watts, M.J. 1994. Climate change and social vulnerability: Towards a sociology and geography of food insecurity. *Global Environmental Change* 4: 37–48.

Brooks, N. 2003. Vulnerability, risk and adaptation: A conceptual framework. Working Paper 38. Norwich, U.K.: Tyndall Centre for Climate Change Research.

Brooks, N., Adger, W.N., and Kelly, P.M. 2005. The determinants of vulnerability and adaptive capacity at the national level and the implications for adaptation. *Global Environmental Change* 15: 151–163.

Downing, T.E., Butterfield, R., Cohen, S., Huq, S., Moss, R., Rahman, A., Sokona, Y., and Stephen, L. 2001. *Climate Change Vulnerability: Linking Impacts and Adaptation*. Nairobi, Kenya: The Governing Council of the United Nations Environment Programme.

Downing, T.E. and Patwardhan, A. 2004. Assessing vulnerability for climate adaptation. In: Lim, B. and Spanger-Siegfried, E. (eds.), *Adaptation Policy Frameworks for Climate Change: Developing Strategies, Policies, and Measures*, pp. 69–89. Cambridge, U.K.: Cambridge University Press (Chapter 3).

Eakin, H. and Luers, A.L. 2006. Assessing the vulnerability of social–environmental systems. *Annual Review of Environment and Resources* 31(1): 365–394.

Eriksen, S.H. and Kelly, P.M. 2007. Developing credible vulnerability indicators for climate adaptation policy assessment. *Mitigation and Adaptation Strategies for Global Change* 12(4): 495–524.

Füssel, H.M. 2007. Vulnerability: A generally applicable conceptual framework for climate change research. *Global Environmental Change* 17: 155–167.

Füssel, H.-M. and Klein, R.J.T. 2006. Climate change vulnerability assessments: An evolution of conceptual thinking. *Climatic Change* 75: 301–329.

Glick, P., Staudt, A., and Stein, B. 2009. *A New Era for Conservation: Review of Climate Change Adaptation Literature*. Reston, VA: National Wildlife Federation.

Handmer, J.W., Dovers, S., and Downing, T.E. 1999. Societal vulnerability to climate change and variability. *Mitigation and Adaptation Strategies for Global Change* 4: 267–281.

Hinkel, J. 2011. Indicators of vulnerability and adaptive capacity: Towards a clarification of the science–policy interface. *Global Environmental Change* 21: 198–208.

Huang, Y. and Cai, M. 2008. *Methodologies Guidelines: Vulnerability Assessment of Freshwater Resources to Environmental Change*. Beijing, China: Developed jointly by United Nations Environment Programme and Peking University, in collaboration with Asian Institute of Technology and Mongolia Water Authority.

IPCC. 2007a. Climate change 2007: Impacts, adaptation and vulnerability. In: Parry, M.L., Canziani, O.F., Palutikof, J.P., van der Linden, P.J., and Hanson, C.E. (eds.), Contribution of Working Group II to the Fourth Assessment Report of the Intergovernmental Panel on Climate Change (IPCC). Cambridge, U.K.: Cambridge University Press.

IPCC. 2007b. Climate change 2007: Mitigation. In: Metz, B., Davidson, O.R., Bosch, P.R., Dave, R., and Meyer, L.A. (eds.), Contribution of Working Group III to the Fourth Assessment Report of the Intergovernmental Panel on Climate Change (IPCC). Cambridge, U.K.: Cambridge University Press.

Iyengar, N.S. and Sudarshan, P. 1982. A method of classifying regions from multivariate data. *Economic and Political Weekly* 17(51): 2048–2052.

Kasperson, J.X. and Kasperson, R.E. 2001. International workshop on vulnerability and global environmental change: SEI Risk and Vulnerability Programme Report 2001-01. Stockholm, Sweden: Stockholm Environment Institute (SEI).

Kasperson, J.X., Kasperson, R.E., Turner II, B.L., Schiller, A.M.D., and Hsieh, W. 2005. Vulnerability to global environmental change. In: Kasperson, J.X. and Kasperson, R.E. (eds.), *Social Contours of Risk*, Vol. II: *Risk Analysis Corporations and the Globalization of Risk*, pp. 245–285. London, U.K.: Earthscan (Chapter 14).

Kelly, P.M. and Adger, W.N. 2000. Theory and practice in assessing vulnerability to climate change and facilitating adaptation. *Climatic Change* 47: 325–352.

Liverman, D.M. 1990. Vulnerability to global environmental change. In: Kasperson, R.E., Dow, K., Golding, D., and Kasperson, J.X. (eds.), *Understanding Global Environmental Change: The Contributions of Risk Analysis and Management*, pp. 27–44. Worcester, MA: Clark University (Chapter 26).

Luers, A.L., Lobell, D.B., Sklar, L.S., Addams, C.L., and Matson, P.A. 2003. A method for quantifying vulnerability, applied to the Yaqui Valley, Mexico. *Global Environmental Change* 13: 255–267.

Metzger, M.J., Leemans, R., and Schröter, D. 2005. A multidisciplinary multi-scale framework for assessing vulnerabilities to global change. *International Journal of Applied Earth Observation and Geoinformation* 7: 253–267.

Moss, R.H., Malone, E.L., and Brenkert, A.L. 2001. *Vulnerability to Climate Change: A Quantitative Approach.* Prepared for the US Department of Energy. Richland, WA: Pacific Northwest National Laboratory.

Nitschke, C.R. and Innes, J.L. 2008. Integrating climate change into forest management in South-Central British Columbia: An assessment of landscape vulnerability and development of a climate-smart framework. *Forest Ecology and Management* 256: 313–327.

O'Brien, K., Eriksen, S., Nygaard, L.P., and Schjolden, A. 2007. Why different interpretations of vulnerability matter in climate change discourses. *Climate Policy* 7: 73–88.

O'Brien, K., Eriksen, S., Schjolen, A., and Nygaard, L. 2004. What's in a word? Conflicting interpretations of vulnerability in climate change research. CICERO Working Paper 04 Oslo, Norway.

Olmos, S. 2001. Vulnerability and adaptation to climate change: Concepts, issues, assessment methods. Foundation Paper Prepared for Climate Change Knowledge Network (www.cckn.net).

Pandey, V.P., Babel, M.S., and Kazama, F. 2009. Analysis of a Nepalese water resources system: Stress, adaptive capacity and vulnerability. *Water Science and Technology/Water Supply* 9(2): 213–222.

Pandey, V.P., Babel, M.S., Shrestha, S., and Kazama, F. 2010. Vulnerability of freshwater resources in large and medium Nepalese river basins to environmental change. *Water Science and Technology* 61(6): 1525–1534.

Pandey, V.P., Babel, M.S., Shrestha, S., and Kazama, F. 2011. A framework to assess adaptive capacity of the water resources system in Nepalese river basins. *Ecological Indicators* 11(2): 480–488.

Refsgaard, J.C., van der Sluijs, J.P., Højberg, A.L., and Vanrolleghen, P.A. 2007. Uncertainty in the environmental modeling process: A framework and guidance. *Environmental Modeling & Software* 22: 1543–1556.

Saaty, T. 1980. *The Analytical Hierarchy Process.* New York: McGraw-Hill.

Timmermann, P. 1981. Vulnerability, resilience and the collapse of society. *Environmental Monograph*, Vol. 1. Toronto, Ontario, Canada: Institute for Environmental Studies, University of Toronto.

Tol, R.S.J. and Yohe, G.W. 2007. The weakest link hypothesis for adaptive capacity: An empirical test. *Global Environmental Change* 17(2): 218–227.

Turner II, B.L., Kasperson, R.E., Matson, P.A. et al. 2003. A framework for vulnerability analysis in sustainability science. *Proceedings of the National Academy of Sciences of the United States of America* 100: 8074–8079.

US Fish and Wildlife Service. 2011. Scanning the conservation horizon. Available at http://training.fws.gov/EC/Resources/climate_change/vulnerability.html (accessed 13 April 2013).

Vogel, C., Moser, S.C., Kasperson, R.E., and Dabelko, G.D. 2007. Linking vulnerability, adaptation, and resilience science to practice: Pathways, players, and partnerships. *Global Environmental Change* 17: 349–364.

Walker, W.E., Harremoës, H., Rotmans, J. et al. 2003. Defining uncertainty: A conceptual basis for uncertainty management in model-based decision support. *Integrated Assessment* 4: 5–17.

World Economic Forum. 2002. Environmental sustainability index. *World Economic Forum.* Available at http://sedac.ciesin.columbia.edu/data/collection/esi/.

# 8 Climate Change Adaptation in Water

*S.V.R.K. Prabhakar, Binaya Raj Shivakoti, and Bijon Kumer Mitra*

## CONTENTS

8.1 Introduction ...................................................................................................209
8.2 Decision Support Resources ......................................................................... 211
8.3 Considerations for Integrating Climate Change into Decision Support Systems ........................................................................................................... 215
    8.3.1 Projected Climate Change Impacts ................................................... 215
    8.3.2 Differentiating Adaptation Actions ................................................... 215
    8.3.3 Managing Uncertainty ....................................................................... 217
    8.3.4 Multistakeholder Engagement ........................................................... 217
8.4 Adaptation Options for Water ....................................................................... 217
    8.4.1 Structural Adaptation Options ........................................................... 218
    8.4.2 Nonstructural Options and Approaches ............................................ 219
        8.4.2.1 Community-Based Water Resources Management .......... 222
        8.4.2.2 Water Budgeting and Allocation ....................................... 222
        8.4.2.3 Water Pricing ..................................................................... 223
        8.4.2.4 Water Scheduling .............................................................. 223
        8.4.2.5 Water Rationing ................................................................ 224
        8.4.2.6 Water Trading .................................................................... 225
    8.4.3 Integrated Approaches ....................................................................... 225
        8.4.3.1 Integrated Water Resources Management ........................ 225
        8.4.3.2 Upstream and Downstream Integration ............................ 226
        8.4.3.3 Climate Cobenefits and Water Resources ......................... 227
8.5 Barriers for Mainstreaming Adaptation ........................................................ 229
8.6 Conclusions ................................................................................................... 231
References ............................................................................................................. 233

## 8.1 INTRODUCTION

Water resources play an important role in the sustainable development and are central to all activities of human beings and to the health of the natural ecosystems. For the past several decades, water resources have been subjected to two kinds of pressures, that is, developmental pressures and climate change–related pressures. Among the developmental pressures, the primary demand has been from the rapidly growing population resulting in competing uses from agriculture, industry, and domestic sectors. The additional developmental pressures are also from rapidly changing lifestyles, including eating habits, and growing nexus between water and energy production.

Out of total water withdrawals in the world, agriculture accounts for majority of water use followed by industrial and municipal uses. The other form of water use, which is often less documented, is the evaporation from steadily growing reservoir construction. The regional disparities in water withdrawals are large due to differences in socioeconomic and developmental factors. Asia leads in total water withdrawals and consumption followed by North America and Europe (Shiklomanov 1999). On an average, per capita water withdrawal has been increasing over the years and currently stands between 610 $m^3$/capita/year (World Energy Council 2010) and 652 $m^3$/capita/year (calculated from the world water withdrawal of 4500 billion $m^3$ as reported in 2030 World Water Resources Group [2009]) with considerable difference between developing and developed economies. The continuation of business as usual (BAU) practice may lead to greater freshwater stress and scarcity putting more than 2.8 billion people in 48 countries in Africa at risk by 2025 (Rekacewicz 2005).

In addition to the aforementioned developmental pressures, water resources are put to enormous pressure from climate change–related impacts. In water sector, disasters and climate change deserve special attention since both can undermine the decades of development achieved in the sector. The historical data clearly show that several continents have been vulnerable to several hydrometeorological disasters and the number of these disasters has been steadily increasing over the decades with a clear trend in drought-related disasters. Many cities and towns are already vulnerable to floods and other water-related risks. Climate change is known to bring additional dimension to these risks. Climate change can impact water through disturbing the water cycle leading to hydrometeorological disasters, in terms of excessive rainfall leading to floods or reduced rainfall leading to droughts, which could in turn disrupt natural ecosystems that help maintain the water quality and quantity. The Intergovernmental Panel on Climate Change (IPCC)-led models have suggested that an estimated 75–250 million and 350–600 million additional people will be exposed to water stress by 2025 and 2050, respectively, putting pressure on already depleted water resources in Africa alone. The depletion of freshwater resources is expected to exacerbate in Central, South, East, and Southeast Asia particularly in large river basins by 2050s (Kundzewicz et al. 2007). Changes in runoff in major river basins are particularly projected to be effected with interseasonal disparities, increased runoff during rainy season, and water shortages in dry seasons, affecting the livelihoods across the river basins. Deficit in technology and infrastructure in water sector would further heighten the countries' vulnerability to climate change and variability.

The concept of green growth has gained popularity during recent years as a holistic approach to development that values human, social, and natural capital, efficiently and sustainably using the ecosystem services and building resilience in an increasingly changing world (African Development Bank 2013). As a common denominator, water is necessary for green growth and is impacted by the green growth. Hence, water serves as an important single entry point for implementing green growth strategies. Within the water sector, two important considerations stand out, that is, increasing access to safe drinking water for public consumption including meeting economic demands, which can be achieved through a combination of approaches that increases the quantity of available water, and efficient use of existing water resources. The earlier broader context calls for supply-side strategies that increase the water availability and demand-side approaches that efficiently use the limited available water resources. Overarching efficiency measures on both ends could contribute to both greening the sector and increasing the resilience of livelihoods. Integrated river basin development approaches could provide an easy entry point for such approaches among many others discussed in this chapter.

Keeping the aforementioned background in view, this chapter elicits various tools for prioritizing adaptation actions within water sector, evaluates different adaptation options including structural and nonstructural adaptation, and identifies barriers that could undermine the rapid expansion of these adaptation options. Managing precious water resources involve decision making by various stakeholders at various levels, and these stakeholders are often challenged to take decisions in short-, medium-, and long-term time scales in rapidly changing global and local conditions necessitating simple and scalable decision-making tools that go hand in hand with the multiple objectives these stakeholders are expected to achieve. While many integrated approaches have been in vogue for several years and have been implemented as good developmental practices, this chapter reiterates and finds evidence that these measures could have climate change adaptation benefits as well. Though most cases have been derived from Asia and the Pacific, examples from other parts of the world were also included wherever deemed necessary.

## 8.2 DECISION SUPPORT RESOURCES

Water resources management involves decision making by various developmental practitioners, water managers, and policy makers at various time and geographical scales in increasingly challenging conditions for achieving certain outcomes (discussed in Table 8.1). Hence, it is important that the decision-making resources satisfy the expectations of different stakeholders engaged in water and related sectors. There are several decision support resources these stakeholders resort to for helping meeting their decision-making needs (Gibbs et al. 2012; Palaniappan et al. 2008; Serrat-Capdevila et al. 2011). These resources could be classified as evaluation tools, process guides, technical briefs, technical references, and policy papers (Palaniappan et al. 2008). One of the limitations of the existing traditional decision support resources has been that they often tend to be administrative (i.e., providing more information on water resources, physical assets) and do not help in answering "what if" questions such as what specific infrastructure is suitable for a given project location or what kinds of capacities are required to achieve water management objectives under given conditions. They also lack ability to consider the changing factors with implications for water users. This is particularly of a limitation with the nondynamic and nonevaluative category of decision support resources. To overcome this issue, there have been developments particularly among evaluation tools; both simulation models and optimization models are being employed where simulation models come in handy to answer "what if" question while optimization models help

## TABLE 8.1
## Mapping of Various Stakeholders in the Water Sector at National and Subnational Levels That the Decision Tools Should Be Able to Address

| Level | Stakeholders Involved | Type of Decisions Made | Relevant Climate Change Adaptation Questions |
|---|---|---|---|
| Villages | Communities including farmers, water user associations, and community-based organizations | Mostly decisions have immediate impacts but may also contribute medium to long term (such as construction of water storage tanks, setting use and maintenance rules). | What water practices are suitable to village adaptive capacity? What are the social, economic, and environmental costs and benefits accrued from these practices? What costs and benefits mean to their own well-being and to the local water resources? |
| District and subdistrict | Community-based organizations, water resources and irrigation departments, and water user associations | Decisions ranging between immediate and long term (such as development of programs/projects and its implementation, management of rivers and wetlands, fund-raising). | What water projects and programs are suitable in the (sub)district? What are the technical infrastructure and capacity needs to implement the projects and programs? What is the time scale at which adaptation should take place? What are the social, economic, and environmental costs and benefits accrued from these practices? What do these costs and benefits mean in terms of socioeconomic development? |
| State and country | Local and national governments and ministries such as from environment, water resources, agriculture, and forestry | Most decisions ranging between medium and long term (drafting water policies and legislation, allocating adaptation fund, introducing nationwide programs). | What water programs, policies, legislations, and institutional arrangements are necessary for climate change adaptation? What is the time scale at which adaptation should take place? What are the additional technical, infrastructure, and capacity needs? What are the social, economic, and environmental costs and benefits accrued from these programs and policies, and what do these costs and benefits mean in terms of national development and poverty alleviation? |

## TABLE 8.2
## Some Examples of Decision Support Resources Employed in Water Resources Management

| Decision Support Resource Employed | Nature | Location | Reference |
|---|---|---|---|
| Potomac Reservoir and River Simulation model | Simulation models | Washington, DC Metropolitan Area | Loucks (2008) |
| Multidisciplinary models | Simulation models integrating the Everglades Screening model, the South Florida Water Management model, the Natural System model, and the Across Trophic Level System Simulation model | Greater Everglades, South Florida, United States | Loucks (2008) |
| Bayesian networks | Simulation model in combination with community participation | Upper Guadiana Basin, Spain | Zorilla et al. (2009) |
| Collaborative decision making | Participation among various ministries and transboundary | Langat River Basin, Malaysia | Elfithri et al. (2008) |
| Multicriteria decision-making tool | Optimization model | Generic | Atoyev (2007) |
| Stakeholder-oriented valuation | Stakeholder participation | Generic | Hermans et al. (2006) |

address the questions such as "what should be" (Loucks 2008). Table 8.2 provides a summary of decision support resources available for decision makers in water resources management. However, most of these support resources do not consider implications of climate change in decision making. Some of these support resources provide opportunity for continuous evaluation and improvement and hence are considered to address the issue of uncertainty and adaptive management. There is still a need for developing new decision support resources for integrating the climate change adaptation concerns into the decision-making process (Purkey et al. 2007) for the reason that most of the current decision-making resources are ignorant of climate change impacts and the projected water demand and supply could only be precisely estimated when climate projections are taken into consideration.

Despite their complexity, integrated decision support systems have greater capability to provide answers to complex questions, such as integrating climate change into decision support, than simpler single-model-based decision support systems.

Food and water security integrated system (FAWSIM) is one such integrated decision support system integrating modules from water and food sectors for addressing the climate change concerns emerging from the relevant stakeholders (Li et al. 2011). The model is driven by climate change scenarios and employs a statistical downscaling method for providing impacts at the scale at which the model is applied for decision making. Another example for such integrated decision support systems is the "decision support system for optimal agriculture production under global environmental changes"; developed by the Research Program on Climate Change Adaptation, University of Tokyo, the model integrates climate, agriculture, water, and soil models to provide optimum crop cultivation options for stable farm incomes (Ninomiya 2012).

### TABLE 8.3
### Approaches for Integrating Climate Change Adaptation into Decision Making in Water Resources Management

| Decision Support Resource Employed | Nature of Integrating Climate Change Concerns | Motivation for Integration | Location | Reference |
|---|---|---|---|---|
| Integrated hydrology/water allocation framework constructed on water evaluation and planning platform | Sensitivity, significance, and stakeholder support as determining criteria leading to gap identification | Uncertainty and relevance of impacts to water stakeholders | Sacramento Valley, California, United States | Purkey et al. (2007) |
| Covers the entire the country and hence relevant to most decision support resources employed in the country | Promoting adaptive management principles in combination with use of climate projection scenarios | Assumption that future climate change would be continuation of historical trends | United States | Brekke et al. (2009) |
| Integrated regional water management plans already established in the state of California, United States | Use of downscaled climate change for flooding and ecosystem-related impacts in combination with adaptive management | Realization on possible impacts of climate change on relevance of regional water management plans | California, United States | Conrad (2012) |

There is a need for thorough understanding of factors to be considered for integrating climate change adaptation into the existing or development of new decision support resources. Table 8.3 presents a list of approaches reported in the published literature for integrating climate change adaptation into decision support systems. Evaluation of these experiences suggests the following: (1) The access to dependable climate projection information is of paramount importance to make water resources management decisions climate proof, and (2) the principles such as adaptive management appears to be the second most important prerequisite to deal with the uncertainty component involved in the climate-integrated decision making. The other factors that contribute to making decision support resources climate proof are elaborated in the following.

## 8.3 CONSIDERATIONS FOR INTEGRATING CLIMATE CHANGE INTO DECISION SUPPORT SYSTEMS

From the foregone discussion, it is clear that stakeholders engaged in water resource management need to take into several considerations for integrating climate change adaptation into the decisions taken. These include considering projected climate change impacts, being able to differentiate adaptation actions from those BAU actions, being able to recognize and address the uncertainty involved in climate change impacts, and being able to engage multiple stakeholders in the decision-making process.

### 8.3.1 Projected Climate Change Impacts

Obtaining and integrating the information on projected climate change impacts on water resources is the first and foremost prerequisite for making a decision support resource climate proof. Figure 8.1 depicts the difference between a traditional water resources management approach and an approach where projected climate change information is used. Traditional decision-making approaches are mostly driven by historical data and experiences, while integrating climate change requires using available projected impacts of climate change on water resources and other elements of the project or decision boundary.

### 8.3.2 Differentiating Adaptation Actions

Two seemingly mutually contradictory but relevant notions could be found on how climate change adaptation should be compared to BAU practices. One notion says that climate change adaptation needs to be completely different from BAU practices and that the practices would have to be invented for the future climate. The second notion says that climate change adaptation practices could be a selection among BAU practices but implemented under different circumstances (different socioeconomic, geographical, and climate contexts) leading to radically different outcomes. These could challenge water managers while deciding the adaptation actions in water sector. Decision-making tools should be able to help resolve this conflict and be

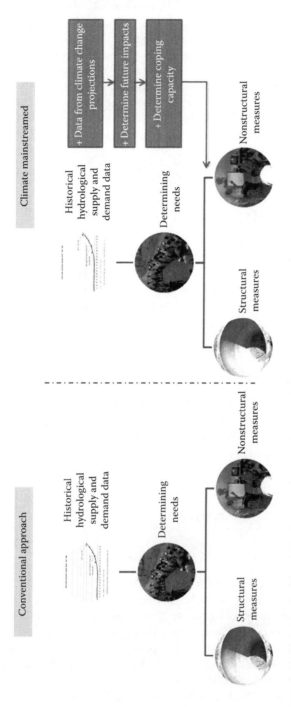

**FIGURE 8.1** Comparing conventional approach with climate-mainstreamed decision-making process.

# Climate Change Adaptation in Water 217

able to help water managers identify adaptation benefits accrued from individual adaptation practices irrespective of whether they are traditional practices or are latest innovations.

### 8.3.3 Managing Uncertainty

Among the challenges faced by the decision makers is the uncertainty factor associated with the nature and degree of impacts of climate change (World Water Assessment Program 2012) in the medium to long time scales. Taking into consideration the uncertainty factor is of paramount importance since most water-related decisions involve infrastructure investments with long shelf life and long gestation period for returning adaptation benefits. Decision-making tools should be able to incorporate the uncertainty factor or be able to provide the decision maker sufficient information based on which uncertainty could be addressed. More details on climate change uncertainties are provided in Chapter 4.

### 8.3.4 Multistakeholder Engagement

Water is the single most important entry point for most developmental activities where several stakeholders are engaged. Recognizing this fact is even more important especially given that the nexus between water, energy, and food production are increasingly being identified as an important area of intervention for resource efficiency and effectiveness. From the multistakeholder engagement point of view, the simplicity and relevance of tools are two important criteria since a range of decision-making tools could envisage from complex to simple, which could either become too much complex for some stakeholders or be too much simple for others. Developing a decision-making tool that is relevant for most stakeholders is a challenge to be overcome while not jeopardizing the relevance of the tool. As a result, these tools should be able to recognize and resolve the nexus between different water-dependent sectors and activities at least at macro level.

## 8.4 ADAPTATION OPTIONS FOR WATER

A variety of structural and nonstructural measures are discussed in this section with a range of advantages that could help achieve water management objectives from the point of view of climate change adaptation. Several structural and nonstructural interventions have been implemented throughout the history of water resources management. The most significant form of interventions is related to structural options, for the reason that controlling and managing water has been the priority for most part of the past developmental history, which was primarily done through land manipulation to harvest runoff, erecting dams for storing water, and construction of canals for channel water from where it is stored to the place needed. Structural options made up most of the interventions when water scarcity was to be dealt during the early and mid-1900s. For most part, the structural options worked well wherever strict guidelines and regulations were put in place. The failure of structural interventions, mostly in terms of maintenance and benefit sharing, has led to the advent of nonstructural interventions primarily to make sure that the ownership of structural interventions is passed on from a centralized agency to decentralized entities such as local communities and user associations that have greater incentives to safeguard the structural interventions being primary benefactors from these interventions. This has resulted in better management of structural interventions while also building the capacities of water users. As a result, one could see gaining importance of nonstructural measures over the past several decades. Today, most water managers and practitioners give equal or even greater importance to nonstructural measures, as situation demanded, in their planning processes.

### 8.4.1 Structural Adaptation Options

Modern history of water resources management is analogous to engineering interventions such as construction of multipurpose dams/reservoirs, water conveyance structures, extensive network of irrigation canals, water supply and sewer facilities, and flood prevention dikes. Also known as structural measures, most of the existing water infrastructures were designed based on historical hydroclimatic observations and accumulated experiences. An underlying basic assumption was that hydroclimatic events tend to follow a uniform pattern in the medium term, such as 50 or 100 years, and it served as a reliable guide for decision making. This convention of designing water infrastructure is being challenged recently due to climate change (Bates et al. 2008), which will modify the statistics of climatic variables at a quicker pace than we are used to (Hallegatte 2009). Existing structural or new measures may not be able to withstand a large range of hydroclimatic shocks and new extremes striking on a frequent basis, thereby making the design more difficult and the construction too expensive (Hallegatte 2009). Additional reinforcing to cope with increasingly uncertain water regimes needs to be identified and developed soon.

A number of factors need to be considered to devise structural measures of adaptation in the water sector because existing water infrastructures could have limitations such as inflexible and single-purpose design and less adaptive operation and management. Strategies like scenario analysis, no-regret/low-regret options, and/or good practices need to be trialed to improve our learning-by-doing capacity in the short term. The use of multipurpose infrastructure could minimize the risk of failure even after exposure to a climate shock in the future. By enabling diversified use of a measure, it could be possible to switch or reverse functions to cope with entirely distinct events that could unfold in series at a place such as use of dams to control floods and mitigate droughts, alternative use of irrigation canal for drinking water supply, and use of flood retention area to recharge groundwater. As structural measures require relatively high upfront investment, assessments from different angles could assist in singling out measures that are effective, productive, efficient, and proven as low risk. Safer strategies such as win-win, no-regret, low-regret, reversible, or higher safety margins could be explored initially (Asia Pacific Water Forum 2012; Hallegatte 2009; UN-Economic Commission for Europe 2009).

Structural measures of adaptation could be on the hard side involving major construction. On the other hand, soft practices take advantage of natural infrastructures and its manipulation or combination of both (Glibert and Vellinga 1990;

Linham and Nicholls 2010; Stalker 2006). While it is not a prerequisite that all adaptation measures should be novel and use advanced technology, in fact they could be among the already well-established and fairly simple measures being practiced at different locations or during specific occasions (Füssel 2007). Here, the basic distinction is in segregating "adaptation relevancy" on how a candidate measure leads to increased adaptive capacity of the people, minimize/escape chances of exposure, or decrease vulnerability from future impacts. However, as already indicated, existing measures may need additional customization, reinforcement, or innovations to improve their robustness and resilience to tackle with a range of uncertain scenarios. This is an area where more research, knowledge transfer, and capacity building are required.

Adaptation to too much water intends to reduce vulnerability from hazards, such as floods and cyclones, by improving combination of four categories of capacity: threshold (or preventing damage), coping (or reducing damage), recovery (or damage reactions or resilience), and adaptation (or damage anticipation) (Graaf 2008). Structural measures have an important function under too-much-water situation because they could protect from direct exposure to the impacts, accommodate impact without significant losses/damages, or facilitate (planned) retreat (Linham and Nicholls 2010). Structures like drainage canals, dams, and dikes along the river bank, multipurpose flood and cyclone relief shelters, and floating agriculture fall under this category. Structural measures under the situation of too much water need to consider postevent needs such as continuity of basic services like safe drinking water and sanitation, health services, food supply, and safe shelters. Hence, supplementary structures for emergency situation should also be introduced alongside (Sinisi and Aertgeerts 2010).

Adapting to too little water, such as acute water shortages or drought, is rather a complex issue because impacts are mostly nonstructural and insidious. It is often extremely difficult to forecast the initiation and cessation of an event such as drought (Pereira et al. 2009). Under the supply side, the primary target would be to expand capture and storage of water such as rainwater harvesting; diverting excess runoff to storage dams/ponds or canals; water transfers; and building overhead or underground tanks, irrigation cum drinking water dams/ponds, etc. Table 8.4 highlights the supply-side measures to deal with too-much-water situation that could be adopted as adaptation measures, where and when found appropriate.

In addition to such supply-side interventions, more can be done through structural measures intended for improving access, distribution, and application at the point of use. These strategies are useful where further scope for increasing volume of water supply has terminated. Examples are improving network delivery and minimizing leaks/evaporation, use of low water-consuming application methods (such as drip irrigation), and promoting water recycling and reuse.

### 8.4.2 Nonstructural Options and Approaches

Structural (hard) measures are recognized as essential part of future climate change adaptation in water sector. However, decades of experience in water resources management revealed that hard approaches alone cannot solve water problems due to

## TABLE 8.4
## Structural Measures of Adaptation

| Specific Measures | Adaptation Functions and/or Benefits | Example Cases |
|---|---|---|
| *Capture* | | |
| Infiltration (such as into canals, river bank, spreading) and injection wells | Could be no regret/low regret<br>Source diversification<br>Could be applied as a distributed system (will increase redundancy)<br>Prevent loss of excess water (rain and runoff water, used water) | Multiple cases (Steenbergen and Tuinhof 2010; World Bank 2010) |
| Controlled flooding/water diversion (spate irrigation) | Retain excess runoff for groundwater storage or direct use<br>Effective in flood prevention | Pakistan, Iran, North, Africa, Sudan, and Yemen (Steenbergen and Tuinhof 2010) |
| Rooftop rainwater harvesting (small scale) | Source diversification<br>Effective for household water supply<br>Could be used for groundwater recharge | Multiple cases (Elliot et al. 2011; UN-HABITAT 2005) |
| *Storage* | | |
| Dams/reservoirs | Supply security<br>Multiple functions (irrigation, drinking water, hydro-energy/cooling, flood prevention, recreation, aquaculture, auto-recharge of groundwater) | Multiple cases (Bates et al. 2008; Kabat and van Schaik 2003) |
| Overhead/underground storage tanks (preferably coupled with rainwater harvesting) | Easy access<br>Increase reliability of supply during water shortages<br>Suitable for domestic uses and microirrigation systems | Multiple cases (Steenbergen and Tuinhof 2010; Vulnerability and Adaptation Programme 2009; WOCAT 2011) |
| Constructed/conservation ponds, small reservoirs, natural wetlands | Easy access<br>Ideal to cope with water scarcity<br>Could be used for floating agriculture and aquaculture | Bangladesh (Selvaraju et al. 2006), India (Vulnerability and Adaptation Programme 2009), Ndiva (Enfors and Gordon 2008), and other cases (Elliot et al. 2011) |
| Groundwater recharge/subsurface dam | Captures excess runoff and reuse water<br>Contributes drought proofing<br>Cheaper compared to surface dams<br>Effective in controlling land subsidence, which increases vulnerable to flooding, and salt water intrusion from sea-level rise | Multiple cases (Elliot et al. 2011; Faurès et al. 2012; Steenbergen and Tuinhof 2010; World Bank 2010) |

various challenges of social, economic, and political issues discussed in the previous section. Generally, hard measures are costly and rigid in nature and may have negative environmental impact (Doswald and Osti 2011). Challenges associated with hard measures can be addressed by combining them with soft measures based on community participation, institutional reforms, incentives, and behavioral change. The soft-path approach targets water demand and supply by behavioral changes of water users and reform water governance systems. Soft options constitute a range of nontraditional elements that include empowering communities in water management, water pricing, water allocation, water scheduling, water restriction, water marketing, and water recycling (Table 8.5). The soft options are often comparatively cost-effective than hard options. Irrigation scheduling and public education were found the least cost adaptation options (MacNeil 2004). Construction of reservoir as a structural option costs almost double that of implementing irrigation scheduling. Proper mix of hard and soft options is very important for future water security. Initial investment on hard infrastructure development may return more benefit, but their sustainability can only be ensured when they are combined with the soft options. When infrastructure is enough to meet basic demand, water investment on soft measures returns

### TABLE 8.5
### Examples of Existing Practices of Soft Measures in Water Resources Management

| Soft Options | Nature | Sector | Location | Reference |
|---|---|---|---|---|
| Community-based water management | Groundwater management | Agriculture | Andhra Pradesh, India | Garduño et al. (2009) |
| Water pricing | Water demand management in agriculture during drought | Agriculture | Australia | Gill (2011) |
| Water restriction | Water demand management in drought | Municipalities | Some municipalities in Colorado | Kenney et al. (2004) |
| Water rationing | Household water demand management during drought emergency | Municipalities | Municipalities in Pennsylvania | Pennsylvania Department of Environmental Protection (2013) |
| Irrigation scheduling | Adapting irrigated agriculture to drought | Agriculture | San Joaquin Valley, California | Ayar (2010) |
| Water trading | Water trading as an approach to climate change, persistent drought | — | Australia | RICS (2011) |

more benefits than hard options. However, the challenge that remains is to achieve a balance between hard and soft options. Some of the nonstructural approaches are discussed hereunder.

### 8.4.2.1 Community-Based Water Resources Management

Weather-dependent farming community is the most vulnerable to climate change. It is expected that climate change–related water disasters such as droughts and floods will hit more frequently and exacerbate rural poverty in developing countries unless suitable adaptation interventions are taken (Institute for Security Studies 2010; International Fund for Agriculture Development 2010). Although farmers have much experience of coping with unexpected climatic events, climate change events may push situation beyond their ability. It is difficult to grow agricultural crops due to uncertainty of rainfall, which affects crop selection, sowing time, irrigation time, and harvesting time.

Community-based water resources management (CBWM) is an approach that enables individuals, groups, and institutions to participate in identifying and addressing water-related local issues (Ali 2011). It is led by the local communities that empower local people for coping with climatic vagaries. In this system, local priorities, knowledge, needs, and capacities are key factors for making an adaptation plan, which offer opportunity of interactions between decision makers and stakeholders. CBWM promotes civic participation in decision making, implementation, and evaluation of water resources management practices and enhances sense of ownership, resulting in successful outcome of water resources management (Marks and Davis 2012). Successful examples of CBWM exist in different parts of the world. CBWM has become popular in the global south, particularly in the rural water with sifting from centralized and technocratic top-down water resources management system (Mehta 1997). Often, top-down approaches fail to meet objective of resource management due to limited reflection of local realities. In Kenya, more than 30% of rural water supply schemes are ran by community-based organization (Rukunga et al. 2006). The community-based groundwater management has been successful in the Andhra Pradesh state of India for almost 20 years covering 638 villages (Mallants 2013). Similar successful cases also exist in Jeppes Reef area of South Africa, where community tap water supply improved lives of people (Thwala 2010). Despite the increasing popularity of CBWM among the state governments, international donors, and NGOs, there are several challenges to promote it due to lack of enabling political environment, institutional arrangements, human resource capacity, heterogenic interest of water users, and financial mechanisms.

### 8.4.2.2 Water Budgeting and Allocation

Water budgeting is a basic means of understanding how much water is entered, stored, evaporated, and drained out of the basin. Water budget is a tool that evaluates water availability in nature and the ability of the water supply to meet demand of water users. Water budget is a foundation for management planning of efficient water resource including water release from the reservoirs, agriculture area selection, cropping pattern, irrigation methods, and identifying required management options. Based on the water allocation plans, farmers can select economically viable cropping

Climate Change Adaptation in Water

pattern under water scarcity (Reddy and Kumar 2008). Furthermore, historical water budgets provide trend of water stress with climate change variability, which can be useful for strategic planning to cope with climate change impact in water sector. It is used as a basic indicator for good water management planning particularly for dry seasons. For example, the Royal Irrigation Department of Thailand, the designated organization for water allocation plan development for the dry season, uses reservoir storage level and projected inflow for water allocation planning in drought, putting priority on human basic needs, irrigation, and industrial uses (Modernization of Water Management System Project 2003). An experiment in Andhra Pradesh of India shows that in dry-season minimum allocation of water during flowering stage and grain formation stage of rice, the total yield in irrigation command area increased by 50%, although per hectare yield decreased by 10% (Bergkamp et al. 2003).

### 8.4.2.3 Water Pricing

Demand-side management is crucial to cope with water scarcity with expected climate change variability for sustainable development. Underprice of water has caused wasteful use of this resource, which intensified water scarcity (Bithas 2008). Water pricing is an economic instrument that can control wastage of water by behavioral changes of users. Furthermore, water pricing works to increase awareness on water scarcity. Water pricing system is subject to regulatory restriction, perception of water right, and political willingness (Dinar 2000; Dinar and Saleth 2005; Le Blanc 2008). A number of successful cases of water pricing together with other measures have been reported in different regions such as Denmark, Madrid of Spain, Bangkok of Thailand, and Bogor of Indonesia (European Communities 2004; Global Water Intelligence 2008; Institute for Global Environmental Strategies 2007). Water pricing should consider climatic context together with social, environmental, and economic cost (Duke and Ehemann 2004). Melbourne City has established variable water price as a water resource risk management tool, which provides direct economic incentive for water conservation (London Climate Change Partnership 2006).

Agriculture is the highest water consumer in developing countries, and improving irrigation water use efficiency would be a sensible point of start to combat water scarcity (Sophie 2013). It is expected that the irrigation water demand will increase due to increase of drought risk from climate change. Since in most countries water is considered as a common resource, farmers have been putting priority on optimal yield and have often overused this limited resource. Water pricing rewards less water use and could encourage farmers to adopt water conservation technological options. Water pricing has been imposed on agricultural water use in different parts of the world, resulting in improvement of water use efficiency in the farms. For example, the irrigation authority of Morocco induced high water price in surface water to discourage wasting of water, which helps to balance the demand in the dry season (Cornish et al. 2004).

### 8.4.2.4 Water Scheduling

Judicious water scheduling in agricultural production can improve water productivity and profitability by reducing nonbeneficial loss due to runoff, deep percolation, and evaporation (Raghuwanshi and Wallender 1998). Water scheduling reduces water demand for irrigation and ensures farmer income under water scarcity. Irrigation scheduling is reported as an effective water management practice, which helps to determine appropriate time of irrigation and amount of water needed for irrigation. Using local climate, soil, and crop information, farmers can irrigate precisely to meet crop requirement under a limited supply of water. Raghuwanshi and Wallender (1998) found that optimum irrigation schedule could reduce irrigation water requirement by 48%–63% in California compared to farmer's full irrigation practices. Water scheduling improvement is particularly effective during dry years when water demand is greater and water conflict is increased. Therefore, using this measure, farmers can reduce their vulnerability to water scarcity. The government of British Columbia of Canada has developed an irrigation schedule calculator, which has proved very useful for the farmers. Such interventions could be useful for farmers of developing countries but need technical capacity building to adopt water scheduling in their climate smart strategies because irrigation scheduling relies on various technical indicators such as soil water content, critical period for water stress, crop water production function, and good weather forecast (Ragab 1996).

Canal irrigation networks have been developed in different parts of the world, which are mainly operated with rigid operation schedule (Kaplesh and Patel 2013). Rigid operation schedule of canal irrigation causing water use efficiency in command area is very low due to mismatching of canal water delivery schedule to critical water periods for growing crops (Devi et al. 2012; Rajput and Patel 2006). Optimum irrigation scheduling of canal can improve climate change adaptation for agriculture (Kalpesh and Patel 2013; Prabhakar et al. 2013). A vast number of planning models have been developed for optimum canal irrigation scheduling; however, irrigation scheduling is only at inception level in most of the developing countries (Ramesh et al. 2009).

### 8.4.2.5 Water Rationing

There is a long tradition of imposing water restriction in to cope with water scarcity particularly during drought and emergency. It is adopted in the drought management plan of urban areas in developed countries such as the United States, Australia, Canada, Singapore, and the United Kingdom, often imposed by national or local governments. The aim of water restriction is to reduce water consumption in nonessential activities such as lawn watering, car washing, and recreational use, which can compensate basic needs of water during water scarcity situation (Kenney et al. 2004; Willis et al. 2013). Water restriction can be voluntary and mandatory, depending on severity of drought period. Mandatory restriction is effective in reducing water consumption than voluntary restriction. Mandatory restriction can reduce net water use by 53% in Lafayette Municipality of Colorado, whereas voluntary water restriction can reduce up to 13% of net water use (Kenney et al. 2004). Successful cases of water restriction in the developed world suggested that water restriction could compensate basic water requirement significantly during climatic events.

### 8.4.2.6 Water Trading

Water trading is the process of shifting water rights, which offers opportunity of judicious water allocation between competitors (Griffin 2006). Water trading has been identified as an effective and economic way to handle the water scarcity situation in different parts of the world, which can reduce required investment for construction of water storage capacity (Becker et al. 1996; Landry 1998). Adoption of water trading will be essential to cope with climate-induced water scarcity in the coming decades (Adam et al. 2013). A study on the creek watershed of Canada reported that the trading scheme would save 1.5 million m$^3$ of water without having any economic loss in comparison with the nontrading mechanism (Luo et al. 2010). Water trading scheme exists in the developed world such as in the United States and Australia (Adam et al. 2013; Brewer et al. 2007). Some countries have set up water trading as an approach to coping climate change and drought; Australia is one of them (RICS 2011). Water trading also exists in the developing countries like India, which tend to be local and informal in nature (Mohanty and Gupta 2002).

### 8.4.3 INTEGRATED APPROACHES

#### 8.4.3.1 Integrated Water Resources Management

Integrated water resources management (IWRM) has gained wider attention since the Earth Summit 1992 at Rio and, subsequently, after Johannesburg Plan of Implementation in 2002 when countries were requested to formulate IWRM plans. Despite ambiguity over the operational aspect of IWRM, its sound concept has continued to attract governments and water experts alike as an infallible solution to solve increasingly complex and heterogeneous water resources management problems and issues that are multisectoral, multiregional, multi-interest (e.g., upstream–downstream), multiagenda, and multicause (Biswas 2008; Grigg 2008). The main attraction point of IWRM lies on its multidimensional scope that aims to integrate nearly all aspects, including cross-sectoral linkages, of water resources management, while recognizing interdependency and trade-off from multiple uses and functions provided by water. Successful implementation of IWRM includes, among others, capturing society's views, reshaping planning processes, coordinating land and water resources management, recognizing water quantity and quality linkages, conjunctive use of surface water and groundwater, and protecting and restoring natural systems (Bates et al. 2008). Ideally, IWRM is a pathway toward sustainable water resources management wherein the principles of equity, efficiency, and environmental integrity are well balanced.

IWRM is considered effective in planning and implementation of climate change adaptation too (Bates et al. 2008; Cap-Net 2009). There could be several reasons behind that assumption. Climate change adaptation involves multiple cross-linkages across sector, administrative levels, stakeholders, and regions. IWRM, as a holistic framework, could be a converging point where those linkages could be better coordinated to result into effective implementation of adaptation actions by overcoming barriers and balancing trade-offs. IWRM could be a common ground for formulating response strategies to manage climatic and nonclimatic stressors as it could enable multidisciplinary, multi-institutional, and multistakeholder coordination (Biswas 2008). Similarly, a mix of structural and nonstructural measures and other multisectoral approaches of adaptation could be synchronized under IWRM. For instance, soft structural measures are best practiced under integrated approaches that intend to harmonize with the way nature functions such as the role of wetlands in absorbing excess runoff and assimilating pollutants. Planning horizon is another dimension to consider. IWRM is a continuous process involving planning at different time scales, so it could be capitalized to accommodate adaptation process as well. IWRM could also be effective in mainstreaming adaptation into national water policies because IWRM is increasingly followed worldwide as the main approach to deal with water resources management issues (Bates et al. 2008).

Both IWRM and adaptation processes are time bound and location specific. Heterogeneity and complexity of water resources management cases often hamper in scoping the right level of integration and scale of adaptation action because the scope may not match well with available resources and capacity. Until now, river basins act as a preferred spatial unit for integration to happen, and numerous river basin organizations (RBOs) have been already established in different parts of the world (lists could be accessed from the International Network of River Basin Organizations, www.inbo-news.org). RBOs are increasingly found inextricably linked to IWRM, but very few RBOs have been found to internalize IWRM principles into real practices (Pangare et al. 2012). In this context, the RBOs also need certain transformations into their practices to make IWRM relevant to cope with challenges from climate change. Adaptive water management (AWM) is a new prerequisite to deal with climate change (Cap-Net 2009; Mysiak et al. 2010). AWM is a value-added extension to IWRM that explicitly incorporates uncertainty arising from climate change and other nonclimatic pressures (Mysiak et al. 2010). AWM is a social and participatory process in which iterative learning from the outcomes of implementation strategies is used to formulate more robust and flexible management practices to cope with the uncertainties and inevitable surprises. For doing that, it parallels two complementary cycles for classifying and analyzing problems: conventional planning cycle and learning cycle. Those problems that could not be managed through conventional planning cycle are subject to learning cycle, which aim to instigate learning experiments.

There are no such universal roadmaps for shifting toward more adaptive IWRM. Hence, models of adaptive IWRM need to be devised within the alternative approaches such as through managed aquifer recharge and buffer management (Steenbergen and Tuinhof 2010), conjunctive uses of surface and groundwater (Foster et al. 2010), or other landscape design, ecosystem-based soft structural approaches.

### 8.4.3.2 Upstream and Downstream Integration

Among integrated approaches, upstream and downstream deserve special attention apart from the IWRM approaches for the significant social implications it has. Upstream–downstream integration is vital for planning and implementation of adaptation measures specifically in the context of river basins. There is an inherent physical relationship of water flow between upstream and downstream. Any adaptation implemented at the upstream could have beneficial, neutral, or nonbeneficial outcomes.

Beneficial adaptation occurs when the flow services and benefits resulting from the management of water at the upstream could be shared downstream such as productivity gains achieved by supplying irrigation water from upstream dams. Neutral adaptation occurs when actions in the upstream have negligible impact to the downstream such as return flow after nonconsumptive use of water. Nonbeneficial adaptation occurs when actions taken at the upstream cause high environmental externalities, competition for water, and transfer of risk such as downstream risk from the construction of flood prevention walls in the upstream, discharge of pollutants, and diversion of water sources. This latter part is of highest concern for climate change adaptation because it could increase vulnerability and conflicts.

Adaptation action involving upstream–downstream linkages needs to coordinate a number of interacting factors. The scale of impacts should be analyzed from both short and long term such as seasonal diversion of water sources versus permanent holding of water. Similarly, there could be cascade of impacts arising from an action either directly or indirectly. For instance, adaptation approaches aimed at storing water could reduce available water for downstream leading to decreased food production, increased malnutrition, or increased food prices. Forward–backward transfer of risk and benefits could be another dimension to evaluate. Payment of ecosystem services to the upstream for supplying water or preserving quality, revenue from selling electricity to the downstream, or export of food items could be some of the examples of benefit transfers. On the risk side, there are numerous cases of upstream to downstream transfer risk due to the manipulation of flow and quality of water sources. Though seemingly rare, the opposite could be true where lack of coordination and advance capture of water use rights could curtail upstream users from accessing water sources leading to potential conflicts (Salman 2010).

Developing a symbiotic upstream–downstream relationship is essential for planning, coordinating actions, and equitable sharing of risk and benefits. Especially the role of nonstructural measures, such as institutional and legal, could be effective while choosing structural measures of adaptation. The role of institution could be effective in distributing risks, shape incentive structures, and mediate (external) interventions (Agrawal and Perrin 2009).

### 8.4.3.3 Climate Cobenefits and Water Resources

The cobenefit approach constitutes identifying and recognizing certain kinds of benefits that have not been widely recognized but are important for meeting several environmental and developmental benefits accrued from them. The idea of cobenefits arises from the realization that certain policies are designed by keeping in mind only certain kinds of benefits to be obtained from and accounted for leaving out other benefits unrecognized for the purpose of evaluating the impact of the policy. IPCC defines climate cobenefits as "The benefits of policies that are implemented for various reasons at the same time—including climate change 'mitigation' —acknowledging that most policies designed to address greenhouse gas mitigation also have other, often at least equally important rationales (e.g., related to objectives of development, sustainability, and equity)" (Intergovernmental Panel on Climate Change 2007). In a way, recognizing cobenefits is also about capturing holistic benefits accrued from a policy or practice as against capturing only few select benefits.

Recognizing the cobenefits can incentivize the project developers to promote certain types of approaches as against others.

Table 8.6 presents a range of direct and cobenefits associated with selected projects. It is clear from the table that water projects can lead to several climate cobenefits and climate projects can lead to a range of water cobenefits. For example, in a project on IWRM, the planned benefits from the project could often be related to sustainable development and resilience. However, it is also possible that such integrated resource management may bring climate benefits such as mitigation of greenhouse gas (GHG) emissions from reduced energy consumption and emissions resulting from crop irrigation practices. Cobenefit approach in water resources management becomes even important with the growing recognition of the nexus between water resources, energy, and food where promoting cobenefits of managing one resource across in other resources makes even greater sense especially when systems are managed for multiuse (Hoff 2011). The cobenefits often go unrecognized for several reasons and hence do not provide sufficient impetus to promote practices that provide several cobenefits over other options available leading to partial capturing of benefits in the decision-making process.

### TABLE 8.6
### Benefits and Cobenefits Identified in Various Water Resources Management Projects

| Focus of the Project | Direct Benefits | Cobenefits | Reference |
|---|---|---|---|
| REDD programs | Reduced forest degradation, promotion of native species, $CO_2$ mitigation | Improvement in aquatic environment, clean drinking water, flood protection | Coe et al. (2010) |
| Wastewater treatment with methane capture funded as a CDM project | Reduced methane emissions, improving urban sanitation infrastructure | Reduced water pollution | Ministry of Environment, Japan (2008); Verified Carbon Standard (2012) |
| Water-quality trading program | Reduced water pollution | GHG sequestration, reduced nutrient pollution in runoff, and promoting ecosystem restoration | Gasper et al. (2012) |
| A combination of afforestation of pastures, nitrogen budgeting, and waste treatment methods | GHG emission reduction | Water-quality benefits in terms of improved physicochemical properties | Wilcock et al. (2008) |
| Water conservation project | Water conservation | GHG reduction and energy conservation | Mass (2009) |

Climate Change Adaptation in Water 229

This happens primarily for the fact that practices and policies are chosen based on to what extent they address a selected problem and the benefits that are not directly related to the objectives of the project tend to get unnoticed. Secondarily, the part of the problem also lies with the limited understanding and recognition among different practitioners due to traditional fragmented approach to problem solving over integrated approaches.

Originally, the concepts of cobenefits have been introduced to provide fillip to GHG mitigation practices that also have developmental benefits. In the context of climate change adaptation and water resources, promoting cobenefits means identifying and recognizing those water-related cobenefits accrued from projects that are not directly targeting water resources as their primary objective. For example, in a project that uses a combination of afforestation of pastures, nitrogen budgeting, and waste treatment methods (fourth example in Table 8.6), the primary objective is to reduce GHG emissions. However, the project delivers an important water-related cobenefit in the form of improving the water quality that could in turn promote available quality water for various uses including the ecosystem restoration and improving the resilience in the natural and social systems. One important hurdle in promoting cobenefits has been the methodological limitations in quantifying and integrating the benefits into the decision-making process and recognition of cobenefits by project donors and developers.

## 8.5 BARRIERS FOR MAINSTREAMING ADAPTATION

There are manifold barriers at different stages in areas of understanding the problems and impacts, identifying and developing options, implementing options, or assessing the effectiveness of options (Moser and Ekstrom 2010). Among the barriers, the critical one is the absence of information about magnitude and timing of future climate variability and its damage potential that will hamper in developing a clear process for identifying and introducing robust adaptation measures. In addition, some measures require relatively higher upfront investments that are costly to revert back once interventions are initiated. Another barrier is the time because design and construction could consume longer time. Similarly, gestation period and operational life of water infrastructure could last for years and decades so that its threshold capacity (Graaf 2008) could become obsolete at any point of time in future when climate impacts are likely to intensify. Meanwhile, it could take even more time for upscaling and diffusion because any adopted measures may need testing and retesting and often involve continuous learning process, before its maturity could be verified. Along with these general barriers, there could be specific barriers for each adapted measure. Tables 8.7 and 8.8 identify some of the specific barriers to the adaptation measures discussed in this chapter.

For simplicity, identified barriers have been categorized into financial, technical, and socioecological. Financial barriers are inherent to large-scale hard structural interventions particularly those falling under too-much-water and supply-side (too-little-water) measures. However, the same ideology may not apply to small-scale interventions as well as those employing soft structural measures unless investment is not supported or subsidized by the state. In addition to these two factors, operation

## TABLE 8.7
## Potential Barriers for Introducing Adaptation in Hard Measures

|  |  | Too Little Water |  |
|---|---|---|---|
| Barriers | Too Much Water | Supply Side | Users' Side |
| *Financial* |  |  |  |
| High investment capital | √ | √ (for large scale) | √ (for large scale) |
| High operation and management cost | ○ | √ (desalination, wastewater treatment) | ○ |
| *Technical* |  |  |  |
| Competition for space | √ | √ | × |
| Weather (windy/extreme temperature) | √ (floating structures, GLOF prevention) | √ (mist harvesting could be affected by increased temperature) | √ (sprinkler irrigation) |
| Remoteness/topography | √ (GLOF prevention) | √ (hilly areas) | × |
| High-tech and sensitive design | √ | √ | √ |
| Clogging | √ (water way from solid wastes, debris) | √ (managed groundwater recharge) | √ (sprinkler irrigation) |
| *Socioecological* |  |  |  |
| High environmental impact | √ (huge dikes) | √ (big storage dams) | × |
| High GHG footprint | × | √ (desalination, wastewater treatment) | ○ (groundwater pumping) |
| Off-site risk/risk transfer | √ | √ | × |
| Health risk and public acceptance | × | √ (wastewater reuse) | ○ (from use of contaminated water) |

*Note:* Barriers are in relative sense (√ is a barrier; ○ could be a barrier; × is not a barrier). Each parenthesis represents specific case(s).

and management cost also dictates the feasibility of adopting structural measures, in particular, at users' side such as operation of desalination plants. Decision criteria such as cost–benefit analysis could be used to determine the financial viability of a measure. In case of technical barriers, there could be several aspects to consider such as the competition for the space to implement soft or hard structural measures, unfavorable weather, remoteness, clogging of pipes, high-tech designs, etc. Although these financial and technical barriers could be overcome at some stage of planning or implementation, social and environmental externalities could be a key factor in determining the long-term sustainability. For instance, groundwater pumping and desalination could be energy intensive, thereby contributing to GHG emission. Similarly, construction of large-scale hard infrastructures such as dams or dikes could have a negative environmental footprint. In particular, hard infrastructure may not be a permanent solution because they tend to transfer risk from one location to another (Jha et al. 2012).

# Climate Change Adaptation in Water 231

**TABLE 8.8**
**Issues to Be Considered for Promoting Soft Measures in the Water Resources Management**

| Sectors | Soft Measures | Issues to Be Considered |
|---|---|---|
| Municipal | Water pricing | Institutional framework |
| | Water restriction | Institutional framework and potential political impact |
| | Standard for water appliances | Potential political impact |
| | Water recycling | Additional investment |
| | Reduce leakage loss | Expensive option for old system |
| Industrial | Increase water use efficiency and water recycling | Additional investment to upgrade |
| Agriculture | Community-based irrigation management | Institutional arrangement, human resource capacity, heterogenic interest of water users |
| | Irrigation water use efficiency | Pricing or technological intervention |
| | Irrigation scheduling | Require active involvement of farmers |

In addition to the aforementioned barriers, often overlooked is the issue of pollution and associated health risk. Any efforts toward adaptation to too much or too little water could be easily dampened by "too polluted" water. In fact, climate change could also exacerbate pollution of water bodies (Bates et al. 2008). In particular, in developing countries, the issue of pollution of water bodies from the point and nonpoint source has resulted in the decrease in usability of water, increase in health risk, and rise in the cost of water treatment. Therefore, construction of appropriate water treatment/purification systems, centralized, clustered, or on-site, should be considered.

## 8.6 CONCLUSIONS

Integrating climate change adaptation options into water resources management needs to consider not only climatic changes but also other global changes brought out during recent decades that have bearing on water resources. The role of decision support systems has been widely advocated as a go-to tool for water resources managers and policy makers. Integrating climate change elements into these support resources would make sense than developing independent tools. There are already several efforts to consider climate information in the decision making. However, most of these tools are at the research and development phase to piloting phase. Actual implementation of these tools on the ground and most importantly in deciding the long-term decisions is still a gray area. Understanding and addressing various uncertainties involved in climate decision making would help overcome the cautious approach being followed in this area.

Several structural and nonstructural adaptation options have been advocated for climate change adaptation benefits. The review of literature suggests that most of these practices have been in vogue even before climate change adaptation has become a concern. Though most of these practices were found to be effective in the current and historical contexts, it was entirely not clear if they hold good even for the future climates. Most literature that evaluates the effectiveness of these practices either does not differentiate between climate variability and change or does not necessarily talk about adaptation but resilience. Ability of the practices to hold good for the future climate could be seen as a litmus test for screening climate change adaptation practices. However, in absence of clear yardstick to measure how these practices would fare well in the future climates, it is difficult to conclude what measures may hold well to what extent in the future.

In general, to date, most part of water resources management has been dominated by structural interventions. The importance of nonstructural interventions has slowly become known with the advent of participatory approaches in water resources management. The research has shown that in fact the benefits from nonstructural measures could outweigh the benefits from structural measures. When sustainability of structural measures is important, it was found that combining structural and nonstructural measures would make the project interventions long-lasting than implementing only one kind of measure independently. Specifically, involvement of local communities' right from the beginning of the project design and implementation has led to longer-lasting measures with greater sharing of responsibilities and benefits among stakeholders. Their association with each other and the similarity in benefits equate top-down approaches to nonstructural measures and bottom-up measures to nonstructural measures. It was also found that most bottom-up measures are in compatibility with promoting the multiuse system approaches when compared to top-down measures.

While it is possible that traditional knowledge can still play a vital role in climate change adaptation, it requires a great deal of innovation at various levels, which includes innovation in how the traditional knowledge is identified and used for the context-specific circumstances. Sometimes, this could mean being able to apply in the contexts for which those practices and knowledge were not originally meant for. Literature also clearly states that adaptation requires integration of various facets of decision making. Integration should happen across disciplines, domains of knowledge, convergence of top-down and bottom-up approaches, and integration of upstream and downstream areas. In water resources management, often the dominant form of integration has been on geographical scales due to the spread of water resources across geographical scales and political boundaries. Such transboundary water resources management though has provided greater flexibility in how the water resources are managed and often has addressed the upstream and downstream concerns; the approach still falls short of achieving climate change adaptation for the reason that integrating climate information into such geographically integrated approaches remains a challenge for the reason that climate information has not been effectively downscaled to that level of decision making. This signifies the need to identify and address sources of uncertainty in decision making, and climate change impact projections comprise a large proportion of uncertainty that water managers have to deal with in addition to uncertainties associated with the future developmental trends and associated water supply and demand projections that are influenced by the rapidly changing developmental patterns.

## REFERENCES

2030 World Water Resources Group. 2009. *Charting Our Water Future: Economic Frameworks to Inform Decision-Making*. Washington, DC: 2030 World Water Resources Group.

Adam, L., Sarah, W., Henning, B. et al. 2013. *The Role of Water Markets in Climate Change Adaptation*. Gold Coast, Queensland, Australia: National Climate Change Adaptation Research Facility.

African Development Bank. 2013. *Enabling Sustainable Development Pathways in Africa: AfDB Green Growth Framework. First Order Draft*. Tunis, Tunisia: African Development Bank.

Agrawal, A. and Perrin, N. 2009. Climate adaptation, local institutions and rural livelihoods. In: Adger, W.N., Lorenzoni, I., and O'Brien, K.L. (eds.), *Adapting to Climate Change: Thresholds, Values, Governance*. Cambridge, U.K.: Cambridge University Press, pp. 350–367.

Ali, M.H. 2011. *Practices of Irrigation and On-Farm Water Management*. London, U.K.: Springer.

Asia Pacific Water Forum. 2012. *Framework Document on Water and Climate Change Adaptation for Leaders and Policy-Makers in the Asia-Pacific Region*. Tokyo, Japan: Asia Pacific Water Forum (APWF).

Atoyev, K. 2007. Multiple criteria decision-making tool for optimal water resource management. Wastewater reuse–risk assessment. In: Bobyley, B. (ed.) *Multiple Criteria Decision-Making and Environmental Security*. NATO Science for Peace and Security Series. Dordrecht, the Netherlands: Springer, pp. 131–144.

Ayar, J.E. 2010. Adapting irrigated agriculture to drought in the San Joaquin valley of California. In: *International Drought Symposium*, University of California, Riverside, CA.

Bates, B.C., Kundzewicz, Z.W., Wu, S., and Palutikof, J.P. (eds.). 2008. Climate change and water. Technical paper of the Intergovernmental Panel on Climate Change (IPCC). Geneva, Switzerland: IPCC Secretariat, 210pp.

Becker, N., Zeitouni, N., and Shechter, M. 1996. Reallocating water resources in the Middle East through market mechanisms. *Water Resources Development* 12(1): 17–32.

Bergkamp, G., Orlando, B., and Burton, I. 2003. *Change: Adaptation of Water Resource Management to Climate Change*. Cambridge, U.K.: IUCN.

Biswas, A.K. 2008. Integrated water resources management: Is it working? *International Journal of Water Resources Development* 24(1): 5–22.

Bithas, K. 2008. The sustainable residential water use: Sustainability, efficiency and social equity. The European experience. *Ecological Economics* 68: 221–229.

Brekke, L.D., Kiang, J.E., Olsen, J.R. et al. 2009. Climate change and water resource management: A federal perspective, U.S. Department of Interior and U.S. Geological Survey: Circular 1331. Reston, VA: U.S. Geological Survey.

Brewer, J., Robert, G., Alan, K., and Gary, L.D. 2007. Water markets in the west: Prices, trading, and contractual forms. Arizona legal studies discussion paper no. 07-07. Tucson, AZ: The University of Arizona.

Cap-Net. 2009. IWRM as a tool for adaptation to climate change. Training Manual and Facilitator's Guide. Cap-Net/UNESCO-IHE. Pretoria, South Africa.

Coe, M.T., Stickler, C.M., and Lefebvre, P.A. 2010. *Ecological Co-benefits: Pan-Amazon Deforestation, Regional Climate, and Water Resources*. Falmouth, MA: The Woods Hole Research Center.

Conrad, E. 2012. *Climate Change and Integrated Regional Water Management in California: A Preliminary Assessment of Regional Approaches*. Berkeley, CA: University of California, Department of Environmental Science, Policy and Management.

Cornish, G., Bosworth, B., Perry, C., and Burke, J. 2004. *Water Charging in Irrigated Agriculture: An Analysis of International Experience*. Rome, Italy: Food and Agriculture Organization.

Devi, B.S., Ranghaswami, M.V., and Mayilswami, 2012. Performance evaluation of water delivery system in canal command area of Pap basin, Tamilnadu. *India Water Week 2012— Water, Energy and Food Security: Call for Solutions*, April 10–14, New Delhi, India.

Dinar, A. 2000. *The Political Economy of Water Pricing Reforms*. New York: The World Bank and Oxford University Press.

Dinar, A. and Saleth, R.M. 2005. Issues in water pricing reforms: From getting correct prices to setting appropriate institutions. In: Folmer, H. and Tietenberg, T. (eds.), *The International Yearbook of Environmental and Resource Economics*. Northampton, U.K.: Edward Elgar. pp. 1–51.

Doswald, N. and Osti, M. 2011. *Ecosystem-Based Approaches to Adaptation and Mitigation— Good Practice Examples and Lessons Learned in Europe*. Bonn, Germany: Federal Ministry of Environment and Nature Conservation and Nuclear Safety.

Duke, J.M. and Ehemann, R. 2004. An application of water scarcity pricing with varying threshold, elasticity, and deficit. *Journal of Soil Water Conservation* 59: 59–65.

Elfithri, R., Mokhtar, M.B., Shah, A.H.H., and Idrus, S. 2008. Collaborative decision making within integrated water resources management: Tool for transboundary waters management. In: *IV International Symposium on Transboundary Waters Management*, Thessaloniki, Greece. pp. 1–7.

Elliot, M., Armstrong, A., Lobuglio, J., and Bartram, J. 2011. In: De Lopez, T. (ed.), *Technologies for Climate Change Adaptation—The Water Sector*. Roskilde, Denmark: UNEP Risoe Centre.

Enfors, E.I. and Gordon, L.J. 2008. Dealing with drought: The challenge of using water system technologies to break dryland poverty traps. *Global Environmental Change* 18(4): 607–616.

European Communities. 2004. *Sustainable Production and Consumption in the European Union*. Luxembourg, Europe: Office for Official Publications of the European Communities.

Faurès, J.M., Hoogeveen, J., Winpenny, J., Steduto, P., and Burke, J. 2012. Coping with water scarcity—An action framework for agriculture and food security. FAO water report 38. Rome, Italy: Food and Agriculture Organization of the United Nations.

Foster, S., Steenbergen, F.V., Zuleta, J., and Garduño, H. 2010. Conjunctive use of groundwater and surface water from spontaneous coping strategy to adaptive resource management. GW-MATE Strategic Overview Series No. 16. Washington, DC: World Bank.

Füssel, H.M. 2007. Adaptation planning for climate change: Concepts, assessment, approaches and key lessons. *Sustainability Science* 2: 265–275.

Garduño, H., Foster, S., Raj, P., and Steenbergen, F. 2009. *Addressing Groundwater Depletion Through Community-Based Management Actions in the Weathered Granitic Basement Aquifer of Drought-Prone Andhra Pradesh—India*. Washington, DC: The World Bank.

Gasper, R.R., Selman, M., and Ruth, M. 2012. Climate co-benefits of water quality trading in the Chesapeake Bay watershed. *Water Policy* 14: 758–765.

Gibbs, M.S., Maier, H.R., and Dandy, G.C. 2012. Development of decision support frameworks for water resource management in the South East. Goyder Institute for Water Research Technical Report Series No. 12/3. Adelaide, South Australia, Australia: Goyder Institute for Water Research.

Gill, R. 2011. Drought and flooding rains: Water provisions for a growing Australia. *CIS Policy Monographs* 115: 1–28.
Glibert, J. and Vellinga, P. 1990. Coastal zone management. In: Berenthal, F., Dowdeswell, E., Luo, J., Attard, D., Vellinga, P., and Karimanzira, R. *Intergovernmental Panel on Climate Change, Climate Change IPCC Response Strategies*. Geneva, Switzerland: Intergovernmental Panel on Climate Change.
Global Water Intelligence. 2008. Tariff hikes check Spanish water consumption. *Global Water Intelligence* 9(8): 25–40.
Graaf, R.E. 2008. Reducing flood vulnerability of urban lowland areas. In: *11th International Conference on Urban Drainage*, Edinburgh, Scotland, U.K. pp. 1–10.
Griffin, R. 2006. *Water Resource Economics: The Analysis of Scarcity, Policy and Projects*. Cambridge, MA: The MIT Press.
Grigg, N.S. 2008. Integrated water resources management: Balancing views and improving practice. *Water International* 33(3): 279–292.
Hallegatte, S. 2009. Strategies to adapt to an uncertain climate change. *Global Environmental Change* 19: 240–247.
Hermans, L., Renault, D., Emerton, L., Perrot-Maitre, D., Nguyen-Koha, S., and Smith, L. 2006. *Stakeholder-Oriented Valuation to Support Water Resources Management Processes Confronting Concepts with Local Practice*. Rome, Italy: Food and Agriculture Organization.
Hoff, H. 2011. Water, energy and food security challenges and opportunities. In: *The Water, Energy and Food Security Nexus: Solutions for the Green Economy*, November 16–18, 2011, Bonn, Germany. pp. 55–84.
Institute for Global Environmental Strategies. 2007. Sustainable groundwater management in Asian cities: A final report of research on sustainable water management policy. Hayama, Japan: IGES.
Institute for Security Studies. 2010. *The Impact of Climate Change in Africa*. Pretoria, South Africa: Institute for Security Studies.
International Fund for Agriculture Development (IFAD). 2010. Climate change exacerbates struggles of rural poor amidst growing global competition for natural resources. Press release no.: IFAD/79/2010. Rome, Italy: IFAD.
Intergovernmental Panel on Climate Change. 2007. Climate Change 2001. Working Group 3 Contribution to the Third Assessment Report of the Intergovernmental Panel on Climate Change. Cambridge, UK.: Cambridge University Press.
Jha, A.K., Bloch, R., and Lamond, J. 2012. *Cities and Flooding: A Guide to Integrated Urban Flood Risk Management for the 21st Century*. Washington, DC: World Bank.
Kabat, P. and van Schaik, H. 2003. *Climate Changes the Water Rules: How Water Managers Can Cope with Today's Climate Variability and Tomorrow's Climate Change*. Wageningen, the Netherlands: Dialogue on Water and Climate.
Kalpesh, K.A.P. and Patel, N.G. 2013. Literature review on canal irrigation scheduling using Ga. *Indian Journal of Research* 3(4): 154–156.
Kenney, D.S., Klein, R.A., and Clark, M.P. 2004. Use and effectiveness of municipal water restrictions during drought in Colorado. *Journal of American Water Resource Association* 40(1): 77–87.
Kundzewicz, Z.W., Mata, L.J., Arnell, N.W. et al. 2007. Freshwater resources and their management. In: Parry, M.L., Canziani, O.F., Palutikof, J.P., van der Linden, P.J., and Hanson, C.E. (eds.), *Climate Change 2007: Impacts, Adaptation and Vulnerability. Contribution of Working Group II to the Fourth Assessment Report of the Intergovernmental Panel on Climate Change*. Cambridge, U.K.: Cambridge University Press. pp. 173–210.
Landry, C. 1998. Market transfers of water for environmental protection in the Western United State. *Water Policy* 1: 457–469.

Le Blanc, D. 2008. A framework for analyzing tariffs and subsidies in water provision to urban households in developing countries. DESA working paper no. 63. New York: United Nations, Department of Economic and Social Affairs.

Li, Y., Ye, W., and Yan, X. 2011. Development of a co-evolutionary decision support system—Food and water security integrated model system (FAWSIM). *APN Science Bulletin* 1: 23–28.

Linham, M.M. and Nicholls, R.J. 2010. *Technologies for Climate Change Adaptation: Coastal Erosion and Flooding*. TNA Guidebook Series. Roskilde, Denmark: UNEP Risø Centre on Energy, Climate and Sustainable Development.

London Climate Change Partnership. 2006. *Adapting to Climate Change: Lessons for London*. London, U.K.: Greater London Authority.

Loucks, D.P. 2008. Water resource management models. *Technologies for Clean Water* 38: 24–30.

Luo, B., Maqsood, I., and Gong, Y. 2010. Modeling climate change impacts on water trading. *Science of the Total Environment* 408: 2034–2041.

MacNeil, R. 2004. Costs of adaptation options. In: Cohen, S., Neilsen, D., and Welbourn, R. (eds.), *Expanding the Dialogue on Climate Change and Water Management in the Okanagan Basin: Final Report*. British Columbia, Ontario, Canada: Environment Canada. pp. 161–163.

Mallants, D. 2013. Bringing public participation to the water table. Available at: *ECOS*. http://www.ecosmagazine.com. Accessed on 20 June 2013.

Marks, J.S. and Davis, J. 2012. Does user participation lead to sense of ownership for rural water systems? Evidence from Kenya. *World Development* 40(8): 1569–1576.

Mass, C. 2009. Greenhouse gas and energy co-benefits of water conservation. Ontario Water Conservation Alliance. Available at http://www.conserveourwater.ca/?p=120.

Mehta, L. 1997. Social difference and water resources management: Insights from Kutch, India. *IDS Bulletin* 28: 79–88.

Ministry of Environment Japan. 2008. The Co-benefits Approach for GHG Emission Reduction Projects. Tokyo, Japan: Ministry of Environment, Government of Japan. Available at: http://www.env.go.jp/en/earth/ets/icbaghserp081127.pdf. Accessed on 10 June 2013.

Modernization of Water Management System Project. 2003. *The Guideline for Water Management Planning and Operation*. Osaka, Japan: The Modernization of Water Management System Project, RID, DOAE, JICA.

Mohanty, N. and Gupta, S. 2002. *Breaking the Gridlock in Water Reforms through Water Markets: International Experience and Implementation Issues for India*. New Delhi, India: Julian L. Simon Centre for Policy Research.

Moser, S.C. and Ekstrom, J.A. 2010. A framework to diagnose barriers to climate change adaptation. *Proceedings of the National Academy of Sciences of the United States of America* 107(51): 22026–22031.

Mysiak, J., Henrikson, H.J., Sullivan, C., Bromley, J., and Pahl-Wostl, C. 2010. *The Adaptive Water Resource Management Handbook*. London, U.K.: Earthscan.

Ninomiya, S. 2012. *Development of Decision Support System for Optimal Agricultural Production under Global Environment Changes*. Tokyo, Japan: Research Program on Climate Change Adaptation, Tokyo University.

Palaniappan, M., Lang, M., and Gleick, P.H. 2008. *A Review of Decision-Making Support Tools in the Water, Sanitation, and Hygiene Sector*. Oakland, CA: Woodrow Wilson International Center for Scholars, Pacific Institute.

Pangare, G., Das, B., Lincklaen, W.A., and Makin, I. 2012. *Water Wealth? Investing in Basin Management in Asia and the Pacific*. New Delhi, India: Academic Foundation.

Pennsylvania Department of Environmental Protection. 2013. *Drop by Drop: Use Water Wisely During a Drought*. Available at: http://www.elibrary.dep.state.pa.us/dsweb/Get/Document-95168/3940-FSDEP4230.pdf. Accessed on 10 July 2013.

Pereira, L.S., Cordery, I., and Iacovides, I. 2009. *Coping with Water Scarcity: Addressing the Challenges*. Dordrecht, the Netherlands: Springer.

Prabhakar, S.V.R.K., Kotru, R., Pradhan, N. et al. 2013. *Adaptation Effectiveness Indicators for Agriculture in the Gangetic Basin*. Hayama, Japan: Institute for Global Environmental Strategies.

Purkey, D.R., Huber-Lee, A., Yates, D.N., Hanemann, M., and Julius, S.H. 2007. Integrating a climate change assessment tool into stakeholder-driven water management decision-making processes in California. *Water Resources Management* 21: 315–329.

Ragab, R. 1996. Constraints and applicability of irrigation scheduling under limited water resources, variable rainfall and saline conditions. In: Smith, M., Pereira, L.S., Berengena, J., Itier, B., Goussard, R., Ragab, R., Tollefson, P., and van Hofwegen, L. (eds.), *Irrigation Scheduling from Theory to Practice*. Rome, Italy: Food and Agriculture Organization. pp. 149–165.

Raghuwanshi, N.S. and Wallender, W.W. 1998. Optimal furrow irrigation scheduling under heterogeneous conditions. *Agricultural System* 58(1): 39–55.

Rajput, T.B.S. and Patel, N. 2006. Determination of the optimal date for sowing of wheat in canal irrigated areas using FAO CROPWAT model. In: *The 27th Annual International Irrigation Show*, November 5–7, San Antonio, TX. pp. 432–442.

Ramesh, B.R., Venugopal, K., and Karunakaran, K. 2009. Zero-one programming model for daily operation scheduling of irrigation canal. *Journal of Agriculture Science* 1(1): 13–20.

Reddy, G.M.J. and Kumar, D.N. 2008. Evolving strategies for crop planning and operation of irrigation reservoir system using multi-objective differential evolution. *Irrigation Science* 26: 177–190.

Rekacewicz, P. 2005. Freshwater stress and scarcity in Africa by 2025. UNEP and GRID-Arendal. Arenda, Norway. Available at: http://www.grida.no/graphicslib/detail/freshwater-stress-and-scarcity-in-africa-by-2025_4036. Accessed on 7 August 2013.

RICS. 2011. Water management-drought. Available at: http://www.rics.org/gh/about-rics/what-we-do/influencing-policy/policy-positions/water-management-drought/. Accessed on 5 August 2013.

Rukunga, G., Kioko, T., and Kanyangi, I. 2006. Operation and maintenance of rural water services in Kenya: Suitable solutions. Well Country Brief 15.1. Available at: http://www.lboro.ac.uk/well/resources/Publications/Country%20Notes/CN15.1.htm. Accessed on 20 July 2013.

Salman, M.A.S. 2010. Downstream riparians can also harm upstream riparians: The concept of foreclosure of future uses. *Water International* 35(4): 350–364.

Selvaraju, R., Subbiah, A.R., Baas, S., and Juergens, I. 2006. Livelihood adaptation to climate variability and change in drought-prone areas of Bangladesh developing institutions and options. Institutions for Rural Development 5. Rome, Italy: Food and Agriculture Organization of the United Nations.

Serrat-Capdevila, A., Valdes, J.B., and Gupta, H.V. 2011. Decision support systems in water resources planning and management: Stakeholder participation and the sustainable path to science-Based decision making. In: Jao, C. (ed.), *Efficient Decision Support Systems—Practice and Challenges from Current to Future*. Rijeka, Croatia: InTech Publications. pp. 51–69.

Shiklomanov, I.A. 1999. *State Hydrological Institute and United Educational*. St. Petersburg, Russia: St. Petersburg and Scientific and Cultural Organization.

Sinisi, L. and Aertgeerts, R. 2010. Guidance on water supply and sanitation. In: *Extreme Weather Events*. Copenhagen, Denmark: WHO Regional Office for Europe.

Sophie, W. 2013. To combat scarcity, increase water-use efficiency in agriculture. Available at: http://www.worldwatch.org/combat-scarcity-increase-water-use-efficiency-agriculture-0. Accessed on 20 July 2013.

Stalker, P. 2006. *Technologies for Adaptation to Climate Change*. Bonn, Germany: United Nations Framework Convention on Climate Change.

Steenbergen, F.V. and Tuinhof, A. 2010. Managing the water buffer for development and climate change adaptation: Groundwater recharge, retention, reuse and rainwater storage. Bundesanstalt für Geowissenschaften und Rohstoffe, the Co-operative Programme on Water and Climate (CPWC), the Netherlands National Committee IHP-HWRP. Paris, France: International Hydrological Programme (IHP)-UNESCO.

Thawala, W.D. 2010. Community participation is a necessity for project success: A case study of rural water supply project in Jeppes Reefs, South Africa. *African Journal of Agriculture Research* 5(10): 970–979.

UN-Economic Commission for Europe. 2009. *Guidance on Water and Adaptation to Climate Change*. New York: United Nations Publications.

UN-HABITAT. 2005. *Rainwater Harvesting and Utilisation Book 3: Project Managers & Implementing Agencies*. Blue Drop Series. Nairobi, Kenya: UN-HABITAT.

Verified Carbon Standard. 2012. Methane capture, flare and utilization at Tyson wastewater treatment facilities—Lexington, United States of America. Verified Carbon Standard. Available at: https://vcsprojectdatabase2.apx.com/myModule/Interactive.asp?Tab=Projects&a=2&i=16&lat=40.763414&lon=-99.739709. Accessed on 12 June 2013.

Vulnerability and Adaptation Programme. 2009. Vulnerability and adaptation experiences from Rajasthan and Andhra Pradesh-Water Resources Management Project India. Mahbubnagar, India: SDC Vulnerability and Adaptation Programme India.

Wilcock, R., Elliot, S., Hudson, N., Parkyn, S., and Quinn, J. 2008. Climate change mitigation for agriculture: Water quality benefits and costs. *Water Science and Technology* 58: 2093–2099.

Willis, E., Pearce, M., Mamerow, L., Jorgensen, B., and Martin, J. 2013. Perceptions of water pricing during a drought: A case study from South Australia. *Water* 5: 197–223.

WOCAT. 2011. A multiple-use water system (Nepal). *Natural Resource Management Approaches and Technologies in Nepal*. Kathmandu, Nepal: WOCAT.

World Bank. 2010. *Deep Wells and Prudence: Towards Pragmatic Action for Addressing Groundwater Overexploitation in India*. Washington, DC: The World Bank.

World Energy Council. 2010. *Water for Energy*. London, U.K.: World Energy Council.

World Water Assessment Programme. 2012. *The United Nations World Water Development Report 4: Managing Water under Uncertainty and Risk*. Paris, France: UNESCO.

Zorilla, P., Carmona, G., Hera, A.D. et al. 2009. Evaluation of Bayesian networks in participatory water resources management, Upper Guadiana Basin, Spain. *Ecology and Society* 15(3): 12.

# 9 Managing Climate Risk for the Water Sector with Tools and Decision Support

*Julian Doczi*

## CONTENTS

| | |
|---|---|
| 9.1 Introduction | 239 |
| 9.2 Climate Risks on the Water Sector | 240 |
| 9.3 Deep Uncertainty and Its Implications for Water Sector Climate Risk Management | 241 |
| 9.4 Need for Decision Support and Emergence of *Tools* | 244 |
| 9.5 What Is a *Tool* for Climate Risk Management in the Water Sector? | 245 |
| 9.6 What Types of Tools Exist? | 246 |
| 9.7 Some Examples of Tools | 249 |
| 9.8 Tool *Niches* and Their Degree of Overlap | 251 |
| 9.9 Is There Evidence of User Demand for Tools? | 253 |
| 9.10 New Alternatives for Robust, Climate-Sensitive Decision Support? | 256 |
| 9.11 Tools and Robust Decision Support Methods: Complementary or Clashing? | 257 |
| 9.12 Conclusions | 259 |
| Acknowledgments | 260 |
| Appendix | 261 |
| Tool References | 282 |
| Main Text References | 288 |

## 9.1 INTRODUCTION

This chapter aims to review and reflect on decision support approaches for adapting to, and managing the risk of, climate variability and change on all aspects of water management, particularly for the developing world. It focuses mainly on the many so-called tools for managing this climate risk that have been developed in the last few years. Examining these tools in depth, it reviews 137 tools relevant to climate risk management for the water sector that serve a wide variety of different functions and audiences and that are intended to support an equally wide variety of decisions. This process raises important questions about the degree of overlap in these tools across functions and audiences, as well as questions about the degree to which they are being actively used.

The chapter then analyzes these tools from the newer perspective of the need for a paradigm shift toward more *robust* decision making (RDM), due to the *deeply uncertain* context in which water sector climate adaptation operates. It argues that this new body of literature has sharpened the appreciation of the need for more holistic decision making in many circumstances, with a resulting shift toward more scenario-based analysis and away from planning around a specific, *most likely* future. It notes that many existing tools still function around this latter, *science-first* approach, though an increasing number are incorporating elements of the newer, *decision-centric* approach.

## 9.2 CLIMATE RISKS ON THE WATER SECTOR

As the previous chapters in this book have detailed, the water sector is being acutely impacted by climate variability and change, especially in the developing world. One of the key studies on these impacts was the Intergovernmental Panel of Climate Change's (IPCC's) *Climate Change and Water* technical report (Bates et al. 2008). This report focused in particular on how climate change will impact on the frequency, location, and intensity of precipitation events, and the implications of this for the water sector. The implications are indeed enormous, with more extreme precipitation events in some areas increasing the frequency of floods, while reduced precipitation, elevated evapotranspiration, and greater numbers of hot days in other areas will increase the demand for water and the subsequent risk of droughts. Shifts in the timing and seasonality of precipitation are also already widespread, with growing evidence as well for significant shifts in the intensity and severity of tropical cyclones and in the global *climate engines*, such as the El Niño–Southern Oscillation (ENSO) and the North Atlantic Oscillation (NAO), among others.

In monsoon climates, which include many of the developing world's most vulnerable nations, all of those mentioned earlier may occur in the same region. For example, in the Caribbean, rainy seasons are projected to be shorter but with greater intensity of heavy rains (potentially resulting in floods), while dry seasons could be longer and with decreased precipitation overall (potentially resulting in droughts). Both of these extremes pose serious threats to the water sector, by affecting water quality and quantity, which subsequently impact on the operation of infrastructure and delivery of services.

In addition to increases in climate variability, shifts in *mean* climate conditions are occurring as well, with the emergence of novel climate states. These new states will become increasingly widespread and present significant challenges for the development and reliable operation of long-lived water infrastructure (Brown 2010; Matthews et al. 2011). Given the significant investments these represent in many developing countries, maintaining existing services will prove very challenging even without any additional demographic, economic, and environmental pressures.

The scientific confidence in these projections varies widely across the world, requiring a deeply nuanced approach to any discussion of them. As detailed by the previously mentioned authors and in the IPCC's Fourth Assessment Report (2007), climate models generally have greater agreement on the direction and magnitude of temperature change than they do for precipitation change, though confidence is lower in particular on the timing (temporal resolution) of these changes. Confidence levels vary widely by region and country as well. The climate model projections for precipitation trends in southern Africa, for example, are generally in more agreement than they are for the Horn of Africa. However, Bouwer et al. (2013, p. 2) caution that even where climate models show *consensus* in a region, *the level of confidence in these projections should be guided by application*, as this consensus may reflect a convergence in their often similar assumptions, rather than a convergence in a set of independent hypotheses.

## 9.3 DEEP UNCERTAINTY AND ITS IMPLICATIONS FOR WATER SECTOR CLIMATE RISK MANAGEMENT

Within this context, water managers are facing the need to adapt and build resilience to these ongoing and forthcoming climate impacts but are doing so in a complex and uncertain context. While the weight of available data indicates that many aspects of policy and planning do indeed need to change, understanding how to operationalize this for many developing countries at local level—and even country level—is not at all straightforward. This is due at least in part to how climate model projections are considered and used in the decision-making process. As Bouwer et al. (2013) highlight, these models were generally not developed to guide climate adaptation decision making and often frame impacts and uncertainties in ways that make their application to decision making problematic, especially for the small spatial and temporal scales at which water is typically managed.

Uncertainty itself is not inherently a problem. As Ranger (2013) highlights, infrastructure decisions, particularly in the water sector, have always been made under uncertainty in climatic variability. However, until recently, this uncertainty had been considered *quantifiable*, as engineers could rely on historical data about the typical range of variation and optimize the project accordingly, with this *most likely* future in mind. Now though, as Milly et al. (2008) famously declared, this quantifiable uncertainty via *stationarity* is dead with climate change, as this historical data can no longer be relied upon to predict the hydrological future. Instead, climate change has resulted in a situation of *deep uncertainty*, defined by Hallegatte et al. (2012, p. 2) as "a situation where analysts do not know or cannot agree on (1) models that relate key forces that shape the future, (2) probability distributions of key variables and parameters in these models, and/or (3) the value of alternative outcomes." This is particularly the case when attempting to understand how physical climate change translates into impacts on local societies and sectors, where the *answers* to this are often generated only by progressing through a series of *downscaling* questions, each with its own assumptions and degree of uncertainty. Thus arises a *cascade of uncertainty*, as the linked and contingent nature of each of these questions compounds the uncertainty of the subsequent ones (Wilby and Dessai 2010). Two different representations of this *cascade* are illustrated in Figure 9.1, though the number and type of *links* in the cascade will vary for each overarching question and context examined and may not always be so uncertain that decision making is prevented.

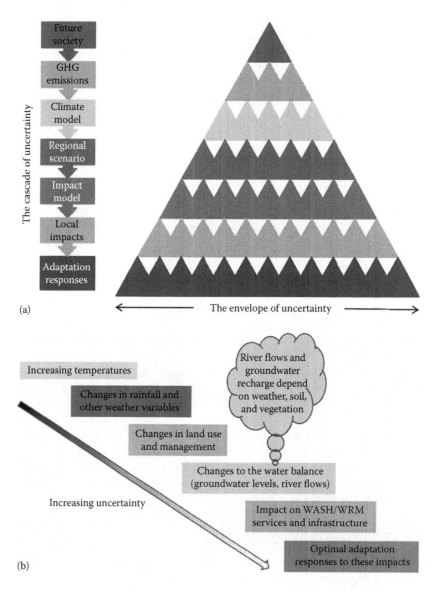

**FIGURE 9.1** Two different depictions of the *cascade of uncertainty*, where each subsequent downward step in this series of projections, impacts, and/or decisions compounds the level of total uncertainty in the outcome estimates. In both cases, though, these are general depictions that are not necessarily applicable in all contexts (a: general image is taken from Wilby, R.L. and Dessai, S., *Weather* 65(7), 181, 2010; b: water sector-specific image is adapted from Richard Carter (unpublished presentation, 2013). With permission.)

This application of the *deep uncertainty* concept to the climate change challenge is relatively new* and potentially presents the need for a paradigm shift by some adaptation planners in the water sector. As authors like Hallegatte et al. (2012), Bouwer et al. (2013), Walker et al. (2013), and Weaver et al. (2012) have argued, the death of stationarity and arising of this deep uncertainty should generally shift adaptation planning away from a *predict then act* paradigm toward a *seek robust solutions* paradigm instead—shifting away from the optimization of interventions for a single *most likely* future toward the design of interventions to *minimize future regret* across a broad range of possible scenarios. This is driven in part by Barnett and O'Neill's (2010) conception of *maladaptation*, whereby—without proper foresight and planning—some actions taken to supposedly adapt to climate impacts could end up actually increasing climate vulnerability, either of the system itself or as a negative externality on neighboring systems.

Some of this challenge can be avoided due to the water sector's so-called development deficit. As Calow et al. (2011), Batchelor et al. (2010), and the Global Water Partnership (GWP) (GWP 2007) argue, the many existing problems facing water supply, sanitation, and hygiene (WASH) and water resource management (WRM) services (i.e., the development deficit) are the primary drivers behind much of their existing vulnerability to climate impacts. This is especially since many existing WASH and WRM services (or lack thereof) often cannot yet even cope with existing climate variability, never mind future impacts. By effectively addressing this existing deficit, the sector would already be well on its way to building climate resilience.†

Likewise, Ranger (2013) argues that the necessary paradigm shift need not be onerous and that, in many cases, this deep uncertainty should not adversely affect *planned adaptation* either (i.e., beyond just addressing the development deficit). She argues that this is because

1. Short- and medium-term climate impact projections are not as uncertain as long-term impact projections, especially when placed alongside existing ranges of climate variability.
2. Most adaptation-relevant decisions are short-lived/focused on the near future and so will not actually be directly affected by forthcoming climate change impacts.
3. For those decisions that do require a long-term perspective (e.g., large infrastructure), there are many well-known approaches that can reduce risks and minimize future regret.
4. Not all adaptation needs to be done as a discrete solution right now—adaptation should instead be a process where decisions and interventions can be updated and improved as the future unfolds.

---

* Though, as Ranger (2013) notes, it has been recognized and studied for other phenomena with similarly cascading challenges, such as population growth, exchange rates, and economic growth.
† Nevertheless, additional, planned adaptation is still needed for the sector. A focus solely on resilience to current climate variability would overlook issues such as the need for transformation in sectoral decision making, in light of the new risks and opportunities presented by systemic climate change (e.g., Brown [2011]).

The rest of her paper then goes on to detail methods and strategies for making this possible, focusing on methods for more robustly appraising potential adaptation options, which will be discussed in more detail in the succeeding text.

## 9.4 NEED FOR DECISION SUPPORT AND EMERGENCE OF *TOOLS*

While this body of literature on the nuances of climate risk management under deep uncertainty is relatively new, the idea of providing decision support for various stakeholders to adapt to climate change is not. *Decision support* is a broad concept but can be thought of here as the result of any human or material resource that has assisted in making a climate-sensitive decision for the water sector. Obviously, the vast majority of decisions of any significance are made with at least some degree of support, be it from stakeholder dialogues, a mentor, a tool, an instruction from a higher authority, a financial incentive, or anything else that can plausibly be said to have influenced the decision—so the term itself is not particularly useful. Nevertheless, the term and its derivatives have been abundantly used in the climate adaptation and water management discourses (as evidenced in nearly every article cited in this chapter), especially surrounding decision-making *tools* for these topics.

As briefly discussed in Chapter 8, a very wide variety of the so-called tools for climate risk management have been created in the last decade, generally with the stated aims of helping users to navigate the complexity of climate-sensitive decision making and avoid maladaptation. Indeed, from a practitioner's perspective, it is rare for a month to pass without learning about some sort of new, so-called tool/toolbox/toolkit that claims to fill a particular niche and add new *decision support* value for its desired stakeholders.

Several previous academic reviews of these tools have already been undertaken and will be discussed in further detail in the following, but none have yet considered these (generally older) tools alongside this (generally newer) discourse of deep uncertainty and the need to seek robust solutions to climate impacts. Likewise, most previous reviews have not focused specifically on climate risk management for the water sector. This chapter thus focuses on the following questions:

- What is a *tool* anyway?
- What types of tools exist that are relevant for climate adaptation and risk management in the water sector, especially in terms of their function, audience, and purpose?
- How are these tools similar to/different from each other, in terms of their degree of overlap?
- Is there evidence of strong user demand for these tools in general?
- To what extent do existing tools incorporate elements compatible with a paradigm shift toward less prescriptive, more robust climate risk management interventions, especially for issues of water management? Are there better alternatives?

Section 9.5 will discuss the first question. Sections 9.6 and 9.7 will discuss the second. Section 9.8 will discuss the third. Section 9.9 will discuss the fourth. Sections 9.10 and 9.11 will then discuss the final questions, followed with a brief conclusion in Section 9.12.

## 9.5 WHAT IS A *TOOL* FOR CLIMATE RISK MANAGEMENT IN THE WATER SECTOR?

Much like *decision support*, the term *tool* is a similarly broad concept and has been interpreted in many different ways, even within the climate change and water management sector. To discuss them in a more structured manner, though, we define the term more specifically here than decision support was.

Based on previous definitions, the possible spectrum of so-called tools apparently ranges from technology all the way to broad political frameworks. The one end of this spectrum includes authors like Palaniappan et al. (2008, pp. 3–4), who, in their review of WASH-related tools, define the term as *the technologies, financing strategies, and approaches that are being used in the WASH sector.* That is, this definition would include technologies like pit latrines and activities like microcredit lending as *tools*. Meanwhile, at the other end of this spectrum, authors like UNDP Cap-Net (2009) define integrated water resources management (IWRM)—a broad political concept or ideal, at best—similarly as a *tool* for climate adaptation.

In our opinion, neither of these extremes of the spectrum are *tools*, as they both lack any clear practicality to individual users. Defining a tool this broadly ends up including almost anything within the definition and defeats the purpose of using the term at all. A better definition should therefore strike a more appropriate balance between specificity and generality, aiming to only include a useful range of *practical* resources. Indeed, p*racticality* and *usefulness for a particular function* are the traits most commonly associated with *general* tools (e.g., hammers and screwdrivers) in standard English dictionaries (e.g., Pearsall 1999).

With this in mind, we instead define *tools* in this context as *documents, computer programs, or websites that clearly and thoroughly operationalize a set of principles or practices that could build the resilience of the water sector to current climate variability or future climate impacts, preferably in an engaging and user-friendly manner.* This definition builds on those proposed by the UNFCCC (2008) and Hammill and Tanner (2011), though it focuses more directly on the need for these tools to have a practical operational purpose—thus reducing the total number of possible tools to a more manageable and useful amount. Recalling the previous definition spectrum from technologies to political frameworks, note that while this definition excludes static technology, it *can* include tools designed to assist in technology choice, as these are specific operationalizations of a set of recommendations. It could likewise include tools designed to provide guidance on how to specifically implement IWRM principles.

## 9.6 WHAT TYPES OF TOOLS EXIST?

The popularity of tools in the climate adaptation discourse is illustrated by the number of previous reviews of them that have already been undertaken. The related research from which this chapter is adapted identified at least 11 existing reviews of climate risk management tools, authored by Garg et al. (2007), UNFCCC (2008), Palaniappan et al. (2008), GIZ (2009), OECD (2009), Olhoff and Schaer (2010), Hammill and Tanner (2011), Traerup and Olhoff (2011), CCCCC (2013), SEI (2013), and ALM (2013). However, it is important to note that none of these previous reviews defined a *tool* as rigorously as we do here, meaning that we excluded many of their reviewed resources for not being practical or relevant enough for our purposes. Nevertheless, after drawing out the relevant tools from these existing reviews and after also undertaking an extensive literature search, we identified a total of at least 137 tools with practical relevance for climate adaptation and risk management for the water sector (though there are likely many more). The full list of these is displayed in the Appendix to this chapter.

As might be predicted, this large number of tools displays a similarly large degree of diversity. To better understand their similarities/differences, it is useful to discuss ways to classify and organize these tools. In the Appendix, we classify them based on *function* and *target sectoral audience*, as these were the simplest and least ambiguous methods, though one could also classify based on the tools' temporal/physical scale, desired users, methodologies used/activities undertaken, or outputs generated, to name a few.* Together, though, this combination of function and sectoral audience also allows some simple conclusions to be made on the types of decisions that these tools can support (i.e., a reflection of their purpose).

To classify based on function, we use the categories from Hammill and Tanner (2011, pp. 17–20), who define three types of tool function (which we label 1, 2, 3):

1. Process guidance tools, which *guide users through the implementation of one or several steps in the climate risk management process*, including (i) communications and engagement with stakeholders, (ii) screening of the development activities for climate risks, (iii) assessing in greater depth the climate risks and potential adaptation/response options for those activities identified via screening, (iv) assisting with the implementation of the selected options, and/or (v) assisting with the monitoring and evaluation of the selected options.
2. Data and information provision tools, which *generate or present information that can be used as inputs for implementing one or several of these process steps.*
3. Knowledge-sharing tools, which *allow users to share knowledge and experience to inform, support and refine the implementation of these process steps.*

---

* These latter classification methods were more ambiguous because many tools were designed to be versatile: across several different scales, targeting a variety of different users, using a variety of methods, etc.

Managing Climate Risk for the Water Sector    247

To classify based on target sectoral audience, we define three different audience categories within this water sector climate risk management context (which we label A, B, C):

A. Tools designed for *general* climate risk management that do not necessarily mention water sector issues anywhere but that may still be usefully applied to this context.
B. Tools designed for water sector best management that do not necessarily mention climate change but whose use could indirectly improve the resilience of WASH and WRM to climate impacts (e.g., by tackling the development deficit).
C. Tools designed *specifically* for climate risk management on WASH and/or WRM.

Combining these two classification schemes creates nine categories (e.g., 1-A, 2-B, 3-C), allows the list of 137 tools to become better organized, and begins to illustrate the areas where many/few tools exist. The summary of this is displayed in Table 9.1. As visible in this table, the number of tools of each type varied significantly. Looking first at the functional typology, the vast majority were of the type 1, *process guidance* category. This is mainly due to a positive bias in our definition toward this type of tool, focusing on those tools with a clear operational aspect for addressing water sector climate risk management. Many of the potential type 2 and 3 tools that we examined were simply less capable of supporting this narrative and were thus excluded. For example, there are a large number of potential type 2 tools that undertake climate impact projections in various different ways, but almost none of them were included here, as their clear and direct operational links to WASH/WRM resilience were simply too weak.

Looking next at the target sectoral audience typology, the numbers are more evenly spaced. Type A tools designed for general climate risk management and type B water sector best management tools were approximately equivalent, while there were fewer type C tools designed specifically for water sector climate risk management.

### TABLE 9.1
### Summary of the Different Types of Tools Present within the Overall List of 137

Total number of unique climate risk management tools for the water sector = 137

| | | |
|---|---|---|
| Number of 1s = 107 | Number of As = 58 | |
| Number of 2s = 17 | Number of Bs = 52 | |
| Number of 3s = 15 | Number of Cs = 29 | |
| Number of 1-A's = 50 | Number of 2-A's = 1 | Number of 3-A's = 7 |
| Number of 1-B's = 38 | Number of 2-B's = 11 | Number of 3-B's = 5 |
| Number of 1-C's = 20 | Number of 2-C's = 7 | Number of 3-C's = 3 |

*Note:* Each set of numbers does not add up to 137, due to a few tools exhibiting more than one each category.

From a water manager's perspective though, these type C tools are likely the most useful for this context, so are worth special attention by potential tool users.

As will be further illustrated in the following section (where examples of tools are given), these nine categories essentially reflect nine different *purposes* for tools, via the types of decisions that the different combinations of function and sectoral audience can support. These nine types of decisions will obviously be generalizations but may nonetheless be helpful for differentiating between the categories. Firstly, we propose the intended decisions for each general category as follows:

- Type A tools: support decisions on non-sector-specific issues related to general climate adaptation
- Type B tools: support decisions on building general water sector resilience and tackling the development deficit
- Type C tools: support decisions on planned adaptation interventions specifically for the water sector
- Type 1 tools: support decisions on practical, project-/program-/policy-level topics
- Type 2 tools: support decisions based on climatic and/or hydrological modeling and scenario planning topics
- Type 3 tools: support decisions on *bigger-picture* strategic topics

These are then brought together in Table 9.2 to distinguish between the supported decisions in each of these categories of tools.

### TABLE 9.2
### Summary of the General Types of Decisions That Are Supported by Each of the Nine Categories of Tools Defined Here

| Type | A | B | C |
|---|---|---|---|
| 1 | Support practical, project-/program-/policy-level decisions on non-sector-specific issues related to general climate adaptation | Support practical, project-/program-/policy-level decisions on building general water sector resilience and tackling the development deficit | Support practical, project-/program-/policy-level decisions on planned adaptation interventions specifically for the water sector |
| 2 | Support climatic and/or hydrological modeling and scenario planning decisions on non-sector-specific issues related to general climate adaptation | Support climatic and/or hydrological modeling and scenario planning decisions on building general water sector resilience and tackling the development deficit | Support climatic and/or hydrological modeling and scenario planning decisions on planned adaptation interventions specifically for the water sector |
| 3 | Support *bigger-picture* strategic decisions on non-sector-specific issues related to general climate adaptation | Support *bigger-picture* strategic decisions on building general water sector resilience and tackling the development deficit | Support *bigger-picture* strategic decisions on planned adaptation interventions specifically for the water sector |

Managing Climate Risk for the Water Sector 249

## 9.7 SOME EXAMPLES OF TOOLS

We now discuss a few examples of tools from the different categories, to illustrate their typical characteristics, the decisions they can support, and the reasons why we judged them to fit within our definition. We will very briefly discuss one example from each of the nine categories.

### Category 1-A Example: Community-Based Risk Screening Tool—Adaptation and Livelihoods

The Community-Based Risk Screening Tool—Adaptation and Livelihoods (CRiSTAL) tool, developed by IISD and partners, is probably one of the best known and most widely used general climate risk management tools. As a tool that targets local project planners, it focuses on identifying and prioritizing climate risks for projects at the local level while also helping users to identify and make decisions on the most important local livelihood resources for use in designing adaptation interventions. It was originally piloted in 2004 and is continually updated, now on its fifth version. It has been applied in over 20 countries across Asia, Africa, and Latin America, including in the water sector. Its broad array of existing, practical resource material, and focus on general climate risk management, qualifies it as a type 1-A tool here.

### Category 1-B Example: Water Safety Planning

The water safety planning (WSP) concept is an initiative of the World Health Organization (WHO), which created a detailed guidebook for the process in 2009. WSPs have the main aim of consistently ensuring the safety and acceptability of a drinking water supply and are versatile in their ability to do this for anything from a simple groundwater well to a complex treatment and distribution system. The strength of this tool has been its comprehensive and detailed risk assessment and management methodology, which takes stakeholders through all the steps in the water supply chain from source to consumer, to proactively identify risks before they occur and to make decisions on how to avoid or manage them. This clearly categorizes the tool as type 1, while its general focus on water supply then categorizes it as type B.

### Category 1-C Example: Rapid Climate Change Adaptation Assessment for WASH Providers in Informal Settlements

A newer tool developed by Heath et al. (2012) is the rapid climate change adaptation assessment (RCAA). This tool was developed specifically to assess climate vulnerabilities and to help local stakeholders choose appropriate adaptations for the WASH sector in urban slum communities and has been trialled in three such slums in Africa. Its method consists of initial risk screening, impact modeling, and subsequent risk analysis and management. While the modeling portion of the tool might suggest a type 2 classification, because its purpose was still focused on the risk assessment and management, we classified it as type 1. Its specific design for climate risk management on WASH likewise classifies it into the type C category.

### Category 2-A Example: SimCLIM

SimCLIM is an integrated computer modeling system that can be used to assess climate impacts and adaptations. With a wide variety of available data, plus a high degree of customizability, the software is versatile enough to support climate-related decision making and risk management across many different locations and sectors, including the water sector. Developed by CLIMsystems Ltd., the tool requires purchasing but, after this point, arrives with detailed documentation that qualifies it as a tool here. Its nature as a computer model classifies it into the type 2 category, while its general climate focus—though with stated applicability to the water sector—likewise classifies it into the type A category.

### Category 2-B Example: Water Security Index

The recently developed Water Security Index by the Asian Development Bank (ADB)/Asia-Pacific Water Forum (APWF) illustrates that not all type 2 tools are pieces of computer software. Located in the appendix of an accompanying report for policymakers (ADB and APWF 2013), this detailed methodology is described for how to measure national progress toward water security. *Water security* in this report is composed of five parts: household water security, economic water security, urban water security, environmental water security, and resilience to water-related disasters, all of which received numerical scores for each country assessed and with each score based on a variety of supporting subordinate data. While presented here at the country level, the described methodology could be easily downscaled for decision making at the subnational or local level, contingent on data availability. The clear comparative usefulness of the resulting scores, as well as the usefulness of gathering the relevant supporting data, gives the tool a strong policy relevance and its results could likely inspire policy change to achieve better water security and climate resilience. Its focus on data analysis for water sector best management categorizes it as a type 2-B tool.

### Category 2-C Example: Water Evaluation and Planning System

Returning to computer software, the Water Evaluation and Planning System (WEAP) software, developed by the Stockholm Environment Institute (SEI), is similar to SimCLIM but focused on water resource planning. It can build a model of the current water resource conditions in an area and then can explore various scenarios related to overall supply and demand, such as new reservoirs, population growth, or water use patterns. It can do so from a water balance perspective or from a policy perspective, and although it was not designed specifically for climate adaptation, it nonetheless has seen a lot of useful application to decision making around water sector adaptation planning, with some users using it specifically for this purpose. It is one of the most widely used tools, available in 22 languages and with nearly 12,000 members on its discussion forum as of July 2013. Its modeling focus for the water sector, including climate adaptation aspects, categorizes it as a type 2-C tool.

Managing Climate Risk for the Water Sector 251

### Category 3-A Example: weADAPT

The weADAPT website focuses on all things climate adaptation and includes related reports, organizations, tools, projects/programs, and various data. Its moderated wiki format also allows anyone to sign up and add their own content to the database, such as new tools or projects. Its content includes many water-related tools, reports, data, and projects. This type of knowledge-sharing environment categorizes it well as a type 3 tool, with its adaptation focus—including applicability to water issues—categorizing it as a type A tool. Its inclusion of a wide variety of practical guidance documents and its focus on adaptation and risk management qualifies it as a tool here, with this abundant and concentrated guidance capable of supporting *bigger-picture* strategic decision making, such as inspiring the choice to invest in adaptation in the first place.

### Category 3-B Example: Global Water Partnership Toolbox for Integrated Water Resources Management

The GWP toolbox on IWRM serves to share knowledge on 61 different methods for better accomplishing an IWRM process, along with a variety of case studies, example IWRM plans and policies, and links to partners and other relevant publications. Resources on climate adaptation are also available. A case could thus be made that this tool fits in the type C category, but since its driving focus is mainly on IWRM, we judged it more suitable in type B. Its nature as a content-heavy website, with focus on practical guidelines, likewise gives the type 3 category and could support strategic decision making around IWRM investments.

### Category 3-C Example: Technologies for Climate Change Adaptation—The Water Sector

Lastly, the UN Environment Programme's (UNEP) handbook on technologies for climate adaptation in the water sector is an example of a type 3-C tool. This handbook presents a variety of WASH-related technologies and discusses when, where, and how they can be best adapted against climate variability and change. Note that the technologies themselves are not tools, but a strategic decision support manual of this nature can indeed qualify as a type 3 tool. Because the focus is specifically on WASH technology climate resilience, the handbook fits a type C categorization.

## 9.8 TOOL *NICHES* AND THEIR DEGREE OF OVERLAP

With these examples of tools and an associated classification system in place, we can now probe a bit more deeply into the variety of tools within these categories and the number of unique *niches* available for them to operate in without overlapping (in terms of users/functions). In other words, do many of these tools overlap, with all possible niches already full, or are most of them unique, with many more potential niches remaining to be filled? If we rely only on the data of the nine possible categories in Table 9.1, a relative abundance of type 1 tools is clearly visible, but this alone does not really satisfy the question. A better answer could be achieved if each tool was given further subdividing classifications, such as classifications based on their scale, methodology used, desired users, and possible outputs.* This would create many more overall categories of tools and allow the results to better indicate how evenly distributed the tools were across all these categories. As mentioned earlier, though, these other classification systems were not used, due to their greater levels of ambiguity from overlap across multiple categories.

Instead, we rely on our own best judgment to discuss this topic of overlap, though this still allows us to note several interesting points from the various categories. Most notably, we observe significant overlap occurring between the 50 type 1-A tools, with the other categories also experiencing some overlap, though to a lesser extent. Within these 50, there are tools that cover all aspects of Hammill and Tanner's functional type 1, from communication tools (e.g., World Wide Fund for Nature [WWF]'s Climate Witness Community Toolkit), to risk screening and assessment tools (e.g., CRiSTAL, among many others), to implementation tools (e.g., UNDP's Designing Climate Change Adaptation Initiatives toolkit) and M&E tools (e.g., AdaptME). Likewise, these tools cover the range of potential audiences, from communities, to local and national governments, to researchers, NGO workers, and staff of multilateral organizations.

More broadly, the biggest strength—or weakness, in terms of tool overlap—of many of these tools is their versatility. Many present themselves as a comprehensive, start-to-finish decision support guidebook, often beginning literally at the *what is climate change* level and proceeding all the way to adaptation option implementation (either specifically for the water sector or more generally) and follow-up for their desired audience (e.g., USAID's Adapting to Coastal Climate Change guidebook). Inevitably, then, a significant amount of overlap exists between tools like these, especially since they are generally obtaining their guidance from similar sources: for example, previous UNFCCC/IPCC agreements on best practice in adaptation/risk management/climate communication. For example, the tools that focus on general community-based vulnerability and adaptation assessments (including CRiSTAL, CARE's Climate Vulnerability and Capacity Analysis [CVCA] and Community-Based Adaptation toolkits, Nakalevu [2006], FAO and IPG [2009], and ENDA and SEI [2013]) take quite similar approaches to these activities.† This is due to the fact that these tools have had to sacrifice specific contextual detail in order to be versatile across these many different community contexts. Likewise, while there are some tools designed for specific contexts (e.g., the BalticClimate toolkit), their layout, methods, and outputs are usually quite similar to the more general tools, though they may introduce some unique regional data or cultural insights.

This also returns to the issue of deep uncertainty and the shift being advocated toward more robust, scenario-based decision making. Given the compounding

---

* For example, if we added spatial scale as a third typology, and if we defined I/II/III/IV for local, state, national, and transboundary level, then we would have 36 overall tool categories, such as type 1-A-II.
† This is not to disparage their approach but rather to just point out that more than one tool is discussing and recommending it.

uncertainties that will underlie many adaptation decisions for the water sector (but by no means all), it makes sense that tools should focus on more confident activities like general resilience building, tackling the development deficit, *no regrets* adaptation measures, robust decision frameworks, and risk management. That said though, many of these (often simply *common sense*) principles are broad enough to be summed up effectively in just a few focused tools, rather than the hundreds that currently exist. While we would agree with the argument of Hammill and Tanner (2011) that having many tools tailored to specific local needs is more desirable than having a few general tools, we disagree that this is what is currently happening. While deep uncertainty likely contributes to this, another important contributor could be related to the political economy of tool development itself, with tool developers potentially viewing versatile, cross-contextual tools as more valuable, due to their wider potential audience. That said, because of the degree of overlap that already exists, a potentially better use of tool developer time would be to focus instead on synthesizing and harmonizing some of the most similar groups of these tools, rather than making new ones. As well, there are notably few tools that attempt to delve deeply into *local context* beyond a climate lens—considering, for example, the political economy of a specific city or country in its activities and recommendations. This is another weakness, as building more resilient political/economic environments at the local level could contribute significantly to building sectoral climate resilience (again, by tackling the development deficit).

## 9.9  IS THERE EVIDENCE OF USER DEMAND FOR TOOLS?

Tool diversity and overlap aside, is anyone actually using them? The presence of at least 137 tools, while encouraging, does not necessarily imply inherent user demand for them. For example, tools could have been created in advance of user demand, to then be able to *capture* latent demand with successful marketing (i.e., a *build it and they will come* approach) but were then never able to actually accomplish this. Obviously, someone is using at least a few of these tools, but hard evidence on their degree of popularity, for developing country users in particular, is fairly scarce.

It is first useful to clarify who the *users* and *developers* are. As mentioned earlier, potential users (audiences) of tools could include donors, NGO workers, businesses, and national/local governments—both in the developed and developing world. That said, any of these actors could also be a developer. For example, a donor like the World Bank could be the user of a water project portfolio risk screening tool, but they could also be the developer of a water adaptation planning tool targeted at local governments in developing countries.

This question of who is using tools is thus quite nuanced and needs to be considered carefully. One of the main general pieces of evidence on tool use to date comes from the study by Hammill and Tanner (2011). They surveyed various tool developers and users to conclude that there was strong interest in these tools in the international development communities of NGO practitioners and donors (both as users and developers)—and that users were often engaged in their design (be they local governments in developing countries or the developed world NGO practitioners themselves). However, they also noted that the initial impetus for the development of most tools was a largely developed world-driven process. That is, most existing tool developers were mainly developed world donors, businesses, or NGOs/academics. The other 10, previously mentioned, tool reviews gave similar implications (especially the review by GIZ [2009]), though without a formal study. In our opinion, this indicates three broad points: (1) most users—whether in the developed world or developing world—see value in *tools* as a concept; (2) developed world actors are actively demanding, developing, and using tools (e.g., many of the major donors and large NGOs now use tools to screen their projects for climate risks); but (3) there is less evidence of this demand, development, and usage activity by developing world actors. Point (3) does not imply that these activities are nonexistent but simply that, overall, there is less evidence of them as compared to the equivalent evidence base for developed world tool activity.

Beyond this, we had to rely solely on anecdotal methods of assessing popularity of individual tools, such as social media hits, search engine results, and relevant website data, where available. We fully acknowledge, though, that these are often very poor proxies for measuring user demand, so the earlier evidence from the previous tool reviews should be prioritized. That said, certain tools, such as the SEI weADAPT and UNDP Adaptation Learning Mechanism (ALM) websites, have created various social media profiles for themselves, such as a Facebook page, a LinkedIn discussion group, and/or a YouTube channel. These had rarely received more than a few hundred *likes*/members/video views though—even if they had been in existence for several years. Although many users may still be using the tool without using its (entirely optional) social media profiles, the figures were nevertheless not very impressive for supposedly global tools that are targeted at a very broad range of users.

In terms of website data, very few tools publicly reported any of their direct usage data, which is unfortunate, as these would allow a much better indication of the tools' popularity. The only tool that fully reported its day-to-day website hits, total number of downloads, and other relevant data was the SEI WEAP tool. These data gave quite a positive view of this tool, with tens of thousands of total downloads and several hundred unique website hits every month. That said, WEAP is well known as one of the most popular tools, so its data—although encouraging—are not necessarily representative as an *average* value.

Another way to check tool popularity was to search for them in Google Scholar, both for total search hits and for citations of the tool publication itself (applicable only when the tool consisted of a single publication). This again was not a particularly useful proxy though, as it seemed to generally underestimate popularity, probably because of the various challenges that arise for Google when trying to locate citations for each hit (e.g., if the citing author referenced the tool incorrectly, it might not appear in the citation list). For example, searching Scholar for the WEAP tool returned only a few hundred total hits and very few citations on the publications that appeared in the search, which clearly underestimates the popularity of the tool in light of its aforementioned website usage data. Nevertheless, the search hits and number of citations on most of the rest of these tools were much lower, which does not suggest a very high overall citation impact of these tools so far, regardless of the search engine's underestimate.

The sum of this anecdotal evidence does not do much to change the initial three arguments we made earlier. Two other, theoretical, pieces of evidence can also be observed, though both of these also essentially support the three points. The first of these theoretical pieces of evidence of demand is the near total lack of active development and social media presence for most of these existing tools. Rather, we observe a large number of one-off reports or websites that are essentially just static publications. If these publications had generated a strong user interest and ongoing dialogue, this would likely be reflected in follow-up reports or other accompanying documentation, unless the targeted user group was comparatively small and already well informed. For example, we would not anticipate that tools developed by donors for themselves or for other donors would need much external documentation or engagement with the Internet at all, while we would expect tools aimed at wider audiences, like CRiSTAL, to do this much more. For these latter tools though, this level of active development was rarely observed, except in computer software-based tools and some notable exceptions like CRiSTAL, which were more reliably updated (though this alone still does not necessarily indicate their usage by a large audience).

The second theoretical note arises if we consider these tools from an *industry* perspective. Examining this list of 137 tools, we observe that the vast majority were created by nonprofit organizations (donors, NGOs, academics, etc.), with only about 10 tools operating on a profit-driven model of payment for access.* In our opinion, this small proportion of profit-driven tools potentially signifies a lack of business interest developing tools for various user groups. Given the profit-seeking motive of business, we could only assume that a lack of business interest in something implies either a real or perceived lack of potential profit opportunity. While this could be due to business perception of low user demand, it could also be due to high market entry costs (i.e., high tool development and marketing costs, in order to create something that users would be interested in purchasing). However, we can argue with near certainty that, if global user demand (i.e., willingness to pay) for tools was high enough, we would expect to see the development of more profit-driven tools to meet that demand, regardless of market entry costs, as nonprofit funding for tools is limited. For example, if, tomorrow, every single local government in every single developing country began demanding unique, tailored tools for their specific context, there would simply not be enough nonprofit funding (e.g., from their respective national governments, from donors, from NGOs) to meet this demand.† Businesses would likely then seize on this opportunity to develop tools, as the real and perceived user demand (and resulting profit opportunity) for tools would be high. Thus, this dearth of profit-driven tools does not support this scenario of a world currently clamoring for tools, though we would only expect to notice a trend like this for certain, larger user groups, like these local developing country governments.

---

* Note that payment for access is not necessarily the only private sector model that tools could take. Some private companies may choose to adopt a free-with-advertising model, or a variety of other business models, though this has, to our knowledge, not yet happened here.
† Assuming no significant change to global aid flows and national budget allocations.

## 9.10 NEW ALTERNATIVES FOR ROBUST, CLIMATE-SENSITIVE DECISION SUPPORT?

So, tools have been popular for donors as a form of decision support, a wide variety have been created, but overlap is now common and user demand seems to similarly vary—but are they nevertheless the best decision support option for robustly adapting the water sector to climate change? This question can be assessed by first discussing the recommendations by Ranger (2013) for *best practice* in robust climate risk management under deep uncertainty. Ranger (2013, pp. 34–36) lays out three general principles for robust and adaptive development interventions, which apply particularly well to the water sector:

1. *Progressive, forward-looking, adapting incrementally over time*—Meaning that climate-sensitive development programming should be a *continuous, forward-looking process of planning, implementation, learning, and adjustment*, with a strong monitoring and evaluation component that feeds back into the decision process as new information is learned about climate impacts and the effectiveness of adaptation options.
2. *Building flexibility into interventions, keeping options open*—Meaning that climate-sensitive programming should avoid inflexible measures: *those that are suitable only over a narrow range of climate conditions and are costly and difficult to adjust*, which can be avoided (especially for infrastructure) by building in safety margins, making the interventions adjustable, investing in options now that could increase the range of available adaptation options in the future, and/or implementing low-cost, short-lifespan measures that could be easily replaced or abandoned if necessary.
3. *Incorporating low-regrets measures*—That is, those measures with outcomes that are *relatively insensitive to climate uncertainties* and that have low costs relative to their benefits, both today and in the future, which could include reactive measures with short lifetimes, measures that reduce vulnerability to current climate variability, measures that reduce other stresses/risks that could increase climate vulnerability, measures with strong cobenefits for multiple stakeholders, measures to reduce general vulnerability and to increase resilience to shocks, and/or measures to remove barriers to autonomous adaptation.

Many elements of these principles are also interpreted specifically for the water sector in the UN's most recent World Water Development Report (UNESCO 2012), which focused entirely on managing water under uncertainty and risk, though this more recent work by Ranger provides a more succinct and thematically equivalent overview.

These principles have also been incorporated into a variety of adaptation option appraisal methods that have been growing in popularity as *robust decision frameworks*. Ranger (2013, pp. 54–62) reviews several of these, of varying function and complexity. For cases where exact probabilities of future events are known, then the simple conventional economic appraisal tools like cost–benefit analysis (CBA), cost-effectiveness analysis (CEA), and multicriteria analysis (MCA) work well, though this is obviously not the case in this deeply uncertain climatic context. However, these methods can still be useful under uncertainty through the use of sensitivity analysis, where these tools are run under several different possible scenarios to help determine how sensitive the outputs are to these different potential futures.

For cases where exact probabilities are known, but will change over time, Ranger discusses the more complex approach of real options analysis (ROA). This method is similar to CBA but allows a deeper analysis of the value of timing and flexibility of decisions and the value of options being available in the future. This allows the appraisal of options like waiting and learning before acting and can thus be useful in identifying key decision points in an incremental adaptation strategy.

For cases where exact probabilities are unknown (i.e., the characteristic of many climate-sensitive decisions in the water sector), Ranger reviews several methods, including robustness matrices, RDM, and *climate-informed decision analysis* (CIDA). The first of these ranks the performance of different options against a set of future scenarios, to determine which are most robust to uncertainty. The second, RDM, is similar but much more in-depth (and resource-intensive), testing the interdependencies of scenarios, priorities, options, and objectives in a participatory manner. The third, CIDA, is again similar but uses expert judgment to assess the *plausibility* different climate scenarios to discredit the least likely.

## 9.11 TOOLS AND ROBUST DECISION SUPPORT METHODS: COMPLEMENTARY OR CLASHING?

If we consider these principles and methodologies as current *best practice* for robust, climate-sensitive decision making in the water sector, then to what extent do the tools reviewed here contain similar content? The application/development of these methodologies for climate-sensitive decision making is still a relatively new activity—with the majority of publications having only occurred since 2010. As a result, most of the older tools reviewed here do not include them by name but may use similar logic and activities. Some of the newer tools are also now beginning to focus on these robust approaches more specifically. For example, the *Climate Adaptation Options Explorer* tool takes a similar focus on the broader socioeconomic context for adaptation decisions, while the climate risk tools for the water sector from Ekstrom et al. (2012) and Dessai and Hulme (2007) both use risk matrix/sensitivity analysis methods similar to the ones described earlier but simply without the same terminology. Two tools that do use these new methods directly are IIED's (2013) stakeholder-focused CBA for the water sector tool and UNEP's (Miller 2011) *MCA4Climate* tool, both of which directly employ the CBA and MCA in a more robust and climate-sensitive manner. There are also tools like the SWITCH project's *COFAS* tool (Peters et al. 2010), which assesses future uncertainties (both climate-related and others) in urban drainage systems and incorporates flexibility into the decision-making process via a variety of methods, including robust risk matrices, ROA, and CBA.

That said, for every tool in this list that utilizes the robust principles and scenario-based methods of a *seek robust solutions* paradigm, there are still several others that rely on a narrower approach to planning for a single *most likely* future, via a *predict then act* paradigm. Ranger (2013) adds to this terminology with two similar descriptions of the situation: *science-first* and *decision-centric* approaches, as displayed in Figure 9.2. Ranger distinguishes these terms by the relative degree of effort invested in different stages of the climate-sensitive decision-making process. She notes that the *science-first* approach was the one initially adopted for work on climate change-related problems, where the majority of effort is directed toward attempting to generate and downscale general climate impact projections for a particular future (arguing that this is a generally *resource-inefficient* activity), with the assessment of the broader socioeconomic context and of other possible future scenarios given less focus. The *decision-centric* approach, in contrast, directs the majority of effort toward deeply understanding the broader problems and uncertainty themselves and on identifying robust solutions. Here, climate change is mainstreamed within the decision process as just one of the factors influencing policy decisions, with climate science playing a more focused role to generate plausible scenarios based on this broader context. With its greater attention to the broader socioeconomic context, Ranger argues that this approach is more in line with the project cycle and policymaking processes, thus being able to have greater impact on them.

Since this is a new framework alongside the robust approaches in general, it is not surprising that many of the older tools take a *science-first* approach (e.g., the tools by CARICOM [2003] and UNDP [2003]), but of more concern is the fact that several newer tools are also still advocating this approach. For example, we argue that the newest versions of UKCIP's Adaptation Wizard (2013) and IISD's CRiSTAL (2013) still take this approach to a certain extent. CRiSTAL, for example, has an activity of *identifying opportunities and barriers to project implementation*,

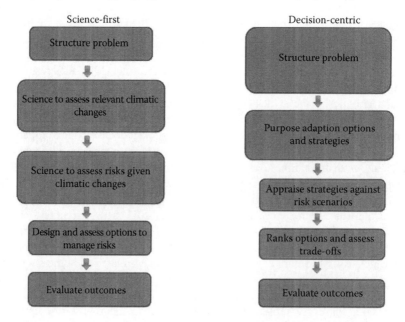

**FIGURE 9.2** Depiction of the *science-first* versus *decision-centric* approaches to climate-sensitive decision making. (Image taken from Ranger, N., *Topic Guide—Adaptation: Decision Making under Uncertainty*, Evidence on Demand, Hertfordshire, U.K., 2013.)

# Managing Climate Risk for the Water Sector

which would be *decision centric*, except that it is relegated at the end of a long process of baseline data collection, climate science, and vulnerability and risk assessments, when, in practice, this might be the most important activity of all for better focusing these climate science assessments onto a number of scenarios and for helping to achieve broader stakeholder buy-in and uptake within the community's particular socioeconomic context. Further examples, to just name a few, include the EU's (2012) Baltic Climate toolkit (with heavy focus on climate science and vulnerability assessment), the US EPA's (2012a) Climate Resilience Evaluation and Awareness Tool for water utilities (CREAT) (a computer software tool that again focuses more on the climate science than on the adaptation options and broader context that could be affecting these water utilities), and Heath et al.'s (2012) RCAA tool (described in Section 9.7 earlier, focusing mainly on assessing climate and hydrological scenarios for impacts and vulnerability, alluding only briefly to the broader problem of a general lack of climate change awareness in the case study trials).

While these *science-first* tools often still offer important activities and guidance, the challenges that have faced effective climate policymaking around the world—coupled with the existing complexity of the water sector—illustrate that the broader socioeconomic context often needs greater focus. There will, of course, continue to be circumstances when a strong, general scientific assessment of climatic/hydrological vulnerabilities and impacts will be the most useful input into a decision process, so tools of this nature will continue to have some value. That said, though, it appears that consensus is emerging on the value of *seeking robust solutions*, via scenario-based planning, which could also help to concretize these scientific assessments into more focused, resource-efficient activities. We thus see more *niches* available for new tools (and other forms of decision support) that guide users through the principles and methods of robust, climate-sensitive decision making for water sector interventions. This is especially the case for those that take a holistic, *decision-centric* approach, as these could also help to engage a wider array of new users/stakeholders from beyond the climate/water sectors, which is absolutely necessary if we hope to effectively address climate change. At the same time, an effort should also be made to modify, combine, or phase out some of the older tools,* especially those still relying on a *stationarity* approach, to avoid potential maladaptation that could be brought about from their well-intentioned use by users who may not be aware of these nuances. This could ideally reduce the total number of tools and their degree of overlap as well.

## 9.12 CONCLUSIONS

This chapter reviewed climate risk management tools for the water sector in light of new arguments for a paradigm shift toward a more robust, *decision-centric* approach to adaptation option appraisal and climate-sensitive decision making. After briefly discussing the latest climate knowledge for the water sector, it highlighted this relatively new body of decision-making literature under climatic *deep uncertainty*, though also emphasized that existing methods of *minimizing regret* were capable

---

* All 137 tools reviewed here were still available online, so new users could still potentially be discovering them via Internet searching.

of addressing most of these challenges. It then focused on *tools* as a decision support solution, first creating a definition for the term and then reviewing the variety in function, audience, and purpose of 137 tools with practical relevance to traditional climate risk management in the water sector. Analysis of these tools found significant degrees of overlap and mixed evidence of user demand. The principles and methods for robust, climate-sensitive decision support were then reviewed and considered alongside these existing tools. Several tools (mainly newer ones) were found to have elements of these robust principles within them, though many more—both old and new—still relied on a largely *science-first* approach, which is less responsive to the idea of minimizing regret via scenario analysis and which devotes less effort to the bigger socioeconomic problems at stake and the role of climate change as just one of many decision factors within them. While *science-first* tools will continue to have value in certain circumstances, the chapter concluded by suggesting that there will be increasing opportunity for *decision-centric* tools to bring new, positive impact to climate-sensitive decision making and, conversely, that tools still relying on *stationary* assessments of a most likely future should be generally phased out.

## ACKNOWLEDGMENTS

This chapter is adapted, with permission, from J. Doczi *Climate risk management tools for the water, sanitation and hygiene sector: an assessment of current practice. ODI Working Paper.* London, U.K.: Overseas Development Institute, 2014. I gratefully acknowledge the funding support from the UK Department for International Development (DFID) to carry out the work that this chapter is based on, under the auspices of a more specific, ODI-led project entitled *Climate risk screening for rural water supply in Ethiopia.* The views expressed here are those of the author and do not necessarily represent the views of ODI or reflect the UK Government's official policies.

## Appendix
### Detailed List of 137 Tools Applicable for Climate Risk Management in the Water Sector

| Tool Name | Developer/Author | Type(s) (1/2/3—A/B/C)[a] | Reference Link (Previous Reviews of the Tool)[b] |
|---|---|---|---|
| Adaptation Database and Planning Tool (ADAPT) | International Council for Local Environmental Initiatives–Local Governments for Sustainability (ICLEI), United States | 1-A | ICLEI (2013) http://www.icleiusa.org/tools/adapt |
| Adaptation Policy Framework for Climate Change (APF) | UNDP | 1-A | UNDP (2003) http://www.eird.org/cd/on-better-terms/docs/UNDP-GEFUsers-Guide-to-the-Adaptation-Policy-Framework.pdf (Reviewed by UNFCCC [2008], Olhoff and Schaer [2010], Hammill and Tanner [2011], and CCCCC [2013]) |
| Adaptation Screening Matrix and Adaptation Decision Matrix | Stratus Consulting | 1-A | Stratus (2013) http://stratusconsulting.com/services/climate/assessing-vulnerability-and-adaptation/ (Reviewed by Garg et al. [2007] and UNFCCC [2008]) |
| Adaptation to Coastal Climate Change: A Guidebook for Development Planners | USAID | 1-A | USAID (2009) http://pdf.usaid.gov/pdf_docs/pnado614.pdf (Reviewed by GIZ [2009], Hammill and Tanner [2011], ALM [2013], and CCCCC [2013]) |
| Adaptation toolkit: Guidebook for researchers and adaptation practitioners working with local communities | Energie, Environnement, Developpement (ENDA) and SEI | 1-A | Ampomah and Devisscher (2013) http://static.weadapt.org/knowledge-base/files/1128/51d4187011529adaptation-toolkit-march-2013.pdf (Reviewed by CCCCC [2013] and weADAPT [2013]) |

(continued)

## Appendix (continued)

### Detailed List of 137 Tools Applicable for Climate Risk Management in the Water Sector

| Tool Name | Developer/Author | Type(s) (1/2/3—A/B/C)[a] | Reference Link (Previous Reviews of the Tool)[b] |
|---|---|---|---|
| Adaptation toolkit: Sea level rise and coastal land use—How governments can use land-use practices to adapt to sea level rise | Georgetown Climate Center | 1-A | Grannis (2011) http://www.georgetownclimate.org/sites/default/files/Adaptation_Tool_Kit_SLR.pdf (Reviewed by CCCCC [2013]) |
| Adaptation Wizard | UK Climate Impacts Programme (UKCIP) | 1-A | UKCIP (2013c) http://www.ukcip.org.uk/wizard/ (Reviewed by UNFCCC [2008], Olhoff and Schaer [2010], Hammill and Tanner [2011], Traerup and Olhoff [2011], and weADAPT [2013]) |
| Adapting to climate change: A planning guide for state coastal managers | National Oceanic and Atmospheric Administration (NOAA) Office of Ocean and Coastal Resource Management | 1-A | NOAA (2010) http://coastalmanagement.noaa.gov/climate/docs/adaptationguide.pdf (Reviewed by CCCCC [2013]) |
| Adapting to climate variability and change: A guidance manual for development planning | USAID | 1-A | USAID (2007) http://pdf.usaid.gov/pdf_docs/PNADJ990.pdf (Reviewed by Olhoff and Schaer [2010], Hammill and Tanner [2011], CCCCC [2013], and weADAPT [2013]) |
| AdaptME (Monitoring and Evaluation) | UKCIP | 1-A | UKCIP (2013d) http://www.ukcip.org.uk/adaptme-toolkit/ |
| Applying climate information for adaptation decision-making | UNDP, UNEP, and the Global Environment Facility (GEF) National Communications Support Programme | 1-A | Lu (2008) http://www.adaptationlearning.net/sites/default/files/ncsp.pdf (Reviewed by ALM [2013]) |

| | | | |
|---|---|---|---|
| BalticClimate toolkit | EU Baltic Sea Region Programme | 1-A | EU (2012)<br>http://www.toolkit.balticclimate.org/index.php<br>*(Reviewed by weADAPT [2013])* |
| Business Area Climate Impacts Assessment Tool (BACLIAT) | UKCIP | 1-A | UKCIP (2013a)<br>http://www.ukcip.org.uk/bacliat/<br>*(Reviewed by UNFCCC [2008] and weADAPT [2013])* |
| Caribbean risk management guidelines for climate change adaptation decision making | The Caribbean Community (CARICOM) | 1-A | CARICOM (2003)<br>http://200.32.211.67/M-Files/openfile.aspx?objecttype=0&docid=2879<br>*(Reviewed by CCCCC [2013])* |
| CEDRIG—Climate, Environment and Disaster Risk Reduction Integration Guidance | Swiss Agency for Development and Cooperation (SDC) | 1-A | SDC (2012)<br>http://www.sdc-drr.net/cedrig<br>*(Reviewed by CCCCC [2013])* |
| CLARA—Climate Adaptation Resource for Advisors | UKCIP | 1-A | UKCIP (2013b)<br>http://www.ukcip.org.uk/clara' |
| Climate Change and Environmental Degradation Risk and Adaptation Assessment (CEDRA) | Tearfund | 1-A | Tearfund (2012)<br>http://tilz.tearfund.org/Topics/Environmental+Sustainability/CEDRA.htm<br>*(Reviewed by GIZ [2009], Hammill and Tanner [2011], Traerup and Olhoff [2011], CCCCC [2013], and weADAPT [2013])* |
| Climate Check | Deutsche Gesellschaft für Internationale Zusammenarbeit (GIZ) | 1-A | Kropp and Scholze (2009)<br>http://www2.gtz.de/dokumente/bib-2009/gtz2009-0175en-climate-change-information.pdf<br>*(Reviewed by GIZ [2009], OECD [2009], and Hammill and Tanner [2011])* |
| Climate Guide | Red Cross/Red Crescent | 1-A | Red Cross/Red Crescent (2007)<br>http://www.climatecentre.org/downloads/File/reports/RCRC_climateguide.pdf<br>*(Reviewed by Olhoff and Schaer [2010] and Hammill and Tanner [2011])* |

*(continued)*

## Appendix (continued)
### Detailed List of 137 Tools Applicable for Climate Risk Management in the Water Sector

| Tool Name | Developer/Author | Type(s) (1/2/3—A/B/C)[a] | Reference Link (Previous Reviews of the Tool)[b] |
|---|---|---|---|
| Climate Proofing for Development | GIZ | 1-A | Hahn and Frode (2011) http://www.preventionweb.net/files/globalplatform/entry_bg_paper~giz2011climateproofing.pdf (Reviewed by Hammill and Tanner [2011], Traerup and Olhoff [2011], and CCCCC [2013]) |
| Climate proofing: A risk-based approach to adaptation | ADB | 1-A | ADB (2005) http://www.adb.org/sites/default/files/pub/2005/climate-proofing.pdf (Reviewed by CCCCC [2013]) |
| CVCA | CARE | 1-A | CARE (2009) http://www.careclimatechange.org/cvca/CARE_CVCAHandbook.pdf (Reviewed by Hammill and Tanner [2011], Traerup and Olhoff [2011], ALM [2013], CCCCC [2013] and weADAPT [2013]) |
| Climate Witness Community Toolkit for the South Pacific | WWF | 1-A | WWF (2009) http://assets.panda.org/downloads/cw_toolkit.pdf (Reviewed by Hammill and Tanner [2011]) |
| Climate-resilient cities: A primer on reducing vulnerabilities to climate change impacts and strengthening disaster risk management in East Asian cities | The World Bank | 1-A | Prasad et al. (2008) http://documents.worldbank.org/curated/en/2008/06/9740566/climate-resilient-cities-primer-reducing-vulnerabilities-climate-change-impacts-strengthening-disaster-risk-management-east-asian-cities (Reviewed by ALM [2013]) |
| CRiSTAL | International Institute for Sustainable Development (IISD) et al. | 1-A | IISD et al. (2013) http://www.iisd.org/cristaltool/ (Reviewed by UNFCCC [2008], GIZ [2009], OECD [2009], Olhoff and Schaer [2010], Hammill and Tanner [2011], Traerup and Olhoff [2011], ALM [2013], CCCCC [2013], and weADAPT [2013]) |

Managing Climate Risk for the Water Sector 265

| | | | |
|---|---|---|---|
| Comprehensive Hazard and Risk Management (CHARM) | Applied Geoscience and Technology Division—Secretariat of the Pacific Community (SOPAC) | 1-A | SOPAC (2001) http://ict.sopac.org/VirLib/DM0044.pdf *(Reviewed by UNFCCC [2008] and CCCCC [2013])* |
| CV&A: A guide to community vulnerability and adaptation assessment and action | Secretariat of the Pacific Regional Environment Programme (SPREP) | 1-A | Nakalevu (2006) http://www.sprep.org/att/publication/000437_cvaguidee.pdf *(Reviewed by ALM [2013])* |
| Designing Climate Change Adaptation Initiatives: A UNDP Toolkit for Practitioners | UNDP | 1-A | UNDP (2010) http://www.undp.org/content/dam/aplaws/publication/en/publications/environment-energy/www-ee-library/environmental-finance/low-emission-climate-resilient-development/designing-adaptation-initiatives-toolkit/Toolkit%20FINAL%20(new%20cover).pdf *(Reviewed by Hammill and Tanner [2011], Traerup and Olhoff [2011], ALM [2013], CCCCC [2013], and weADAPT [2013])* |
| Guidelines for developing a coastal risk management plan | Government of Tasmania—Department of Primary Industries and Water | 1-A | Government of Tasmania (2009) http://www.dpiw.tas.gov.au/inter.nsf/Attachments/KCRE-7PE8CC/$FILE/CRMP_Template_and_Guidelines.pdf *(Reviewed by CCCCC [2013])* |
| ICLEI–ACCCRN urban climate change resilience toolkit | ICLEI and the Asian Cities Climate Change Resilience Network (ACCCRN) | 1-A | ICLEI and ACCCRN (2013) http://www.teriin.org/events/Sunandan-Tiwari.pdf; http://www.scribd.com/doc/144824994/ACCCRN-ICLEI-Replication-Press-Release |
| Identifying Adaptation Options (AdOpt) | UKCIP | 1-A | UKCIP (2012) http://www.ukcip.org.uk/adopt/ *(Reviewed by UNFCCC [2008])* |

*(continued)*

## Appendix (continued)

### Detailed List of 137 Tools Applicable for Climate Risk Management in the Water Sector

| Tool Name | Developer/Author | Type(s) (1/2/3—A/B/C)[a] | Reference Link (*Previous Reviews of the Tool*)[b] |
|---|---|---|---|
| Integrating Climate Change Adaptation into Secure Livelihoods Toolkits | Christian Aid | 1-A | Christian Aid (2010) http://www.adaptationlearning.net/sites/default/files/Adaptation%20toolkit%201.pdf; http://www.adaptationlearning.net/sites/default/files/Adaptation%20toolkit%202.pdf (*Reviewed by Hammill and Tanner [2011], ALM [2013], and CCCCC [2013]*) |
| Local government climate change adaptation toolkit | ICLEI Oceania | 1-A | ICLEI (2008) http://archive.iclei.org/index.php?id=adaptation-toolkit0 (*Reviewed by ALM [2013]*) |
| Mainstreaming climate change adaptation into development planning: A guide for practitioners | UNDP–UNEP Poverty Environment Initiative | 1-A | UNDP and UNEP (2011) http://www.unep.org/pdf/mainstreaming-cc-adaptation-web.pdf |
| Mainstreaming Climate Change Adaptation: A Practitioner's Handbook | CARE International in Vietnam | 1-A | Huxtable and Nguyen (2009) http://www.careclimatechange.org/files/adaptation/CARE_VN_Mainstreaming_Handbook.pdf (*Reviewed by CCCCC [2013]*) |
| Methodology guide for the assessment of total climate risk and resilience measures | Economics of Climate Adaptation Working Group | 1-A | ECA (2009) http://ec.europa.eu/development/icenter/repository/ECA_Shaping_Climate_Resilent_Development.pdf (*Reviewed by CCCCC [2013]*) |
| OECD *Climate Lens* Guidance on Integrating Climate Change Adaptation into Development Projects | Organisation for Economic Co-operation and Development (OECD) | 1-A | OECD (2009) http://www.oecd.org/dac/43652123.pdf (*Reviewed by Olhoff and Schaer [2010] and Hammill and Tanner [2011]*) |

| | | | |
|---|---|---|---|
| Opportunities and Risks of Climate Change and Disasters (ORCHID) and Climate Risk Impacts on Sectors and Programmes (CRISP) | DFID and Institute for Development Studies (IDS), United Kingdom | 1-A | (ORCHID—Multiple publications) IDS (2009) http://www.ids.ac.uk/climatechange/orchid http://www.ids.ac.uk/climatechange/orchid CRISP—Downing et al. (2008) http://www.dewpoint.org.uk/Asset%20Library/DFID/Climate%20Risk%20 Assessment%20Report%20-%20Kenya.pdf Tanner et al. (2008) http://www.ids.ac.uk/files/dmfile/CHinaCLimateSCreeningSynthesisEnglish.pdf (*Reviewed by GIZ [2009], OECD [2009], Olhoff and Schaer [2010], Hammill and Tanner [2011], Traerup and Olhoff [2011], CCCCC [2013], and weADAPT [2013]*) |
| PACT: Helping companies assess their strategic exposure to climate risks | Alexander Ballard Ltd. | 1-A | Alexander Ballard (2013) http://alexanderballard.co.uk/pact/ (*Reviewed by weADAPT [2013]*) |
| Participatory tool on climate and disaster risks[c] | Swiss Interchurch Aid (HEKS) and Bread for All | 1-A | HEKS and Bread for All (2010) http://www.adaptationlearning.net/sites/default/files/CliDR%20Eng_Vers5_0.pdf (*Reviewed by ALM [2013]*) |
| Planning for Community-Based Adaptation to Climate Change | Food and Agricultural Organization (FAO) and the Department of Physical Geography at University of Freiburg (IPG) | 1-A | FAO and IPG (2009) http://www.webgeo.de/fao-webgeo-2-intro/ (*Reviewed by GIZ [2009] and ALM [2013]*) |
| Preparing for Climate Change: A Guidebook for Local, Regional, and State Governments | Center for Science in the Earth System (The Climate Impacts Group) (CIG) et al. | 1-A | CIG et al. (2007) http://www.cses.washington.edu/db/pdf/snoveretalgb574.pdf |

(*continued*)

## Appendix (continued)
### Detailed List of 137 Tools Applicable for Climate Risk Management in the Water Sector

| Tool Name | Developer/Author | Type(s) (1/2/3—A/B/C)[a] | Reference Link (Previous Reviews of the Tool)[b] |
|---|---|---|---|
| Strategic environmental assessment and adaptation to climate change guidebook | OECD | 1-A | OECD (2008) http://www.oecd.org/environment/environment-development/42025733.pdf (Reviewed by Olhoff and Schaer [2010]) |
| The guidance tool: A manual for mainstreaming climate change adaptation into the comprehensive disaster management framework country work program | The Caribbean Disaster Emergency Management Agency (CDEMA) | 1-A | CDEMA (2011) http://unfccc.int/files/adaptation/application/pdf/cca_guidance_tool_e_book_.pdf (Reviewed by CCCCC [2013]) |
| Toolkit for Community-Based Adaptation | CARE | 1-A | CARE (2010a) http://www.careclimatechange.org/files/toolkit/CARE_CBA_Toolkit.pdf (Reviewed by Hammill and Tanner [2011] and ALM [2013]) |
| Toolkit for Integrating Adaptation into Development Projects | CARE | 1-A | CARE (2010b) http://www.careclimatechange.org/files/toolkit/CARE_Integration_Toolkit.pdf (Reviewed by Hammill and Tanner [2011] and CCCCC [2013]) |
| Urban risk assessments: Understanding disaster and climate risk in cities | The World Bank | 1-A | Dickson et al. (2012) http://documents.worldbank.org/curated/en/2012/06/16499064/urban-risk-assessments-understanding-disaster-climate-risk-cities (Reviewed by CCCCC [2013]) |
| Vulnerability and capacity assessment methodology: A guidance manual for the conduct and mainstreaming of climate change vulnerability and capacity assessments in the Caribbean | CCCCC | 1-A | Pulwarty and Hutchinson (2008) http://dms.caribbeanclimate.bz/M-Files/openfile.aspx?objtype=0&docid=2756 (Reviewed by CCCCC [2013]) |

# Managing Climate Risk for the Water Sector

| | | | |
|---|---|---|---|
| Vulnerability and capacity assessment training and toolbox | International Federation of Red Cross and Red Crescent Societies (IFRC) | 1-A | IFRC (various)<br>http://www.ifrc.org/en/what-we-do/disaster-management/preparing-for-disaster/disaster-preparedness-tools/disaster-preparedness-tools/<br>*(Reviewed by CCCCC [2013])* |
| Vulnerability Reduction Assessment (VRA) tool | UNDP | 1-A | UNDP (2008)<br>http://www.seachangecop.org/files/documents/2008_12_CBA_Vulnerability_Reduction_Assessment_Guide.pdf<br>http://www.adaptationlearning.net/sites/default/files/ALM-contribution-VRA-communities-2011.pdf<br>*(Reviewed by ALM [2013])* |
| 10-Step Promotion Program Toolkit for WASH | USAID WaterLinks | 1-B | WaterLinks (2013)<br>http://www.10step-toolkit.org/ |
| A Guide to the Development of On-Site Sanitation | WHO | 1-B | WHO (1992)<br>http://www.who.int/water_sanitation_health/hygiene/envsan/onsitesan.pdf<br>*(Reviewed by Palaniappan et al. [2008])* |
| ASPIRE—A Sustainability, Poverty and Infrastructure Routine for Evaluation | Arup International Development/ Engineers Against Poverty | 1-B | ASPIRE (2013)<br>http://www.oasys-software.com/products/environmental/aspire.html |
| Community-based disaster risk management: Field practitioner's handbook | Asian Disaster Preparedness Center (ADPC) | 1-B | Abarquez and Murshed (2004)<br>http://www.adpc.net/pdr-sea/publications/12Handbk.pdf |
| Community-Based Water Resiliency Tool (CBWR) | US Environmental Protection Agency (US EPA) | 1-B | US EPA (2011)<br>http://water.epa.gov/infrastructure/watersecurity/techtools/cbwr.cfm |

*(continued)*

## Appendix (continued)
### Detailed List of 137 Tools Applicable for Climate Risk Management in the Water Sector

| Tool Name | Developer/Author | Type(s) (1/2/3—A/B/C)[a] | Reference Link (*Previous Reviews of the Tool*)[b] |
|---|---|---|---|
| DFID Guidance Manual on Water Supply and Sanitation Programmes | Water and Environmental Health at London and Loughborough (WELL) and DFID | 1-B | WELL and DFID (1998) http://www.lboro.ac.uk/well/resources/Publications/guidance-manual/guidance-manual.htm (*Reviewed by Palaniappan et al. [2008]*) |
| From source to tap: Guidance on the multi-barrier approach to safe drinking water | Canadian Council of Ministers of the Environment (CCME) | 1-B | CCME (2004) http://www.ccme.ca/assets/pdf/mba_guidance_doc_e.pdf (*Reviewed by Palaniappan et al. [2008]*) |
| Guidelines on Municipal Wastewater Management | UNEP et al. | 1-B | UNEP et al. (2004) http://esa.un.org/iys/docs/san_lib_docs/guidelines_on_municipal_wastewater_english.pdf (*Reviewed by Palaniappan et al. [2008]*) |
| Improving partnership governance in water services through public–private partnerships (PPPs) toolkit | Building Partnerships for Development (BPD) | 1-B | BPD (2011) http://www.partnershipsforwater.net/web/w/www_7_en.aspx (*Reviewed by Palaniappan et al. [2008]*) |
| Just Stir Gently: The Way to Mix Hygiene Education with Water Supply and Sanitation | IRC International Water and Sanitation Centre | 1-B | Boot (1991) http://docs.watsan.net/Scanned_PDF_Files/Class_Code_2_Water/203.2-91JU-8660.pdf (*Reviewed by Palaniappan et al. [2008]*) |
| Linking Technology Choice with Operation and Maintenance in the Context of Community Water Supply and Sanitation | WHO and IRC | 1-B | Brikke and Bredero (2003) http://whqlibdoc.who.int/publications/2003/9241562153.pdf (*Reviewed by Palaniappan et al. [2008]*) |

| | | | |
|---|---|---|---|
| Management Guide: Series of Manuals on Drinking Water Supply | Swiss Resource Centre and Consultancies for Development (SKAT) | 1-B | Frohlich (2001)<br>http://www.rural-water-supply.net/_ressources/documents/default/1_Management-Guide.pdf<br>(Reviewed by Palaniappan et al. [2008]) |
| MIKE by DHI software (various) | DHI Group | 1-B<br>1-C<br>2-B<br>2-C | DHI Group (2013)<br>http://mikebydhi.com/<br>(Reviewed by Garg et al. [2007] and UNFCCC [2008]) |
| Natural disaster mitigation in drinking water and sewerage systems: Guidelines for vulnerability analysis | Pan American Health Organization (PAHO) | 1-B | PAHO (1998)<br>http://www1.paho.org/English/Ped/nd-water_mit.pdf<br>(Reviewed by UNFCCC [2008]) |
| Safe Water Guide for the Australian Aid Program 2005 | Australian Agency for International Development (AusAid) | 1-B | AusAid (2005)<br>http://www.wsportal.org/uploads/IWA%20Toolboxes/WSP/safe_water_guide.pdf<br>(Reviewed by Palaniappan et al. [2008]) |
| Sanitation and Hygiene Promotion—Programming Guidance | Water Supply and Sanitation Collaborative Council (WSSCC) and WHO | 1-B | WSSCC and WHO (2005)<br>http://www.who.int/water_sanitation_health/hygiene/sanhygpromo.pdf<br>(Reviewed by Palaniappan et al. [2008]) |
| Sanitation safety planning | WHO and the International Water Association (IWA) | 1-B | WHO (2010)<br>http://www.who.int/water_sanitation_health/wastewater/sanitation_safety_plans_Concept_NoteV11_4_2_17_092010.pdf |
| Service Delivery Approach for rural water supply | Harold Lockwood and Stef Smits | 1-B | Lockwood and Smits (2011)<br>http://www.waterservicesthatlast.org/publications/multi_country_synthesis |
| Sustainable Community Management of Urban Water and Sanitation Schemes (A Training Manual) | Water and Sanitation Program of the World Bank (WSP-WB) and DAWASA | 1-B | Castro et al. (2009)<br>http://www.wsp.org/sites/wsp.org/files/publications/africa_training_manual.pdf |

(continued)

## Appendix (continued)
### Detailed List of 137 Tools Applicable for Climate Risk Management in the Water Sector

| Tool Name | Developer/Author | Type(s) (1/2/3—A/B/C)[a] | Reference Link (Previous Reviews of the Tool)[b] |
|---|---|---|---|
| Sustainable Wastewater Management: A Handbook for Smaller Communities | Ministry for the Environment (MFE), New Zealand | 1-B | MFE (2003) http://www.mfe.govt.nz/publications/waste/wastewater-mgmt-jun03/# (Reviewed by Palaniappan et al. [2008]) |
| The GEMI Local Water Tool | Global Environmental Management Initiative (GEMI) and the World Business Council for Sustainable Development (WBCSD) | 1-B | WBCSD (2013) http://www.wbcsd.org/work-program/sector-projects/water/localwatertool.aspx |
| The International Benchmarking Network (IBNET) for Water and Sanitation Utilities—Toolkit | IBNET | 1-B | IBNET (2011) http://www.ib-net.org/ (Reviewed by Palaniappan et al. [2008]) |
| The Manager's Non-Revenue Water Handbook: A Guide to Understanding Water Losses | Ranhill Utilities Berhad and USAID | 1-B | Farley et al. (2008) http://www.waterlinks.org/sites/default/files/NRW%20Manager's%20Handbook.pdf |
| The WBCSD Global Water Tool | WBCSD | 1-B | WBCSD (2012) http://www.wbcsd.org/work-program/sector-projects/water/global-water-tool.aspx |
| The WRC Community-Based Health and Hygiene Model and Implementation Kit | South African Water Research Commission (WRC) | 1-B | Onabolu and Ndlovu (2006) http://www.wrc.org.za/Knowledge%20Hub%20Documents/Research%20Reports/TT264-06.pdf (Reviewed by Palaniappan et al. [2008]) |

| | | | |
|---|---|---|---|
| Tools for mainstreaming disaster risk reduction | IFRC and the ProVention Consortium | 1-B | Benson and Twigg (2007) http://www.preventionweb.net/files/1066_toolsformainstreamingDRR.pdf (*Reviewed by Olhoff and Schaer [2010], Traerup and Olhoff [2011], CCCCC [2013], and weADAPT [2013]*) |
| Towards Better Programming: A Manual on Hygiene Promotion | UNICEF and the London School of Hygiene and Tropical Medicine (LSHTM) | 1-B | UNICEF and LSHTM (1999) http://www.unicef.org/wash/files/hman.pdf (*Reviewed by Palaniappan et al. [2008]*) |
| Towards Better Programming: A Sanitation Handbook | UNICEF and USAID | 1-B | UNICEF and USAID (1997) http://www.unicef.org/wash/files/San_e.pdf (*Reviewed by Palaniappan et al. [2008]*) |
| Towards Better Programming: A Water Handbook | UNICEF | 1-B | UNICEF (1999) http://www.unicef.org/wash/files/Wat_e.pdf (*Reviewed by Palaniappan et al. [2008]*) |
| Urban Water Supply and Sanitation Programming Guide | Planning and Development Collaborative International (PADCO) and USAID | 1-B | PADCO (2001) http://www.watersanitationhygiene.org/References/EH_KEY_REFERENCES/WATER/Urban%20Water/Urban%20WatSan%20Guide%20(USAID).pdf (*Reviewed by Palaniappan et al. [2008]*) |
| Urban Water Supply Handbook | Larry W. Mays | 1-B | Mays (2002) http://www.amazon.co.uk/Supply-Handbook-McGraw-Hill-Handbooks-ebook/dp/B000TL2Z6W (*Reviewed by Palaniappan et al. [2008]*) |
| Vulnerability Self-Assessment Tool (VSAT) | US EPA | 1-B | US EPA (2010) http://water.epa.gov/infrastructure/watersecurity/techtools/vsat.cfm (*Reviewed by Garg et al. [2007]*) |
| WASH technology information packages | UNICEF | 1-B | Baumann et al. (2010) http://www.rural-water-supply.net/_ressources/documents/default/1-471-2-1359464901.pdf |

(*continued*)

**Appendix (continued)**

**Detailed List of 137 Tools Applicable for Climate Risk Management in the Water Sector**

| Tool Name | Developer/Author | Type(s) (1/2/3—A/B/C)[a] | Reference Link (*Previous Reviews of the Tool*)[b] |
|---|---|---|---|
| Water and Sanitation for All: A Practitioner's Companion | WSP-WB and Water Utility Partnership (WUP)—Africa | 1-B | WSP-WB and WUP (2003) http://web.mit.edu/urbanupgrading/waterandsanitation/home.html (*Reviewed by Palaniappan et al. [2008]*) |
| Water and sanitation PPP toolkits | The World Bank | 1-B | The World Bank (2011) http://ppp.worldbank.org/public-private-partnership/sector/water-sanitation/toolkits (*Reviewed by Palaniappan et al. [2008]*) |
| Water Health and Economic Analysis Tool (WHEAT) | US EPA | 1-B | US EPA (2012b) http://water.epa.gov/infrastructure/watersecurity/techtools/wheat.cfm |
| Water Operator Partnership Facilitation Guidelines | USAID and WaterLinks | 1-B | WaterLinks (2011) http://www.waterlinks.org/sites/default/files/WOP%20Content%20FINAL.pdf |
| WSP | Various (originally the WHO) | 1-B 3-B | Examples include: Bartram et al. (2009) http://whqlibdoc.who.int/publications/2009/9789241562638_eng.pdf WHO WPRO (2008) http://www.wpro.who.int/publications/docs/TrainingWorkbookonWSPforUrbanSystems.pdf Greaves and Simmons (2011) http://tilz.tearfund.org/webdocs/Tilz/Topics/watsan/Water%20Safety%20Plans/WSP%20for%20communities%20-%20main%20text.pdfWHO (2006) http://www.who.int/wsportal/en/(*Reviewed by Palaniappan et al. [2008]*) |
| Adaptation Tipping Points approach | Deltares | 1-C | Kwadijk et al. (2010) www.deltares.nl/xmlpages/TXP/files?p_file_id=14123 |

Managing Climate Risk for the Water Sector 275

| | | | |
|---|---|---|---|
| AdaptWater | Water Services Association of Australia (WSAA) et al. | 1-C | WSAA (2013) https://www.wsaa.asn.au/WSAAPublications/Corporate/AdaptWater%20Final%20Report.pdf https://www.wsaa.asn.au/WSAAPublications/FactSheets/AdaptWater%20Fact%20Sheet.pdf |
| Assessing Future Uncertainties Associated with Urban Drainage using Flexible Systems—The COFAS Method and Tool | SWITCH Project—UN Educational, Scientific and Cultural Organization (UNESCO) Institute for Water Education (IHE) | 1-C | Peters et al. (2010) http://www.switchurbanwater.eu/outputs/pdfs/W2-1_GEN_MAN_D2.1.4_Assessing_future_uncertainties_urban_drainage_COFAS.pdf |
| Assessing the Robustness of Adaptation Decisions to Climate Change Uncertainties: A Case Study on Water Resources Management in the East of England | Suraje Dessai and Mike Hulme | 1-C | Dessai and Hulme (2007) http://www.sciencedirect.com/science/article/pii/S0959378006000914 |
| Climate Finance Impact Tool (FIT) for Adaptation | Japan International Cooperation Agency (JICA) | 1-C | JICA (2011) http://www.jica.go.jp/english/our_work/climate_change/pdf/adaptation_all.pdf |
| CREAT | US EPA | 1-C | US EPA (2012a) http://water.epa.gov/infrastructure/watersecurity/climate/creat.cfm *(Reviewed by ALM [2013])* |
| Climate safeguards system: Climate screening and adaptation review and evaluation procedures booklet | African Development Bank (AfDB) | 1-C | AfDB (2011) http://www.afdb.org/fileadmin/uploads/afdb/Documents/Generic-Documents/CSS%20Basics-En_def.pdf *(Reviewed by CCCCC [2013] and weADAPT [2013])* |
| Examining climate risk using a modified uncertainty matrix framework for the water sector | Marie Ekström et al. | 1-C | Ekström et al. (2012) http://www.sciencedirect.com/science/article/pii/S0959378012001343 |

*(continued)*

### Appendix (continued)
### Detailed List of 137 Tools Applicable for Climate Risk Management in the Water Sector

| Tool Name | Developer/Author | Type(s) (1/2/3—A/B/C)[a] | Reference Link (Previous Reviews of the Tool)[b] |
|---|---|---|---|
| Guidance on Water and Adaptation to Climate Change | UN Economic Commission for Europe (UNECE): Convention on the Protection and Use of Transboundary Watercourses and International Lakes | 1-C | UNECE (2009) http://www.unece.org/fileadmin/DAM/env/water/publications/documents/Guidance_water_climate.pdf |
| Guidance on water supply and sanitation in extreme weather events | UNECE and WHO Europe | 1-C | Sinisi and Aertgeerts (2010) http://www.unece.org/fileadmin/DAM/env/water/whmop2/WHO_Guidance_EWE_Final_draft_web_opt.pdf (Reviewed by ALM [2013]) |
| Handbook on Methods for Climate Change Impact Assessment and Adaptation Strategies | UNEP | 1-C | Feenstra et al. (1998) http://www.ivm.vu.nl/en/Images/UNEPhandbookEBA2ED27-994E-4538-B0F0C424C6F619FE_tcm53-102683.pdf (Reviewed by UNFCCC [2008] and CCCCC [2013]) |
| How to integrate climate change adaptation into national-level policy and planning in the water sector: A practical guide for developing country governments | Tearfund | 1-C | Venton (2010) http://tilz.tearfund.org/webdocs/Tilz/Topics/watsan/Water%20Adaptation%20Guide_Web.pdf (Reviewed by ALM [2013]) |
| IWRM for climate adaptation—Training manual and facilitator's guide | UNDP Cap-Net | 1-C | UNDP Cap-Net (2009) http://www.cap-net.org/sites/cap-net.org/files/CC&%20IWRM%20English%20manual_.pdf |

Managing Climate Risk for the Water Sector 277

| | | |
|---|---|---|
| Methodology guidebook for the assessment of investment and financial flows to address climate change | UNDP Environment and Energy Group | 1-C | UNDP (2009)<br>http://www.undpcc.org/docs/Investment%20and%20Financial%20flows/Methodology/UNDP_IFF%20methodology.pdf<br>*(Reviewed by CCCCC [2013])* |
| Multi-criteria analysis (MCA) for improving WRM | UNEP MCA4Climate | 1-C | Miller (2011)<br>http://www.mca4climate.info/_assets/files/Water_Management_Final_Report.pdf<br>*(Reviewed by CCCCC [2013])* |
| Stakeholder-focused CBA in the water sector | International Institute for Environment and Development (IIED) | 1-C | IIED (2013)<br>http://www.iied.org/economics-climate-change-adaptation-water-sector |
| Testing an RCAA for Water and Sanitation Providers in Informal Settlements in Three Cities in sub-Saharan Africa | Heath et al. (2012) | 1-C | Heath et al. (2012)<br>http://eau.sagepub.com/content/24/2/619.abstract |
| Water Economy for Livelihoods (WELS) tool | Research-Inspired Policy and Practice Learning in Ethiopia and the Nile Region (RiPPLE)/Overseas Development Institute (ODI) | 1-C | Coulter et al. (2010) and Ludi (2009)<br>http://www.rippleethiopia.org/documents/info/20100521-water-economy-baseline-report; http://www.odi.org.uk/publications/3148-climate-change-water-food-security |
| Water Security and Climate Resilient Development: Strategic Framework & Technical Backgrounder | African Ministers' Council on Water (AMCOW), the Climate and Development Knowledge Network (CDKN), and the GWP | 1-C | AMCOW et al. (2012)<br>http://www.gwp.org/Global/About%20GWP/Publications/CDKN%20publications/SF_WaterSecurity_FINAL.pdf;<br>http://www.gwp.org/Documents/WACDEP/TBD_Final.pdf |

*(continued)*

## Appendix (continued)
### Detailed List of 137 Tools Applicable for Climate Risk Management in the Water Sector

| Tool Name | Developer/Author | Type(s) (1/2/3—A/B/C)[a] | Reference Link (Previous Reviews of the Tool)[b] |
|---|---|---|---|
| SimCLIM | CLIMsystems Ltd. | 2-A | CLIMsystems (2012) http://www.climsystems.com/simclim/index.php (Reviewed by UNFCCC [2008] and weADAPT [2013]) |
| A GIS Data Integration Tool for Assessing Stormwater Management Options: User Guide | SWITCH Project—UNESCO-IHE | 2-B | Viavattene (2009) http://www.switchurbanwater.eu/outputs/pdfs/W2-3_GEN_MAN_D2.3.2a_GIS_data_integration_tool_user_guide.pdf |
| Aquarius | Colorado State University | 2-B | Diaz et al. (2008) http://www.fs.fed.us/rm/value/aquarius (Reviewed by Garg et al. [2007] and UNFCCC [2008]) |
| Geoinformation Internet portal of the Dniester River basin | UNEP/GRID-Arendal and Zoi Environmental Network | 2-B | UNEP and Zoi (2012) http://82.116.78.174/en/about-geoportal |
| Improved Risk Assessment for Water Distribution Systems (IRA-WDS) GIS-based risk analysis tool | Kalanithy Vairavamoorthy et al. | 2-B | Vairavamoorthy et al. (2007) http://www.sciencedirect.com/science/article/pii/S1364815206001538 |
| MODSIM-DSS | Colorado State University | 2-B | Labadie (2013) http://modsim.engr.colostate.edu/version8.shtml |
| RIBASIM | Deltares | 2-B | Deltares (2012) http://www.deltares.nl/en/software/101928/ribasim (Reviewed by Garg et al. [2007] and UNFCCC [2008]) |
| RiverWare | Center for Advanced Decision Support in Water and Environmental Systems (CADSWES), United States | 2-B | CADSWES (2013) http://www.riverware.org/ (Reviewed by Garg et al. [2007] and UNFCCC [2008]) |

| | | | |
|---|---|---|---|
| Water Security Index | ADB and the APWF | 2-B | ADB and APWF (2013) http://www.adb.org/sites/default/files/pub/2013/asian-water-development-outlook-2013.pdf |
| WaterWare | Environmental Software and Services (ESS), Austria | 2-B | ESS (2013) http://www.ess.co.at/WATERWARE/ *(Reviewed by Garg et al. [2007] and UNFCCC [2008])* |
| Assessing GIS-Based Indicator Methodology for Analyzing the Physical Vulnerability of Water and Sanitation Infrastructure | Martin Karlson | 2-C | Karlson (2012) http://liu.diva-portal.org/smash/get/diva2:561073/FULLTEXT01 |
| Statistical Downscaling Model (SDSM) | Loughborough University | 2-C | Dawson (2013) www.sdsm.org.uk *(Reviewed by weADAPT [2013])* |
| WEAP | SEI-US | 2-C | SEI-US (2013) http://www.weap21.org/ *(Reviewed by Garg et al. [2007], UNFCCC [2008], Olhoff and Schaer [2010], and weADAPT [2013])* |
| Water Vulnerability Index, Climate Vulnerability Index, Water Poverty Index[d] | Caroline Sullivan et al. | 2-C  2-C  2-B | Sullivan (2011) http://epubs.scu.edu.au/cgi/viewcontent.cgi?article=1929&context=esm_pubs Sullivan and Meigh (2005) http://hcmus.edu.vn/images/stories/qhqt/hoatdongqhqt/file_2-1.pdf Sullivan et al. (2003) ftp://ftp.fao.org/agl/emailconf/wfe2005/narf_054.pdf |
| Water World | King's College London (KCL) and AmbioTEK | 2-C | KCL and AmbioTEK (2013) http://www.policysupport.org/waterworld |

*(continued)*

## Appendix (continued)

### Detailed List of 137 Tools Applicable for Climate Risk Management in the Water Sector

| Tool Name | Developer/Author | Type(s) (1/2/3—A/B/C)[a] | Reference Link (Previous Reviews of the Tool)[b] |
|---|---|---|---|
| WaterSim | Arizona State University—Decision Center for a Desert City (DCDC) | 2-C | DCDC (2013) http://dcdc.asu.edu/watersim/ |
| ALM | UNDP et al. | 3-A | ALM (2013) http://www.adaptationlearning.net/ (Reviewed by GIZ [2009], Olhoff and Schaer [2010], and weADAPT [2013]) |
| Caribbean Climate Online Risk and Adaptation Tool (CCORAL) | CCCCC | 3-A | CCCCC (2013) http://coral.caribbeanclimate.bz/ (Reviewed by ALM [2013]) |
| ci:grasp 2.0 (Climate Impacts: Global and Regional Adaptation Support Platform) | Potsdam Institute for Climate Impact Research (PIK) and GIZ | 3-A | ci:grasp (2013) http://www.pik-potsdam.de/cigrasp-2/index.html (Reviewed by GIZ [2009]) |
| Climate Adaptation Decision eXplorer (ADx) | weADAPT | 3-A | weADAPT (2013a) http://weadapt.org/knowledge-base/adaptation-decision-making/adaptation-decision-explorer (Reviewed by CCCCC [2013] and weADAPT [2013]) |
| Climate Adaptation Knowledge Exchange (CAKE) | EcoAdapt and Island Press | 3-A | CAKE (2013) http://www.cakex.org/ |
| Climate Change Knowledge Portal | The World Bank | 3-A | World Bank (2013) http://sdwebx.worldbank.org/climateportal/index.cfm (Reviewed by GIZ [2009] and OECD [2009]) |

Managing Climate Risk for the Water Sector   281

| Tool | Organization | Type | Reference |
|---|---|---|---|
| weADAPT | SEI et al. | 3-A | weADAPT (2013b) http://weadapt.org/ (Reviewed by GIZ [2009] and Olhoff and Schaer [2010]) |
| Compendium of Sanitation Systems and Technologies | EAWAG Sandec and WSSCC | 3-B | Tilley et al. (2008) http://www.eawag.ch/forschung/sandec/publikationen/compendium_e/index_EN |
| Cross-border flood risk management simulation game | FLOOD-WISE | 3-B | FLOOD-WISE (2012) http://floodwise.nl/results/the-game/ |
| Election toolkit: How to campaign on water and sanitation issues during an election | End Water Poverty | 3-B | End Water Poverty (2012) http://www.endwaterpoverty.org/sites/endwaterpoverty.org/files/8783_ElectionToolkit_FINAL1_6_1.pdf |
| GWP Toolbox for IWRM | GWP | 3-B | GWP (2008) http://www.gwptoolbox.org/index.php (Reviewed by Palaniappan et al. [2008]) |
| Adapting Urban Water Systems to Climate Change: A Handbook for Decision Makers at the Local Level | ICLEI European Secretariat | 3-C | Loftus et al. (2011) http://www.iwahq.org/ContentSuite/upload/iwa/all/Water%20climate%20and%20energy/SWITCH_Adaption-Handbook_final_small.pdf |
| Climate Change and Urban Water Utilities: Challenges and Opportunities | The World Bank | 3-C | Danilenko et al. (2010) http://www.wsp.org/sites/wsp.org/files/publications/climate_change_urban_water_challenges.pdf |
| Technologies for climate change adaptation: The water sector | UNEP Risoe Centre on Energy, Climate and Sustainable Development | 3-C | Elliot et al. (2011) http://www.zaragoza.es/contenidos/medioambiente/onu/issue06/1149-eng.pdf |

[a] We attempt to classify tools using both functional and sectoral classification, based on our own best judgment, where 1/2/3 represents the three functional types and A/B/C represents the three target sectoral audience types, as described in Section 9.6.

[b] Note that all URLs were current at the time of writing (July 2013) but will inevitably change with time. Citations are thus also provided to the full references below, in the Tool References section. Only tools with active URLs were included.

[c] Note that this tool is based largely on CRiSTAL and CARE's CVCA.

[d] Three similar tools developed by the same authors as semiprogressive iterations, so they are listed here as just one tool to avoid artificially inflating the list.

## TOOL REFERENCES

Abarquez, I. and Murshed, Z. 2004. *Community-Based Disaster Risk Management: Field Practitioner's Handbook*. Pathumthani, Thailand: ADPC.

ADB. 2005. *Climate Proofing: A Risk-Based Approach to Adaptation*. Metro Manila, Philippines: ADB.

ADB and APWF. 2013. *Asian Water Development Outlook 2013: Measuring Water Security in Asia and the Pacific*. Metro Manila, Philippines: ADB and APWF.

AfDB. 2011. *Climate Safeguards System: Climate Screening and Adaptation Review and Evaluation Procedures Booklet*. Tunis, Tunisia: AfDB.

Alexander Ballard. 2013. PACT. Berkshire, U.K.: Alexander Ballard Ltd. (http://alexanderballard.co.uk/pact/) [Accessed July 2013].

ALM. 2013. UNDP adaptation learning mechanism. New York: UNDP ALM (http://www.adaptationlearning.net/) [Accessed July 2013].

AMCOW, CDKN and GWP. 2012. *Water Security and Climate Resilient Development: Strategic Framework and Technical Background Document*. Abuja, Nigeria: AMCOW.

Ampomah, G. and Devisscher, T. 2013. *Adaptation Toolkit: Guidebook for Researchers and Adaptation Practitioners Working with Local Communities*. Oxford, U.K.: ENDA and SEI-UK.

Arup and Engineers Against Poverty. 2013. ASPIRE—A sustainability, poverty and infrastructure routine for evaluation. Newcastle-upon-Tyne, U.K.: Oasys Ltd. (http://www.oasys-software.com/products/environmental/aspire.html) [Accessed July 2013].

AusAID. 2005. *Safe Water Guide for the Australian Aid Program 2005: A Framework and Guidance for Managing Water Quality*. Canberra, Australian Capital Territory, Australia: AusAID.

Bartram, J., Corrales, L., Davison, A. et al. 2009. *Water Safety Plan Manual: Step-by-Step Risk Management for Drinking-Water Suppliers*. Geneva, Switzerland: WHO and IWA.

Baumann, E., Montangero, A., Sutton, S., and Erpf, K. 2010. *WASH Technology Information Packages—For UNICEF WASH Programme and Supply Personnel*. Copenhagen, Denmark: UNICEF and SKAT.

Benson, C. and Twigg, J. 2007. *Tools for Mainstreaming Disaster Risk Reduction: Guidance Notes for Development Organisations*. Geneva, Switzerland: IFRC and the ProVention Consortium.

Boot, M.T. 1991. *Just Stir Gently: The Way to Mix Hygiene Education with Water Supply and Sanitation*. The Hague, the Netherlands: IRC International Water and Sanitation Centre.

BPD. 2011. Tools for public–private partnerships. London, U.K.: BPD (http://www.partnershipsforwater.net/web/w/www_7_en.aspx) [Accessed July 2013].

Brikke, F. and Bredero, M. 2003. *Linking Technology Choice with Operation and Maintenance in the Context of Community Water Supply and Sanitation: A Reference Document for Planners and Project Staff*. Geneva, Switzerland: WHO and IRC International Water and Sanitation Centre.

CADSWES. 2013. Riverware. Boulder, CO: CADSWES (http://www.riverware.org/) [Accessed July 2013].

CAKE. 2013. CAKE—Climate adaptation knowledge exchange. Bainbridge Island, WA: EcoAdapt and Island Press (http://www.cakex.org/) [Accessed July 2013].

CARE. 2009. *Climate Vulnerability and Capacity Analysis—Handbook*. Geneva, Switzerland: CARE International.

CARE. 2010a. *Community-Based Adaptation Toolkit*. Geneva, Switzerland: CARE International.

CARE. 2010b. *Toolkit for Integrating Climate Change Adaptation into Development Projects*. Geneva, Switzerland: CARE International.

CARICOM. 2003. *Caribbean Risk Management Guidelines for Climate Change Adaptation Decision Making*. Georgetown, Guyana: CARICOM.

Castro, V., Msuya, N., and Makoye, C. 2009. *Sustainable Community Management of Urban Water and Sanitation Schemes (A Training Manual)*. Nairobi, Kenya: DAWASA and WSP.

CCCCC. 2013. Caribbean climate online risk and adaptation tool. Belmopan, Belize: CCCCC (http://ccoral.caribbeanclimate.bz/) [Accessed July 2013].

CCME. 2004. *From Source to Tap: Guidance on the Multi-Barrier Approach to Safe Drinking Water*. Winnipeg, Manitoba, Canada: CCME.

CDEMA. 2011. *The Guidance Tool: A Manual for Mainstreaming Climate Change Adaptation into the CDM Country Work Programme*. St. Michael, Barbados: CDEMA.

Christian Aid. 2010. *Integrating Climate Change Adaptation into Secure Livelihoods: Toolkit 1—Framework and Approach, and Toolkit 2—Developing a Climate Change Analysis*. London, U.K.: Christian Aid.

CIG, King County and ICLEI. 2007. *Preparing for Climate Change: A Guidebook for Local, Regional and State Governments*. Seattle, WA: CIG, King County and ICLEI-US.

ci:grasp. 2013. ci:grasp—The climate impacts: Global and regional adaptation support platform. Potsdam, Germany: PIK and GIZ (http://www.pik-potsdam.de/cigrasp-2/index.html) [Accessed July 2013].

CLIMsystems. 2012. SimCLIM. Hamilton, New Zealand: CLIMsystems (http://www.climsystems.com/simclim/index.php) [Accessed July 2013].

Coulter, L., Kebede, S., and Zeleke, B. 2010. *Water Economy Baseline Report: Water and Livelihoods in a Highland to Lowland Transect in Eastern Ethiopia*. Addis Ababa, Ethiopia: RiPPLE.

Danilenko, A., Dickson, E., and Jacobsen, M. 2010. *Climate Change and Urban Water Utilities: Challenges and Opportunities*. Washington, DC: The World Bank.

Dawson, C.W. 2013. SDSM—Statistical downscaling model. Loughborough, U.K.: Loughborough University (http://co-public.lboro.ac.uk/cocwd/SDSM/) [Accessed July 2013].

DCDC. 2013. WaterSim. Tempe, AZ: DCDC (http://dcdc.asu.edu/watersim/) [Accessed July 2013].

Deltares. 2012. Ribasim: River basin planning and management. Delft, the Netherlands: Deltares (http://www.deltares.nl/en/software/101928/ribasim) [Accessed July 2013].

Dessai, S. and Hulme, M. 2007. Assessing the robustness of adaptation decisions to climate change uncertainties: A case study on water resources management in the East of England, *Global Environmental Change* 17(1): 59–72.

DHI. 2013. Modelling the world of water: MIKE by DHI software. Horsholm, Denmark: DHI (http://mikebydhi.com/) [Accessed July 2013].

Diaz, G.E., Brown, T.C., and Sveinsson, O. 2008. Aquarius: A modelling system for river basin water allocation. Fort Collins, CO: USDA and Colorado State University (http://www.fs.fed.us/rm/value/aquarius) [Accessed July 2013].

Dickson, E., Baker, J.L., Hoornweg, D., and Tiwari, A. 2012. *Urban Risk Assessments: Understanding Disaster and Climate Risk in Cities*. Washington, DC: The World Bank.

Downing, C., Preston, F., Parusheva, D., Horrocks, L., Edberg, O., Samazzi, F., Washington, R., Muteti, M., Watkiss, P., and Nyangena, W. 2008. *Final Report: Kenya—Climate Screening and Information Exchange*. London, U.K.: AEA Group.

ECA. 2009. *Shaping Climate-Resilient Development: A Framework for Decision-Making*. San Francisco, CA: ClimateWorks Foundation, GEF, EC, McKinsey & Company, The Rockefeller Foundation, Standard Chartered Bank and SwissRe.

Ekstrom, M., Kuruppu, N., Wilby, R.L., Fowler, H.J., Chiew, F.H.S., Dessai, S., and Young, W.J. 2012. Examination of climate risk using a modified uncertainty matrix framework—Applications in the water sector, *Global Environmental Change* 23(1): 115–129.

Elliot, M., Armstrong, A., Lobuglio, J., and Bartram, J. 2011. *Technologies for Climate Change Adaptation: The Water Sector*. Roskilde, Denmark: UNEP Risoe Centre.

End Water Poverty. 2012. *Election Toolkit: How to Campaign on Water and Sanitation Issues During an Election*. London, U.K.: End Water Poverty.

ESS. 2013. WaterWare: Water resources management information system. Gumpoldskirchen, Austria: ESS (http://www.ess.co.at/WATERWARE/) [Accessed July 2013].

EU. 2012. *BalticClimate—The Toolkit*. Hannover, Germany: EU and the Academy for Spatial Research and Planning.

FAO and IPG. 2009. Planning for community based adaptation to climate (CBA): Introduction to the e-learning tool. Rome, Italy: FAO and IPG (http://www.webgeo.de/fao-webgeo-2-intro/) [Accessed July 2013].

Farley, M., Wyeth, G., Ghazali, Z.B.M., Istandar, A., and Singh, S. 2008. *The Manager's Non-Revenue Water Handbook: A Guide to Understanding Water Losses*. Bangkok, Thailand: USAID and Ranhill Utilities Berhad.

Feenstra, J.F., Burton, I., Smith, J.B., and Tol, R.S.J. 1998. *Handbook on Methods for Climate Change Impact Assessment and Adaptation Strategies*. Amsterdam, the Netherlands: VrijeUniversiteit and UNEP.

FLOOD-WISE. 2012. Cross-border flood risk management: A serious game. Maastricht, the Netherlands: FLOOD-WISE (http://floodwise.nl/results/the-game/) [Accessed July 2013].

Frohlich, U. 2001. *Series of Manuals on Drinking Water Supply, Management Guide*, Vol. 1. St. Gallen, Switzerland: SKAT.

Government of Tasmania. 2009. *Coastal Risk Management Plan: Template and Guidelines*. Hobart, Tasmania, Australia: Government of Tasmania Department of Primary Industries and Water.

Grannis, J. 2011. *Adaptation Tool Kit: Sea-Level Rise and Coastal Land Use*. Washington, DC: Georgetown Climate Center.

Greaves, F. and Simmons, C. 2011. *Water Safety Plans for Communities: Guidance for Adoption of Water Safety Plans at Community Level*. Teddington, U.K.: Tearfund.

GWP. 2008. GWP toolbox for integrated water resources management. Stockholm, Sweden: GWP (http://www.gwptoolbox.org/index.php) [Accessed July 2013].

Hahn, M. and Frode, A. 2011. *Climate Proofing for Development: Adapting to Climate Change, Reducing Risk*. Eschborn, Germany: GIZ.

Heath, T.T., Parker, A.H., and Weatherhead, E.K. 2012. Testing a rapid climate change adaptation assessment for water and sanitation providers in informal settlements in three cities in sub-Saharan Africa, *Environment and Urbanisation* 24(2): 619–637.

HEKS and Bread for All. 2010. *Participatory Tool on Climate and Disaster Risks: Integrating Climate Change and Disaster Risk Reduction into Community-Level Development Projects*. Bern, Switzerland: HEKS and Bread for All.

Huxtable, J. and Nguyen, T.Y. 2009. *Mainstreaming Climate Change Adaptation: A Practitioner's Handbook*. Hanoi, Vietnam: CARE International in Vietnam.

IBNET. 2011. The International Benchmarking Network for Water and Sanitation Utilities (IBNET). Washington, DC: The World Bank (http://www.ib-net.org/) [Accessed July 2013].

ICLEI. 2008. Local government climate change adaptation toolkit. Melbourne, Victoria, Australia: ICLEI-Oceania (http://archive.iclei.org/index.php?id=adaptation-toolkit0) [Accessed July 2013].

ICLEI. 2013. Adaptation database and planning tool. Oakland, CA: ICLEI USA (http://www.icleiusa.org/tools/adapt) [Accessed July 2013].

ICLEI and ACCCRN. 2013. Urban climate change resilience toolkit. Bonn, Germany: ICLEI and ACCCRN (http://www.teriin.org/events/Sunandan-Tiwari.pdf; http://www.scribd.com/doc/144824994/ACCCRN-ICLEI-Replication-Press-Release) [Accessed July 2013].

IDS. 2009. Adaptation screening tools for development cooperation: Piloting ORCHID and other approaches. Brighton, U.K.: IDS (http://www.ids.ac.uk/climatechange/orchid) [Accessed July 2013].

IFRC (various). Vulnerability and capacity assessment (VCA). Geneva, Switzerland: IFRC (http://www.ifrc.org/en/what-we-do/disaster-management/preparing-for-disaster/disaster-preparedness-tools/disaster-preparedness-tools/) [Accessed July 2013].

IIED. 2013. Economics of climate change adaptation in the water sector. London, U.K.: IIED (http://www.iied.org/economics-climate-change-adaptation-water-sector) [Accessed July 2013].

IISD, IUCN, Helvetas, and SEI. 2013. *CRiSTAL—Community-Based Risk Screening Tool—Adaptation and Livelihoods*. Winnipeg, Manitoba, Canada: IISD.

JICA. 2011. *JICA Climate Finance Impact Tool for Adaptation*. Tokyo, Japan: JICA.

Karlson, M. 2012. Assessing GIS-based indicator methodology for analyzing the physical vulnerability of water and sanitation infrastructure, Master's thesis. Linkoping, SE: Water and Environmental Studies—Linkoping University (Supervisor: Wittgren, H.-B.).

KCL and AmbioTEK. 2013. Water World. London, U.K.: KCL and AmbioTEK (http://www.policysupport.org/waterworld) [Accessed July 2013].

Kropp, J. and Scholze, M. 2009. *Climate Change Information for Effective Adaptation: A Practitioner's Manual*. Eschborn, Germany: GIZ.

Kwadijk, J.C.J., Haasnoot, M., Mulder, J.P.M. et al. 2010. Using adaptation tipping points to prepare for climate change and sea level rise, a case study in the Netherlands, *Wiley Interdisciplinary Reviews: Climate Change* 1(5): 729–740.

Labadie, J. 2013. MODSIM-DSS: Water rights planning, water resources management and river operations decision support system. Fort Collins, CO: Colorado State University (http://modsim.engr.colostate.edu/version8.shtml) [Accessed July 2013].

Lockwood, H. and Smits, S. 2011. *Supporting Rural Water Supply—Moving Towards a Service Delivery Approach*. Rugby, U.K.: IRC International Water and Sanitation Centre and Aguaconsult.

Loftus, A.-C., Anton, B., and Philip, R. 2011. *Adapting Urban Water Systems to Climate Change: A Handbook for Decision Makers at the Local Level*. Freiburg, Germany: ICLEI-Europe.

Lu, X. 2008. *Applying Climate Information for Adaptation Decision-Making: A Guidance and Resource Document*. New York: UNDP/UNEP/GEF National Communications Support Programme.

Ludi, E. 2009. *Climate Change, Water and Food Security*. London, U.K.: ODI.

Mays, L.W. 2002. *Urban Water Supply Handbook*. New York: The McGraw Hill Companies, Inc.

MFE. 2003. *Sustainable Wastewater Management: A Handbook for Smaller Communities*. Wellington, New Zealand: MFE.

Miller, K. 2011. *MCA4 Climate: A Practical Framework for Planning Pro-Development Climate Policies—Adaptation Theme Report: Improving Water Resource Management*. Nairobi, Kenya: UNEP.

Nakalevu, T. 2006. *CV&A: A Guide to Community Vulnerability and Adaptation Assessment and Action*. Apia, Western Samoa: SPREP.

NOAA. 2010. *Adapting to Climate Change: A Planning Guide for State Coastal Managers*. Silver Spring, MD: NOAA.

OECD. 2008. *Strategic Environmental Assessment and Adaptation to Climate Change*. Paris, France: OECD DAC Network on Environment and Development Co-operation.

OECD. 2009. *Integrating Climate Change Adaptation into Development Co-Operation: Policy Guidance*. Paris, France: OECD.

Onabolu, B. and Ndlovu, M. 2006. *The WRC Community Based Health and Hygiene Model and Implementation Kit*. Pretoria, South Africa: WRC.

PADCO. 2001. *Urban Water Supply and Sanitation Programming Guide.* Washington, DC: PADCO and USAID.
PAHO. 1998. *Natural Disaster Mitigation in Drinking Water and Sewerage Systems: Guidelines for Vulnerability Analysis.* Washington, DC: PAHO.
Peters, C., Sieker, H., Jin, Z., and Eckart, J. 2010. *Deliverable 2.1.4—Assessing Future Uncertainties Associated with Urban Drainage Using Flexible Systems—The COFAS Method and Tool.* Delft, the Netherlands: SWITCH UNESCO-IHE.
Prasad, N., Ranghieri, F., Shah, F., and Trohanis, Z. 2008. *Climate Resilient Cities 2008 Primer: Reducing Vulnerabilities to Climate Change Impacts and Strengthening Disaster Risk Management in East Asian Cities.* Washington, DC: The World Bank.
Pulwarty, R.S. and Hutchinson, N. 2008. *Vulnerability and Capacity Assessment Methodology: A Guidance Manual for the Conduct and Mainstreaming of Climate Change Vulnerability and Capacity Assessments in the Caribbean Region.* St. Michael, Barbados: CCCCC.
Red Cross/Red Crescent. 2007. *Climate Guide.* Geneva, Switzerland: Red Cross/Red Crescent.
SDC. 2012. CEDRIG. Bern, Switzerland: SDC (http://www.sdc-drr.net/cedrig) [Accessed July 2013].
SEI-US. 2013. WEAP—Water evaluation and planning system. Somerville, MA: SEI-US (http://www.weap21.org/) [Accessed July 2013].
Sinisi, L. and Aertgeerts, R. 2010. *Guidance on Water Supply and Sanitation in Extreme Weather Events.* Geneva, Switzerland: UNECE.
SOPAC. 2001. *Regional Comprehensive Hazard and Risk Management (CHARM) Guidelines.* Suva, Fiji: SOPAC.
Stratus. 2013. Assessing vulnerability and adaptation: Tools for assessing adaptation options. Boulder, CO: Stratus Consulting (http://stratusconsulting.com/services/climate/assessing-vulnerability-and-adaptation/) [Accessed July 2013].
Sullivan, C.A. 2011. *Quantifying Water Vulnerability: A Multi-Dimensional Approach.* East Lismore, New South Wales, Australia: Southern Cross University School of Environment, Science and Engineering.
Sullivan, C.A. and Meigh, J.R. 2005. Targeting attention on local vulnerabilities using an integrated index approach: The example of the Climate Vulnerability Index, *Water Science and Technology* 51(5): 69–78.
Sullivan, C.A., Meigh, J.R., Giacomello, A.M. et al. 2003. The water poverty index: Development and application at the community scale, *Natural Resources Forum* 27: 189–199.
Tanner, T., Jun, X., and Holman, I. 2008. *Screening for Climate Change Adaptation in China: A Process to Assess and Manage the Potential Impact of Climate Change on Development Projects and Programmes in China.* Brighton, U.K.: IDS.
Tearfund. 2012. CEDRA (Climate change and Environmental Degradation Risk and adaptation Assessment). Teddington, U.K.: Tearfund (http://tilz.tearfund.org/Topics/Environmental+Sustainability/CEDRA.htm) [Accessed July 2013].
Tilley, E., Luthi, C., Morel, A., Zurbrugg, C., and Schertenleib, R. 2008. *Compendium of Sanitation Systems and Technologies.* Dubendorf, Switzerland: EAWAG Sandec and WSSCC.
UKCIP. 2012. AdOpt—Identifying adaptation options. Oxford, U.K.: UKCIP (http://www.ukcip.org.uk/adopt/) [Accessed July 2013].
UKCIP. 2013a. BACLIAT—Business Areas Climate Assessment Tool. Oxford, U.K.: UKCIP (http://www.ukcip.org.uk/bacliat/) [Accessed July 2013].
UKCIP. 2013b. CLARA—Climate Adaptation Resource for Advisors. Oxford, U.K.: UKCIP (http://www.ukcip.org.uk/clara/) [Accessed July 2013].
UKCIP. 2013c. UKCIP Adaptation Wizard. Oxford, U.K.: UKCIP (http://www.ukcip.org.uk/wizard/) [Accessed July 2013].
UKCIP. 2013d. UKCIPAdaptME toolkit. Oxford, U.K.: UKCIP (http://www.ukcip.org.uk/adaptme-toolkit/) [Accessed July 2013].

UNDP. 2003. *User's Guidebook for the Adaptation Policy Framework*. New York: UNDP.
UNDP. 2008. *A Guide to the Vulnerability Reduction Assessment*. New York: UNDP.
UNDP. 2009. *Methodology Guidebook for the Assessment of Investment and Financial Flows to Address Climate Change*. New York: UNDP.
UNDP. 2010. *Designing Climate Change Adaptation Initiatives: A UNDP Toolkit for Practitioners*. New York: UNDP.
UNDP and UNEP. 2011. *Mainstreaming Climate Change Adaptation into Development Planning: A Guide for Practitioners*. New York: UNDP and UNEP.
UNDP Cap-Net. 2009. *IWRM as a Tool for Adaptation to Climate Change: Training Manual and Facilitator's Guide*. Pretoria, South Africa: UNDP Cap-Net.
UNECE. 2009. *Guidance on Water and Adaptation to Climate Change*. Geneva, Switzerland: UNECE.
UNEP and Zoi. 2012. Geoportal of the Dniester River Basin. Geneva, Switzerland: UNEP and Zoi Environmental Network (http://82.116.78.174/en/about-geoportal) [Accessed July 2013].
UNEP, WHO, UN-HABITAT and WSSCC. 2004. *Guidelines on Municipal Wastewater Management*. The Hague, the Netherlands: UNEP/GPA Coordination Office.
UNICEF. 1999. *Towards Better Programming: A Water Handbook*. New York: UNICEF.
UNICEF and LSHTM. 1999. *Towards Better Programming: A Manual on Hygiene Promotion*. New York: UNICEF and LSHTM.
UNICEF and USAID. 1997. *Towards Better Programming: A Sanitation Handbook*. New York: UNICEF and USAID.
USAID. 2007. *Adapting to Climate Variability and Change: A Guidance Manual for Development Planning*. Washington, DC: USAID.
USAID. 2009. *Adapting to Coastal Climate Change: A Guidebook for Development Planners*. Washington, DC: USAID.
US EPA. 2010. Vulnerability self-assessment tool (VSAT). Washington, DC: US EPA (http://water.epa.gov/infrastructure/watersecurity/techtools/vsat.cfm) [Accessed July 2013].
US EPA. 2011. Community-based water resiliency (CBWR) tool. Washington, DC: US EPA (http://water.epa.gov/infrastructure/watersecurity/techtools/cbwr.cfm) [Accessed July 2013].
US EPA. 2012a. Climate resilience evaluation and awareness tool (CREAT). Washington, DC: US EPA (http://water.epa.gov/infrastructure/watersecurity/climate/creat.cfm) [Accessed July 2013].
US EPA. 2012b. Water health and economic analysis tool (WHEAT). Washington, DC: US EPA (http://water.epa.gov/infrastructure/watersecurity/techtools/wheat.cfm) [Accessed July 2013].
Vairavamoorthy, K., Yan, J., Galgale, H.M., and Gorantiwar, S.D. 2007. IRA-WDS: A GIS-based risk analysis tool for water distribution systems, *Environmental Modelling and Software* 22(7): 951–965.
Venton, P. 2010. *How to Integrate Climate Change Adaptation into National-Level Policy and Planning in the Water Sector: A Practical Guide for Developing Country Governments*. Teddington, U.K.: Tearfund.
Viavattene, C. 2009. *Deliverable 2.3.2a: A GIS Data Integration Tool for Assessing Stormwater Management Options: User Guide*. Delft, the Netherlands: SWITCH UNESCO-IHE.
WaterLinks. 2011. *Water Operator Partnership Facilitation Guidelines*. Metro Manila, Philippines: USAID WaterLinks and AECOM International Development.
WaterLinks. 2013. 10-Step promotion program toolkit: Helping service professionals improve water, sanitation and hygiene programs. Metro Manila, Philippines: USAID WaterLinks (http://www.10step-toolkit.org/) [Accessed July 2013].
WBCSD. 2012. The WBCSD Global Water Tool©. Geneva, Switzerland: WBCSD (http://www.wbcsd.org/work-program/sector-projects/water/global-water-tool.aspx) [Accessed July 2013].

WBCSD. 2013. The GEMI Local Water Tool™. Geneva, Switzerland: WBCSD (http://www.wbcsd.org/work-program/sector-projects/water/localwatertool.aspx) [Accessed July 2013].
weADAPT. 2013a. Climate adaptation options explorer (ADx). Stockholm, Sweden: SEI (http://weadapt.org/knowledge-base/adaptation-decision-making/adaptation-decision-explorer) [Accessed July 2013].
weADAPT. 2013b. weADAPT homepage. Stockholm, Sweden: SEI (http://weadapt.org/) [Accessed July 2013].
WELL and DFID. 1998. *DFID Guidance Manual on Water Supply and Sanitation Programmes*. London, U.K.: WELL and DFID.
WHO. 1992. *A Guide to the Development of On-Site Sanitation*. Geneva, Switzerland: WHO.
WHO. 2006. WSPortal: Health through water. Geneva, Switzerland: WHO (http://www.who.int/wsportal/en/) [Accessed July 2013].
WHO. 2010. *Concept Note: Sanitation Safety Plans (SSPs)—A Vehicle for Guideline Implementation*. Geneva, Switzerland: WHO.
WHO WPRO. 2008. *Training Workbook on Water Safety Plans for Urban Systems*. Manila, Philippines: WHO WPRO.
World Bank. 2011. Water and sanitation PPP toolkits. Washington, DC: The World Bank (http://ppp.worldbank.org/public-private-partnership/sector/water-sanitation/toolkits) [Accessed July 2013].
World Bank. 2013. Climate change knowledge portal: For development practitioners and policy makers. Washington, DC: The World Bank (http://sdwebx.worldbank.org/climateportal/index.cfm) [Accessed July 2013].
WSAA. 2013. *AdaptWater™: A Climate Change Adaptation and Decision-Making Tool for the Urban Water Industry—Final Report and Fact Sheet*. Melbourne, Victoria, Australia: WSAA.
WSP-WB and WUP. 2003. Water and sanitation for all toolkit: A practitioner's companion. Washington, DC: WSP-WB and WUP (http://web.mit.edu/urbanupgrading/waterandsanitation/home.html) [Accessed July 2013].
WSSCC and WHO. 2005. *Sanitation and Hygiene Promotion: Programming Guidance*. Geneva, Switzerland: WSSCC and WHO.
WWF. 2009. *Climate Witness: Community Toolkit*. Suva, Fiji: WWF South Pacific.

## MAIN TEXT REFERENCES

ALM. 2013. UNDP Adaptation Learning Mechanism—Browse ALM content. New York: UNDP ALM (http://www.adaptationlearning.net/explore) [Accessed July 2013—Tools available by searching for 'tools' on this browse page].
Barnett, J. and O'Neill, S. 2010. Maladaptation, *Global Environmental Change* 20: 211–213.
Batchelor, C., James, A.J., and Smits, S. 2010. *Adaptation of WASH Services Delivery to Climate Change, and Other Sources of Risk and Uncertainty*. The Hague, the Netherlands: IRC International Water and Sanitation Centre.
Bates, B.C., Kundzewicz, Z.W., Wu, S., and Palutikof, J.P. 2008. *Climate Change and Water: Technical Paper of the Intergovernmental Panel on Climate Change*. Geneva, Switzerland: IPCC.
Bouwer, L., Garcia, L., Gilroy, K. et al. 2013. *White Paper 1: Caveat Adaptor—FAQS: The Best Use of Climate Model Simulations for Climate Adaptation & Freshwater Management*. Arlington, TX: Alliance for Global Water Adaptation.

Brown, C. 2010. The end of reliability, *Journal of Water Resources Planning and Management* 136(2): 143–145.

Brown, C. 2011. Decision-scaling for robust planning and policy under climate uncertainty. *World Resources Report Uncertainty Series*. Washington, DC: World Resources Report.

Calow, R., Bonsor, H., Jones, L., O'Meally, S., MacDonald, A., and Kaur, N. 2011. Climate change, water resources and WASH: A scoping study. ODI Working Paper 337. London, U.K.: ODI.

CCCCC. 2013. Caribbean climate online risk and adaptation tool: Toolbox. Belmopan, Belize: CCCCC (http://ccoral.caribbeanclimate.bz/toolbox) [Accessed July 2013].

Garg, A., Rana, A., and Shukla, P.R. 2007. Handbook of current and next generation vulnerability and adaptation assessment tools. BASIC Project Paper 8. Brighton, U.K.: IDS.

GIZ. 2009. *International Workshop on Mainstreaming Adaptation to Climate Change: Guidance and Tools*, May 28–30, 2009. Berlin, Germany: GIZ.

GWP. 2007. Climate change adaptation and integrated water resource management—An initial overview. GWP Technical Committee Policy Brief 5. Stockholm, Sweden: GWP.

Hallegatte, S., Shah, A., Lempert, R., Brown, C., and Gill, S. 2012. *Investment Decision Making under Deep Uncertainty: Application to Climate Change*. Washington, DC: The World Bank.

Hammill, A. and Tanner, T. 2011. Harmonising climate risk management: Adaptation screening and assessment tools for development co-operation. OECD Environment Working Paper 36. Paris, France: OECD.

IPCC. 2007. Climate change 2007: Synthesis report. Contribution of Working Groups I, II and III to the Fourth Assessment Report of the Intergovernmental Panel on Climate Change. Geneva, Switzerland: IPCC.

Matthews, J., Wickel, B.A.J., and Freeman, S. 2011. Converging currents in climate-relevant conservation: Water, infrastructure and institutions, *PLoS Biology* 9(9): e1001159.

Milly, P.C.D., Betancourt, J., Falkenmark, M., Hirsch, R.M., Kundzewicz, Z.W., Lettenmaier, D.P., and Stouffer, R.J. 2008. Stationarity is dead: Whither water management? *Science* 319(5863): 573–574.

OECD. 2009. *Integrating Climate Change Adaptation into Development Co-Operation: Policy Guidance*. Paris, France: OECD.

Olhoff, A. and Schaer, C. 2010. *Screening Tools and Guidelines to Support the Mainstreaming of Climate Change Adaptation into Development Assistance—A Stocktaking Report*. New York: UNDP Environment and Energy Group.

Palaniappan, M., Lang, M., and Gleick, P. 2008. *A Review of Decision-Making Support Tools in the Water, Sanitation and Hygiene Sector*. Oakland, CA: The Pacific Institute and the Environmental Change and Security Program of the Woodrow Wilson International Center for Scholars.

Pearsall, J. 1999. Tool. In *The Concise Oxford Dictionary*, 10th edn. Oxford, U.K.: Oxford University Press.

Ranger, N. 2013. *Topic Guide—Adaptation: Decision Making under Uncertainty*. Hertfordshire, U.K.: Evidence on Demand.

Traerup, S. and Olhoff, A. 2011. *Climate Risk Screening Tools and Their Application: A Guide to the Guidance*. Roskilde, Denmark: UNEP Risoe Centre.

UNDP Cap-Net. 2009. *IWRM as a Tool for Adaptation to Climate Change: Training Manual and Facilitator's Guide*. Pretoria, South Africa: UNDP Cap-Net.

UNESCO. 2012. The United Nations World Water Development Report 4: Managing water under uncertainty and risk, Vol. 1. Paris, France: UNESCO.

UNFCCC. 2008. Compendium on methods and tools to evaluate impacts of, and vulnerability and adaptation to, climate change. Bonn, Germany: UNFCCC.

Walker, W.E., Haasnoot, M., and Kwakkel, J.H. 2013. Adapt or perish: A review of planning approaches for adaptation under deep uncertainty, *Sustainability* 5: 955–979.

weADAPT. 2013. weADAPT homepage. Stockholm, Sweden: SEI (http://weadapt.org/) [Accessed July 2013—Tools available across the website, no single directory page].

Weaver, C.P., Lempert, R.J., Brown, C., Hall, J.A., Revell, D., and Sarewitz, D. 2012. Improving the contribution of climate model information to decision making: The value and demands of robust decision frameworks, *WIRES Climate Change* 4: 39–60.

Wilby, R.L. and Dessai, S. 2010. Robust adaptation to climate change, *Weather* 65(7): 180–185.

# 10 Transboundary River Systems in the Context of Climate Change

*Soni M. Pradhanang and Nihar R. Samal*

## CONTENTS

10.1 Introduction .................................................................................................. 291
10.2 Transboundary River Basins: Examples from around the World ................. 295
    10.2.1 Ganges–Brahmaputra–Meghna (GBM) ........................................... 295
    10.2.2 Danube River Basin .......................................................................... 299
    10.2.3 Colorado River Basin ....................................................................... 310
    10.2.4 Nile River Basin ................................................................................ 315
10.3 Transboundary Water Treaties ...................................................................... 322
    10.3.1 Climate Change and Transboundary River Basins .......................... 322
    10.3.2 International Freshwater Agreements/Declarations ......................... 323
    10.3.3 Regional Accords ............................................................................. 323
    10.3.4 River Basin Treaties ......................................................................... 324
10.4 Conclusions ................................................................................................... 325
References ............................................................................................................ 326

## 10.1 INTRODUCTION

Freshwater is one of the vital resources needed for ecological and societal activities. River basins and their tributaries are crucial sources of freshwater. A river basin or catchment is defined as an interconnected system that transforms natural input of solar energy, atmospheric precipitation, nutrients, and other environmental factors (Burton, 1986). The involvement of many environmental and societal factors and the interconnection between and among systems make river basin management difficult. The level of complexity increases even more for transboundary river basins and their management. A large proportion of freshwater is stored in rivers, lakes, and aquifers shared by two or more countries. There are about 261 river basins that are shared by two or more nations (Table 10.1) (Wolf et al., 1999). Water plays an important role in international conflicts and security. Political borders and river boundaries rarely coincide with borders of watersheds, ensuring that politics inevitably intrude on water policy and water management. Inequities in the distribution, use, and consequences of water management and use have been a source of tension

## TABLE 10.1
### Country Areas in the Ganges–Brahmaputra–Meghna River Basin

| Country | Ganges Basin Basin Area (1000 km$^2$) | Ganges Basin Percentage of Total Area | Brahmaputra Basin Basin Area (1000 km$^2$) | Brahmaputra Basin Percentage of Total Area | Meghna Basin Basin Area (1000 km$^2$) | Meghna Basin Percentage of Total Area |
|---|---|---|---|---|---|---|
| China | 33 | 3 | 293 | 50 | — | — |
| Nepal | 140 | 13 | — | — | — | — |
| Bhutan | 45 | 8 | — | — | — | — |
| India | 861 | 80 | 195 | 34 | 49 | 58 |
| Bangladesh | 46 | 4 | 47 | 8 | 36 | 42 |
| Total | 1080 | 100 | 580 | 100 | 85 | 100 |

*Source:* Joint Rivers Commission Bangladesh (JRCB), Treaty between the government of the People's Republic of Bangladesh and the government of the Republic of India on sharing of the Ganga/Ganges waters at Farakka, Available at: http://www.jrcb.gov.bd/ (accessed on May 23, 2013).

and dispute. In addition, water resources have been used to achieve military and political goals, including the use of water systems and infrastructure, such as dams and supply canals, firefighting as military targets.

Many rivers and water bodies are shared by two or more countries, provinces, or states. International river basins cover about half of the Earth's land surface, generating roughly 60% of global freshwater, and serving around 40% of the world's population (Cooley et al., 2009; Cooley and Gleick, 2011). The first comprehensive collection of information on shared international rivers of the world was first published by the United Nations (UN) in 1958. This early assessment identified 166 major international river basins, but was updated in 1978 to account for an additional 48 river basins (United Nations, 1978). Currently, 263 rivers either cross or demarcate international political boundaries (Figure 10.1). Europe has the largest number of international basins (69), followed by Africa (59), Asia (57), North America (40), and South America (38). The numbers of international river basins, as well as the countries through which they traverse, change over time in response to alterations in the world political map (Giordano, 2003; Giordano and Wolf, 2003). Beyond the sheer number of basins involved, the significance of the world's international waterways is further reflected in their physical extent, abundant resources, and political composition, which make these shared resources vulnerable (Wolf et al., 1999). Climate change is expected to bring a wide range of challenges to freshwater resources, altering water quantity (Pradhanang et al., 2011, 2013), quality, distributions, operation systems, and imposing new governance complications. With a view of global climate change, water scarcity and water stress arise locally, regionally, nationally, and also internationally, which necessitates the implementation of socio-eco-technological methods toward the conservation of water resources (Samal et al., 2006). The conservation of these water resources through water privatization coupled with water pricing both at national and international levels evolves the best management of water systems,

Transboundary River Systems in the Context of Climate Change 293

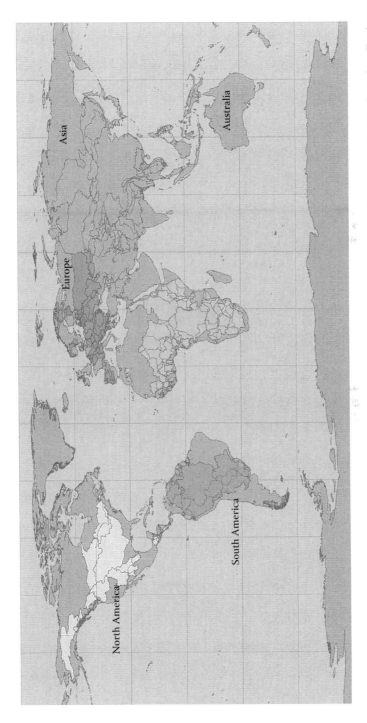

**FIGURE 10.1** Transboundary river basins by continents around the world. International river basins as delineated by the Transboundary Freshwater Dispute Database project, Oregon State University, 2000. (From International River Basins (Wolf, A.T. et al., *Int. J. Water Resourc. Dev.*, 15, 387, 1999), updated 2001.)

such as lakes, rivers, and reservoirs (Samal and Mazumdar, 2005; Samal et al., 2006). Among these local to global challenges, management of river basins, in particular, may be one of the global solutions in the context of worldwide water scarcity and global climate change. Thus, transboundary river basin management under the changing global climatic conditions is the major concern today in the minds of stakeholders, academicians, scientists, and politicians (Figure 10.2). It is the need to investigate how to effectively incorporate information about future hydro-climatological conditions into the politically complex system of transboundary water agreements, including formal treaties, international agreements, and transnational management institutions.

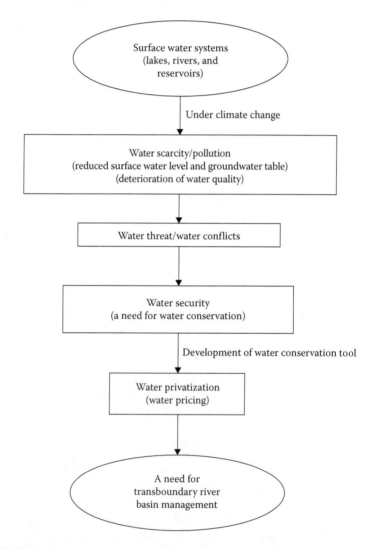

**FIGURE 10.2** Simple conceptual frameworks for flow and development of the transboundary river basin management.

## 10.2 TRANSBOUNDARY RIVER BASINS: EXAMPLES FROM AROUND THE WORLD

### 10.2.1 GANGES–BRAHMAPUTRA–MEGHNA (GBM)

The Ganges–Brahmaputra–Meghna (GBM) River system is the third largest freshwater outlet to the world's oceans and the GBM region consists of river basins of three major river systems that flow through India, Nepal, Bhutan, the Tibet region of China, and Bangladesh (Figure 10.3). GBM is a huge river system with an annual discharge of 1350 billion cubic meters (bcm) and a total drainage area of 1.75 million square kilometers. Of the total annual discharge, the Ganges contributes about 500 bcm, the Brahmaputra 700 bcm, and the Meghna 150 bcm (Faisal, 2002). Not only is each of these three individual rivers big, but each one also has tributaries that are important by themselves in social, economic, and political terms, as well as in terms of water availability, water transport, and water usage. Many of these tributaries are also of transboundary nature (Biswas and Uitto, 2001; Ahmad, 2003). In addition, the system carries up to 1.5 billion tons of sediment per year that originate in the foothills of the Himalayas. The GBM region (Table 10.1) has a combined population of about 600 million, which is growing at a rate of over 2% per year, leading to enormous pressure on the land and water resources throughout the region (Faisal, 2002).

According to the fourth Intergovernmental Panel on Climate Change (IPCC), the increase in precipitation has been assumed to be 13% over the GBM basin for the whole monsoon period, but in reality there will be temporal and spatial variation in the increase of precipitation during the monsoon and over the GBM basin in future (Cell, 2009).

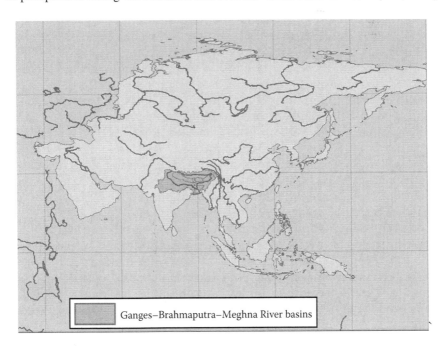

**FIGURE 10.3** Map of the GBM River basins.

Fung and Farquharson (2004) completed a study titled "Impact of Climate and Sea Level Change in Part of the Indian Subcontinent (CLASIC)," in which they used a number of regional climate models (RCMs) (Providing Regional Climates for Impacts Studies [PRECIS] and Hadley Centre Regional Model 2 [HadRM2]) together with a number of general circulation models (GCMs) (Coupled Global Climate Model [CGCM2], Center for Climate System Research/ National Institute of Environmental Studies [CCSRNIES], Geophysical Fluid Dynamics Laboratory [GFDL], Hadley Centre Coupled Model [HadCM3], and CCCma) to assess impacts of climate change in the Ganges, Brahmaputra, and Meghna (GBM) basins. When simulated in the GBM region, the GCMs' results had some uncertainties and inconsistencies among them whereas the RCMs could consistently simulate the effects in the Himalayas. The PRECIS and HadRM2 models were simulated for different climate change scenarios and predicted precipitation for future dry and wet season precipitations. The PRECIS model under SRES A2 scenario showed a 14.8% decrease in dry season precipitation and a 16.1% increase in wet season precipitation in the 2050s. The HadRM2 model with a scenario of 1% per year increase in $CO_2$ from 1990 onward showed a 17.3% decrease in dry season precipitation and a 10.1% increase in wet season precipitation for the same projection period.

Because of the size, complexities, multinational, and transboundary nature of GBM, the planning and management of this basin require a multiple systems approach. Accordingly, following the Ganges Treaty between India and Bangladesh, the main focus of bilateral negotiations between these two countries at present has been on the Teesta River, an important tributary of the Ganges. Currently, these negotiations are ongoing, and no mutually acceptable framework for the management of the Teesta River is in sight. Furthermore, Bangladesh has been so concerned with the Indian plan that it is considering the interlinking of major rivers in recent months, so that any other issue, including a possible treaty on the Teesta, is now receiving somewhat low priority.

The enormity of the development potential of the huge water resources of the GBM region stands out in stark contrast to the region's socioeconomic deprivation (Ahmad, 2003). It is a direct reminder to formulate a long-term vision in order to develop a regional development framework for water utilization. Because of the seasonal availability of water in the Himalayan rivers, harnessing the resource requires that it be stored for meeting year-round demands. Run-of-the-river projects may help, but they cannot store water. Flood control benefits cannot accrue without storage. Thus, good storage schemes are essential for economic and social development of this region. The geographically interlinked character of the major rivers in the GBM region warrants an integrated regional approach to the care and management of the catchments. Sound basin-wide catchment management is an essential long-term strategy to combat the threat of floods and erosion and to preserve the ecosystem apart from the threat of arsenic and fluoride contamination in groundwater in the belt of the GBM basin (Chakraborti et al., 2002). The sediment load in the rivers, which is largely the consequence of geomorphologic processes in the upper catchment areas, tends to increase with the progressive removal of vegetative cover on the slopes. Soil conservation and reforestation in the upper catchments of Nepal and India, and also within Bangladesh, could help in substantially reducing sedimentation and improving groundwater resources.

The GBM rivers create flood problems in their respective basin areas during the monsoon months almost every year. Bangladesh, being the lower riparian country, suffers most from such floods, which cause extensive loss of life and property (Bakker, 2009). Climate change may alter the distribution and quality of the GBM River basin water resources. Some of the impacts include occurrence of more intense rains, changed spatial and temporal distribution of rainfall, higher runoff generation, low groundwater recharge, depletion of groundwater table, melting of glaciers, as well as changes in evaporative demands and water use patterns in agricultural, municipal, and industrial sectors. These impacts lead to severe influences on agricultural production and food security, ecology, biodiversity, river flows, floods, and droughts, water availability and water security, sea level rise, and human and animal health. Spatial and temporal distributions in precipitation are unique characteristics of the GBM River basin; for example, the Ganges River basin is characterized by low precipitation in its northwest region and high precipitation in the coastal areas; the Brahmaputra River basin is characterized by high precipitation zones and dry rain shadow areas; and the world's highest precipitation area is situated in the Meghna River basin (Mirza et al., 1998). The key facts about the river basins are that 92% of the basin areas lie outside Bangladesh (thus, no infrastructural solutions) and that GBM average annual precipitations are 1500, 2500, and 4000 mm, respectively.

The study by De Stephano et al. (2012) reported that the interannual variability in the GBM has been historically low for all country-basin units. The distribution of climate change impacts is somewhat pronounced, with several climate scenarios leading to moderate or high increases in 2030 and 2050. Variability management in the basin has been defined by augmenting dry-season flows and monsoon flood control, and more often operates on intra-annual time scales. If indeed the system transitions to a state of greater interannual variability, then management institutions currently in place will have to adapt to meet fundamentally different challenges that could ensue.

The problems in the GBM River basins are typically conflicting interests of up- and downstream riparian systems. India is both an upstream and a downstream riparian system depending on the portion of the basin under consideration, thus making arrangements of international water treaties on this basin even more complex. India, as one of the upstream riparian systems with respect to Bangladesh, developed plans for water diversions for its own irrigation, navigability, and water supply interests. Initially, Pakistan, and later Bangladesh, had interests in protecting the historic flow of the river for its own downstream uses (Table 10.1). Salman and Uprety (2002) reported that the 1996 Ganges River Treaty may have incorporated important stipulations such as water allocation, yet ignored others, including water augmentation (or variability management) and flood mitigation. Having little recourse to deal with water variability has contributed to political tensions between India and Bangladesh. The potential clash between upstream development and downstream historic use set the stage for attempts at conflict management. Much of the international law that has been signed about the GBM has to do with dividing flow between India and Bangladesh. Agreements signed in 1977 and 1996 and a memorandum of understanding in 1985 regulated flow allocations in the dry season, but have not considered upstream uses of non-signatories, such as Bhutan and Nepal (Table 10.2). Notably, India has used its position of power in the basin to insist on a series of bilateral treaties rather than engaging in multilateral negotiations. This pattern is reflected in the collection of treaties for the basin, which are all bilateral.

## TABLE 10.2
## Chronology of Major Events in the GBM River Basins

| Document Name | Date Signed | Treaty Basin | Country |
|---|---|---|---|
| Agreement between the British government and the State of Jind for regulating the supply of water for irrigation from the western Jumna Canal | NA | Western Jumna Canal | United Kingdom, India |
| Agreement between Great Britain and the Panna State respecting the Ken Canal | 1908 | Ken Canal | United Kingdom |
| Agreement between the government of India and the government of Nepal on the Kosi project | 1954 | Kosi | Nepal |
| Agreement between His Majesty's government of Nepal and the government of India on the Gandak irrigation and power project | 1959 | Gandak, Bagmati | India, Nepal |
| Amended agreement between His Majesty's government of Nepal and the government of India concerning the Kosi project | 1966 | Kosi | India, Nepal |
| Statute of the Indo-Bangladesh Joint Rivers Commission | 1972 | Ganges–Brahmaputra | Bangladesh, India |
| Agreement between His Majesty's government of India and the royal government of Bhutan regarding the Chukkha hydroelectric project. Chukkha Hydroelectric Project; India finances a hydroelectric project (60% grant; 40% low interest loan) to be built in | 1905 | NA | Bhutan, India |
| provisional conclusion of the treaty of April 18, 1975, on the division of the waters of the Ganges | 1975 | Ganges | Bangladesh, India |
| Agreement between the government of the People's Republic of Bangladesh and the government of the Republic of India on sharing of the Ganges waters at Farakka and on augmenting its flows | 1977 | Ganges | Bangladesh, India |
| Agreement between Nepal and India on the renovation and extension of Chandra Canal, Pumped Canal, and distribution of the Western Kosi Canal | 1978 | Kosi | India, Nepal |
| Indo-Bangladesh memorandum of understanding on the sharing of Ganga waters at Farakka | 1982 | Ganges | Bangladesh, India |

## TABLE 10.2 (continued)
## Chronology of Major Events in the GBM River Basins

| Document Name | Date Signed | Treaty Basin | Country |
|---|---|---|---|
| Agreement on ad hoc sharing of the Teesta waters between India and Bangladesh reached during the 25th Meeting of the Indo-Bangladesh Joint Rivers Commission held in July 1983, at Dhaka | 1983 | Teesta/Tista | Bangladesh, India |
| Summary record of discussions of the First Meeting of the Joint Committee of Experts held in Dhaka between January 16 and 18, 1986 | 1986 | Frontier or shared waters | Bangladesh, India |
| Treaty between His Majesty's government of Nepal and the government of India concerning the integrated development of the Mahakali River including Sarada Barrage, Tanakpur Barrage, and Pancheshwar Project | 1996 | Mahakali | India, Nepal |
| Treaty between the government of the Republic of India and the government of the People's Republic of Bangladesh on sharing of the Ganga/Ganges waters at Farakka | 1996 | Ganges | Bangladesh, India |
| Agreement between the British government and the Patiala state regarding the Sirsa branch of the Western Jumna Canal | | Ganges | United Kingdom, India |
| Meeting of the Joint Rivers Commission | 1983 | Ganges | Bangladesh, India |

*Source:* Adapted from Program in water conflict management and transformation, available from: http://www.transboundarywaters.orst.edu/database/interfreshtreatdata.html.

### 10.2.2 Danube River Basin

The Danube River basin (Figure 10.4) is the heart of central Europe and is Europe's second longest river, with a length of 2857 km. The Danube basin covers about 800,000 km² and is rather diverse in terms of geography and climatology including all of Hungary, most of Romania, Austria, Slovenia, Croatia, and Slovakia; and significant parts of Bulgaria, Germany, the Czech Republic, Moldova, Serbia, and Ukraine. Bosnia and Herzegovina, and small parts of Italy, Switzerland, Albania, and Poland are also included in the basin (Sommerwerk et al., 2009) (Table 10.3). The Danube River discharges into the Black Sea through a delta, which is the second largest wetland area in Europe. The 1961–2000 average discharge at the entrance of the delta (CeatalIzmail

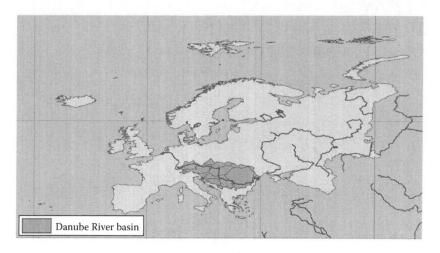

**FIGURE 10.4** Map of the Danube River basin.

### TABLE 10.3
### Country Areas in the Danube River Basin

| Country | Area in DRB (km²) | Percentage of DRB (%) | Percentage of DRB in Country (%) |
| --- | --- | --- | --- |
| Albania | 126 | <0.1 | 0.01 |
| Austria | 80,423 | 10 | 96.1 |
| Bosnia and Herzegovina | 36,636 | 4.6 | 74.9 |
| Bulgaria | 47,413 | 5.9 | 43 |
| Croatia | 34,965 | 4.4 | 62.5 |
| Czech Republic | 21,688 | 2.9 | 27.5 |
| Germany | 56,184 | 7 | 16.8 |
| Hungary | 93,030 | 11.6 | 100 |
| Italy | 565 | <0.1 | 0.2 |
| Macedonia | 109 | <0.1 | 0.2 |
| Moldova | 12,834 | 1.6 | 35.6 |
| Montenegro | 7,075 | 0.9 | 51.2 |
| Poland | 430 | <0.1 | 0.1 |
| Romania | 232,193 | 29 | 97.4 |
| Serbia | 81,560 | 10.2 | 92.3 |
| Slovak Republic | 47,084 | 5.9 | 96 |
| Slovenia | 16,422 | 2 | 81 |
| Switzerland | 1,809 | 0.2 | 4.3 |
| Ukraine | 30,520 | 3.8 | 5.4 |
| Total | 801,463 | 100 | |

*Source:* ICPDR, ICPDR Strategy on Adaptation to Climate Change, International Commission for the Protection of the Danube River Vienna, Austria, 2012.

station, 5 45.22°N, 28.73°E) was about 6500 m³/s (GRDC, 2013). The Alps in the west, the Dinaric–Balkan mountain chains in the south, and the Carpathian mountain bow receive the highest annual precipitation (1000–3200 mm per year) while the Vienna basin, Pannonian basin, Romanian and Prut low plains, and the lowlands and the delta region are very dry (350–600 mm per year) (Lucarini et al., 2007). The upper regions in the west show a strong influence from the Atlantic climate with high precipitation, whereas the eastern regions are affected by a continental climate with lower precipitation and typically cold winters. The dominant land use for the entire Danube basin consists of forest (35%), arable land (34%), and grassland (17%).

The river is shared by a large and ever-growing number of riparian states that, for decades, were in hostile political alliances, some of which are currently locked in intense national disputes. As a consequence, conflicts in the basin tended to be both frequent and intricate, and their resolution especially formidable (Botterweg and Rodda, 1999). Nevertheless, in recent years, the riparian states of the Danube River have established an integrated program for the basin-wide control of water quality, which, if not the first such program, claims to probably being the most active and the most successful of its scale. The Environmental Program for the Danube River is also the first basin-wide international body that actively encourages public and nongovernmental organizations (NGOs) participation throughout the planning process, which, by diffusing the confrontational setting common in planning, may help preclude future conflicts both within countries and internationally.

In the Danube basin, the WaterGAP Global Hydrology Model (WGHM) strongly underestimates observed winter discharges, which might be due to an underestimation of rain or snowmelt in the basin (Döll and Zhang, 2010). According to ECHAM4, climate change will lead to significant increases in river discharge from November to May. Peak flow will shift from May to April, and March and April flows are predicted to more than double by the 2050s. Annual river discharge will increase by more than 35%, and would then be 30% larger than natural flows under the 1961–1990 climates (Döll and Zhang, 2010). Like in the Danube, discharge changes predicted by using HadCM3 input are smaller, but go in the same direction. Higher winter flows and earlier and higher spring flows can be expected to lead to increased sediment transport, disruption of spawning, decreased reproduction and recruitment, and to a change in assemblage structure (Scheurer et al., 2009; Poff and Zimmerman, 2010; Pradhanang et al., 2013).

In 1994, the Danube countries having major shares within the Danube River basin signed the Danube River Protection Convention, defining the three main areas for action, which include the protection of water and associated ecological resources, the sustainable use of water in the Danube basin, and the management of floods and ice hazards. Today, the Danube River Protection Convention has 15 contracting parties: 14 countries and the European Union. Together, they form the International Commission for the Protection of the Danube River (ICPDR). Its permanent secretariat is based in Vienna and started its work in 1998. In order to take the required steps on adaptation to climate change, the ICPDR was asked in the Danube Declaration from 2010 to develop a Climate Change Adaptation Strategy for the Danube River basin by the end of 2012. The detailed chronology of these events is presented in Table 10.4.

## TABLE 10.4
## Chronology of Major Events in the Danube River Basin

| Document Name | Date Signed | Treaty Basin | Country Name |
|---|---|---|---|
| Agreement between Austria and Bavaria on the Inn River | | Inn | Austria |
| Agreement between Austria and Bavaria on the Inn River | | Inn | Germany |
| Treaty between Austria and Bavaria concerning the regime of the frontier line and other territorial relations between Bohemia and Bavaria | | Frontier or shared waters | Austria, Germany |
| Convention between the Austrian and Czechoslovak Republics concerning the delimitation of the frontier between Austria and Czechoslovakia and various questions connected therewith | 1921 | Thaya | Austria, Czech Republic, Slovakia |
| Convention instituting the definitive statute of the Danube | 1921 | Danube | Belgium, Czech Republic France, United Kingdom, Greece, Croatia, Italy, Romania, Serbia, Slovakia, Slovenia |
| Treaty between Germany and Poland for the settlement of frontier questions | 1926 | Frontier or shared waters | Germany, Poland |
| Convention regarding the regime of navigation on the Danube | 1948 | Danube | Bulgaria, Czech Republic, Hungary, Romania, Slovakia, Ukraine, Union of Soviet Socialist Republics, Yugoslavia (former) |
| Protocol between the Federal People's Republic of Yugoslavia and the People's Republic of Romania governing crossing of the frontier by officials of the water control services | 1948 | Danube | Romania, Yugoslavia (former) |

## TABLE 10.4 (continued)
## Chronology of Major Events in the Danube River Basin

| Document Name | Date Signed | Treaty Basin | Country Name |
|---|---|---|---|
| Treaty between the government of the Union of Soviet Socialist Republics and the government of the Romanian People's Republic concerning the regime of the Soviet–Romanian state frontier and final protocol | 1949 | Danube | Romania, Union of Soviet Socialist Republics |
| Agreement between the Austrian Federal government and the Bavarian State government concerning the diversion of water in the Rissbach, Durrach, and Walchen districts | 1950 | Frontier or shared waters; Isar, Rissbach, Durrach, Kesselbach, Blaserbach, Dollmannbach | Hungary, Union of Soviet Socialist Republics, Austria, Germany |
| Agreement between the government of the Polish Republic and the government of the German Democratic Republic concerning navigation in frontier waters and the use and maintenance of frontier waters, signed at Berlin | 1952 | Frontier or shared waters, Oder, NysaŁużycka (Lausitzer Neisse) | Poland |
| Agreement between the government of the Republic of Austria and the government of the Federal Republic of Germany and of the free state of Bavaria concerning the Donaukraftwerk–Jochenstein–Aktiengesellschaft (Danube Power Plant and Jochenstein Joint-Stoch Company) | 1952 | Danube | Austria, Germany |
| Convention between the government of the Union of Soviet Socialist Republics and the government of the Romanian People's Republics concerning measures to prevent floods and to regulate the water regime of the River Prut | 1952 | Prut | Romania, Union of Soviet Socialist Republics |

(*continued*)

## TABLE 10.4 (continued)
## Chronology of Major Events in the Danube River Basin

| Document Name | Date Signed | Treaty Basin | Country Name |
|---|---|---|---|
| Agreement between Czechoslovakia and Hungary concerning the settlement of technical and economic questions relating to frontier water | 1954 | Tisza | Czech Republic, Hungary, Slovakia |
| Convention between the government of the Federal People's Republic of Yugoslavia and the Federal Government of the Austrian Republic concerning water economy questions relating to the Drava, signed at Geneva | 1954 | Drava | Austria, Yugoslavia (former) |
| Agreement between the Federal People's Republic of Yugoslavia and the Republic of Austria concerning water economy questions in respect of the frontier sector of the Mura and the frontier waters of the Mura (the Mura Agreement); and Protocol to the Mura | 1954 | Mura | Austria, Yugoslavia (former) |
| Agreement between Yugoslavia and Romania concerning questions of water control on water control systems and watercourses on or intersected | 1955 | Danube, Tisza | Romania, Yugoslavia (former) |
| Agreement between Yugoslavia and Hungary together with the statute of the Yugoslav–Hungarian water economy commission | 1955 | Mura, Drava, Maros, Tisa, Danube | Hungary, Yugoslavia (former) |
| Treaty between the Hungarian People's Republic and the Republic of Austria concerning the regulation of water economy questions | 1956 | Danube | Austria, Hungary Czech Republic, Hungary, Slovakia, |
| Agreement between the government of the Federal People's Republic of Yugoslavia and the government of the People's Republic of Albania concerning water economy questions, together with the statue of the Yugoslav–Albanian water economic commission and with the protocol concerning fishing in frontier lakes and rivers | 1956 | CrniDrim, BeliDrim, Bojana, Lake Skadar | Albania, Yugoslavia (former) |

## TABLE 10.4 (continued)
## Chronology of Major Events in the Danube River Basin

| Document Name | Date Signed | Treaty Basin | Country Name |
|---|---|---|---|
| Convention between the governments of the Romanian People's Republic, the People's Republic of Bulgaria, the Federal People's Republic of Yugoslavia, and the Union of Soviet Socialist Republics concerning fishing in the waters of the Danube | 1958 | Danube | Bulgaria, Romania, Union of Soviet Socialist Republics, Yugoslavia (former) |
| Agreement concerning water economy questions between the government of the Federal People's Republic of Yugoslavia and the government of the People's Republic of Bulgaria | 1958 | Danube | Bulgaria, Yugoslavia (former) |
| Convention between the government of the Socialist Federal Republic of Yugoslavia and the government of the Romanian People's Republic concerning the operation of the Iron Gates water power and navigation system on the River Danube | 1963 | Danube | Romania, Yugoslavia (former) |
| Agreement between the Socialist Federal Republic of Yugoslavia and the Romanian People's Republic concerning the construction and operation of the Iron Gates water power and navigation system on the River Danube | 1963 | Danube | Romania, Yugoslavia (former) |
| Convention between the Socialist Federal Republic of Yugoslavia and the Romanian People's Republic concerning compensation for damage caused by the construction of the Iron Gates water power and navigation system on the River Danube | 1963 | Danube | Romania, Yugoslavia (former) |

(*continued*)

## TABLE 10.4 (continued)
## Chronology of Major Events in the Danube River Basin

| Document Name | Date Signed | Treaty Basin | Country Name |
|---|---|---|---|
| Convention between the government of the Socialist Federal Republic of Yugoslavia and the government of the Romanian People's Republic concerning the preparation of designs for the construction of the Iron Gates water power and navigation system on the River Danube | 1963 | Danube | Yugoslavia (former), Romania |
| Final act, agreement, and other acts relating to the establishment and operation of the Iron Gates water power and navigation system on the River Danube | 1963 | Danube | Yugoslavia (former) |
| Convention between the government of the Socialist Federal Republic of Yugoslavia and the government of the Romanian People's Republic concerning the determination of the value of investments and mutual accounting in connection with the construction of the Iron Gates water power and navigation system on the River Danube | 1963 | Danube | Romania, Yugoslavia (former) |
| Treaty between the Republic of Austria and the Czechoslovak Socialist Republic concerning the regulation of water management questions relating to frontier waters | 1967 | Danube | Austria |
| Treaty between the Republic of Austria and the Czechoslovak Socialist Republic concerning the regulation of water management questions relating to frontier waters | 1967 | Danube | Czech Republic |
| Treaty between the Republic of Austria and the Czechoslovak Socialist Republic concerning the regulation of water management questions relating to frontier waters | 1967 | Danube | Slovakia |

## TABLE 10.4 (continued)
## Chronology of Major Events in the Danube River Basin

| Document Name | Date Signed | Treaty Basin | Country Name |
|---|---|---|---|
| Agreement between the government of the Czechoslovak Socialist Republic and the government of the Hungarian People's Republic concerning the establishment of a river administration in the Rajka–Gönyü sector of the Danube | 1968 | Danube | Czech Republic, Hungary, Slovakia |
| Agreement between the government of the Socialist Republic of Romania and the government of the Union of Soviet Socialist Republics on the joint construction of the Stinca–Costesti Hydraulic Engineering Scheme on the River Prut and the establishment of the conditions for its operation (with protocol) | 1971 | Prut | Romania, Union of Soviet Socialist Republics |
| Convention concerning the protection of Italo–Swiss waters against pollution, signed at Rome | 1972 | Lake Maggiore, Lake Lugano, Lake Verbano, Lake Ceresio, Doveria, Melezza, Giona, Tresa, Breggia, Maira/Mera, Poschiavino, Spol | Canada, Switzerland, Italy |
| Agreement concerning mutual assistance in the construction of the Gabcikovo–Nagymaros system of locks | 1977 | Gabcikovo–Nagymaros system | Czech Republic, Hungary, Slovakia |
| Treaty between the government of Romania and the government of Hungary on the regulation of water problems in watercourses forming or crossing the boundary | 1986 | Frontier or shared waters | Hungary, Romania |
| Agreement between the Federal Republic of Germany and the European Economic Community, on the one hand, and the Republic of Austria, on the other, on cooperation on management of water resources in the Danube basin, Regensburg | 1987 | Danube | Austria, Germany, European Economic Community |

(*continued*)

## TABLE 10.4 (continued)
## Chronology of Major Events in the Danube River Basin

| Document Name | Date Signed | Treaty Basin | Country Name |
|---|---|---|---|
| Treaty between the Czech Republic government and the Slovak Republic government on mutual relations and principles of cooperation in agriculture, food industry, forestry, and water economy under the conditions of the customs union | 1992 | Not specified | Czech Republic, Slovakia |
| Agreement between the government of the Republic of Croatia and the government of the Republic of Hungary on water management relations | 1994 | Danube, Drava | Croatia, Hungary |
| Convention on cooperation for the protection and sustainable use of the River Danube | 1994 | Danube | Austria, Bulgaria, Germany, European Union |
| Convention on cooperation for the protection and sustainable use of the River Danube | 1994 | Danube | Croatia, Hungary, Moldova, Romania, Slovakia, Ukraine, Moldova |
| Agreement between the government of the Republic of Moldova and the government of Ukraine on the joint use and protection of transboundary waters | 1994 | Dnestr, Danube, Kogilnik, Sarata | Ukraine, Hungary |
| Framework agreement on the Sava River basin | 2002 | Sava | Bosnia and Herzegovina, Croatia, Slovenia, Yugoslavia (former) |
| Framework convention of the protection and sustainable development of the Carpathians | 2003 | General | Czech Republic, Hungary, Montenegro, Poland, Romania, Serbia, Slovakia, Ukraine |
| Agreement between the government of Romania and the government of Ukraine on cooperation in the field of transboundary water management | 1997 | Danube, Tisza, Prut, Siret | Romania, Ukraine |

## TABLE 10.4 (continued)
## Chronology of Major Events in the Danube River Basin

| Document Name | Date Signed | Treaty Basin | Country Name |
|---|---|---|---|
| Draft convention on the cooperation for the protection and sustainable use of the Danube River | 1994 | Danube | Unknown |
| Agreement between the government of Ukraine and the republic of Poland on the cooperation in the field of water management in frontier waters | 1996 | Danube, Dniester, Vistula/Wista | Poland, Ukraine |

*Source:* Adapted from Program in water conflict management and transformation, available from: http://www.transboundarywaters.orst.edu/database/interfreshtreatdata.html.

The "Danube Study—Adaptation to Climate Change" by ICPDR (2012) identified knowledge gaps and requirements for further research besides commonalities in the impact studies. A prerequisite for achieving progress in this regard is having professionals who are able to apply the required activities (institutional adaptation), including the establishment and maintenance of databases, measurement networks, simulation models, analysis software, laboratories, knowledge management systems, and adjusted processes in the concerned institutions. To improve the understanding of ongoing changes and their impacts, better observational data and data access are necessary. Quality assurance and the homogenization of data sets help to improve model projections and are a prerequisite for adaptive management required under conditions of climate change within transboundary regions or catchments. In particular, changes in water availability, as well as changes in water demand, in the Danube basin are of high interest on a monthly or higher temporal resolution scale.

There is a need to compare climate impacts across sectors and to systematically assess climate risks, preferably based on a commonly agreed methodology and database. A basin-wide assessment could guide the selection of regional hot spots for detailed impact studies. An interdisciplinary research team can acquire a multisector impact aggregation and a damage and risk assessment for short-term, medium-term, or long-term applications. The synergies and conflicts between climate change and land use planning need to be clarified. Feedback between land use and climate change should be analyzed more extensively (e.g., by coupled climate and land use change modeling). Furthermore, an evaluation of the water-related consequences of different climate policies and development pathways is also important for a common adaptation strategy. Alongside climate change impacts, socioeconomic and demographic aspects are also crucial for future adaptation measures.

A basin-wide approach that covers all relevant hydrological parameters is valuable to determine the impacts and consequences of climate change in the Danube River basin. Particularly, the middle and lower Danube River basin might benefit from this approach due to the relatively sparse information existing on climate change impacts in these regions. Given the expected increase in water scarcity and drought situations in future summer periods in the southeastern regions, a basin-wide approach can help resolve transboundary environmental crises and in taking suitable adaptation measures. The assessment of water availability for the various major utilities under several future circumstances (e.g., under severe drought and water shortage conditions) can be projected. Upstream–downstream dependencies, taking into consideration socioeconomic and demographic changes, should be clearly presented. Furthermore, model projections for the whole Danube basin with better land-surface properties and interactions as a large-scale climate model can provide suitable information on the catchment scale—the most important scale for water management.

### 10.2.3 COLORADO RIVER BASIN

The Colorado River and its tributaries traverse through the Great Basin, and the Sonoran and Mojave Deserts, providing vital life support to the arid American Southwest (Figure 10.5). The source of the Colorado River is about 3048 m above sea level in the Rocky Mountains of Colorado, the United States, and flows southwest to the Gulf of California in Mexico (Table 10.5). It is the international boundary between the United States and Mexico for 27 km. Before the construction of a number of dams along its route, the Colorado River flowed 128 km through Mexico to the Gulf of California. The 2330 km of its route in the United States makes it the nation's fifth longest river and drains a large portion of the North American continent covering 644,358 km$^2$ in the United States and 10,809 km$^2$ in Mexico. The Colorado River and its tributaries drain southwestern Wyoming and western Colorado, parts of Utah, Nevada, New Mexico, California, and almost all of Arizona. Three quarters of the Basin is federal land devoted to national forests and parks and Indian reservations (Baillat, 2010). The terrain of the Colorado River consists of wet upper slopes, irregular transition plains and hills, deep canyon lands, and dry lower plains.

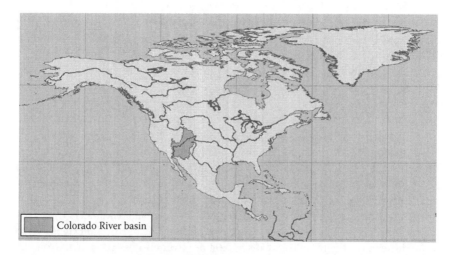

**FIGURE 10.5** Map of the Colorado River basin.

## TABLE 10.5
### Country Areas in the Colorado River Basin

| Country | Total Area of the Country (Million km$^2$) | Area of the Country within the Basin (km$^2$) | Percentage of Total Area of Basin (%) | Percentage of Total Area of Country (%) |
|---|---|---|---|---|
| CRB, the United States | 9.827 | 644,358 | 98.3 | 6.55 |
| CRB, Mexico | 1.973 | 10,809 | 1.7 | 0.55 |
| Total | 11.8 | 655,167 | | 7.10 |

*Source:* Oak Ridge National Laboratory (ORNL), LANDSCAN 2007 global population dataset. Oak Ridge National Laboratory, Oak Ridge, TN, 2008, Available at: http://www.ornl.gov/sci/landscan (accessed on May 24, 2013).

The wet upper slopes consist of numerous streams that feed into the Colorado River from stream-cut canyons and small flat-floored valleys, often occupied by alpine lakes and adjacent steep-walled mountain peaks, and dense forests. Great quantities of sediment are washed into the river and for many years (since the last glacial period, approximately 140,000 years) have been deposited in the lower reaches of the basin, forming marginal sandbars and terraces. These have been accumulating at the river mouth in the Upper Gulf of California, forming what is today known as the Colorado River delta, and constituting the Mexicali and Imperial Valleys. The accumulated sediments formed a land elevation, cutting one arm of the ocean in the Gulf and created the old Lake Cahuilla. This ancient lake, according to botanical studies and geologists, dried up during the Spanish conquest (in the sixteenth century). However, due to the derivation of return flows from the Imperial Irrigation District and flooding periods in 1905, the lake was filled again, forming what is today known as the Salton Sea (Arias et al., 2004).

The hydrology of the Colorado River is largely snowmelt driven, with 70% of the river's annual pre-impoundment flow occurring from May to July (Harding et al., 1995). Much of the basin, especially the border region, is extremely arid, with less than 8 cm of precipitation per year. The basin is suffering from a decade-long drought, with annual flows during this period at about 75% of the average. More than 80% of runoff in the basin originates from less than 20% of the basin area, generally in the Rocky Mountains at elevations above 2500 m (Hoerling et al., 2009). Recent analysis of tree-ring records shows several multiyear droughts that dwarf the current drought, both in duration and severity. By the year 2050, average temperature is estimated to increase by 2°C to 4°C (Christensen and Lettenmaier, 2007; Pierce et al., 2009). Eighteen of nineteen climate models show a drying trend in the lower Mexican portions of the Colorado River basin, with the hydrology becoming consistently drier throughout the century (Seager et al., 2007). However, 80%–85% of the Colorado River runoff originates from precipitation at elevations above 2500 m, where projections of changes in the timing and magnitude of precipitation are less certain. Nonetheless, a recent study projects that greater water losses to evaporation and infiltration to drier soils will likely reduce the Colorado River runoff by 6%–20% by 2050 (Ray et al., 2008).

The two principle reservoirs in the Colorado River are Lake Mead and Lake Powell, each with a usable capacity of greater than 30 km$^3$. Numerous smaller reservoirs include Flaming Gorge, Mohave, Strawberry Reservoir, Lake Havasu, Roosevelt Lake, Taylor Park Reservoir, Blue Mesa Reservoir, McPhee Reservoir, Vallecito Reservoir, and Navajo Reservoir. Historically, the annual flows of the Colorado River at Lee's Ferry have exceeded 29.6 km$^3$ and have been less than 4.6 km$^3$ (USGS, 2004) (Figure 10.4). Most of the flow for the Colorado River originates in the Upper Basin, which encompasses approximately 284,400 km$^2$. About 86% of the annual runoff originates within only 15% of the area, in the high mountains of Colorado.

The Colorado River basin is extremely dynamic with expanding economies and increasing industrialization, especially in the California and Baja California border regions (UNEP, 2004). The population of the Colorado River basin is rapidly growing and urban areas are sprawling. Due to the unmanaged growth in the basin, serious transboundary environmental problems and concerns have developed. Other issues include the impact of urban development on the fauna and flora of already sensitive ecosystems, water security, and storage. About 80%–90% of all water resources are used to irrigate agricultural lands. Considering that the region is characterized mainly by arid and semiarid zones, the problem of freshwater shortage is accentuated in the lower basin. The primary source of water supply in the Colorado River basin states comes from the Colorado River. Groundwater is also an important resource, accounting in some states (e.g., Arizona, California, Baja California, and San Luis) for up to 37% of total water use. As the West's population and need for water have grown, the Colorado River has been tapped through a system of dams and diversions that begin close to its source in the mountains of Colorado and Wyoming. In the United States, water allocation is controlled by state law, with the western and southern states generally relying on prior appropriation systems for surface water allocations, and the northern and eastern states relying mainly on riparian rights systems (Hutchins, 1977). The chronology of major events of treaties, agreements, and cooperation is presented in Table 10.6. Groundwater allocation, which is also under state jurisdiction, is often managed separately from surface water—a perpetual problem in water resources management, given the pervasive interactions between groundwater and surface water. The federal Environmental Protection Agency implements laws to protect the environment, including water quality and aquatic habitat, for which many states have assumed administrative responsibility (CWA, 1972). Under international law, individual states have the right to control territorial resources. Considering the transboundary implications of the Colorado River delta as a shared watershed, the responsibility for its protection relies on both riparian states. To date, both Mexico and the United States and federal government agencies have resisted active binational cooperation to restore the health of the Colorado River delta ecosystem. These agencies instead point to the absence of any formal agreement between the federal governments of the United States and Mexico regarding the allocation of Colorado River water for delta conservation. It is believed that the restoration of the Colorado River delta comes down to all water-consumptive users in the Colorado River basin. There must be a continuity of public participation in policy and management decisions and coordination among the various involved organizations to ensure that efforts are not duplicated.

# TABLE 10.6
## Chronology of Major Events in the Colorado River Basin

| Document Name | Date Signed | Treaty Basin | Country Name |
|---|---|---|---|
| Treaty of Guadalupe Hidalgo in 1848 established borders between the United States and Mexico stipulating the international border along the Rio Grande | | Colorado, Rio Grande | Mexico, United States |
| Boundary Convention between the United States and Mexico | | Colorado, Rio Grande | Mexico, United States |
| Boundary Convention between the United States and Mexico, extending the Convention of March 1, 1889 | | Rio Grande, Colorado | Mexico, United States |
| Boundary Convention between the United States and Mexico, extending the Convention of March 1, 1889 | | Rio Grande, Colorado | Mexico, United States |
| Treaty between the United States and Mexico relating to the waters of the Colorado and Tijuana rivers, and of the Rio Grande (Rio Bravo) from Fort Quitman, Texas, to the Gulf of Mexico, signed at Washington on February 3, 1944, and supplementary | | Colorado, Rio Bravo/Rio Grande, Tijuana | Mexico, United States |
| Exchange of notes constituting an agreement concerning the loan of waters of the Colorado River for irrigation of lands in the Mexicali Valley | 1966 | Colorado | Mexico, United States |
| Treaty to resolve pending boundary differences and maintain the Rio Grande and Colorado River as the international boundary | | Rio Bravo/Rio Grande, Colorado | Mexico, United States |
| Agreement effected by Minute No. 241 of the International Boundary and Water Commission, the United States and Mexico, adopted at El Paso | 1972 | Colorado | Mexico, United States |

*(continued)*

## TABLE 10.6 (continued)
## Chronology of Major Events in the Colorado River Basin

| Document Name | Date Signed | Treaty Basin | Country Name |
| --- | --- | --- | --- |
| Agreement extending Minute No. 241 of the International Boundary and Water Commission, the United States and Mexico, on July 14, 1972, as extended | 1973 | Colorado | Mexico, United States |
| Mexico–US agreement on the permanent and definitive solution to the salinity of the Colorado River basin (International Boundary and Water Commission Minute No. 242) | 1973 | Colorado | Mexico |
| Mexico–US agreement on the permanent and definitive solution to the salinity of the Colorado River basin (International Boundary and Water Commission Minute No. 242) | 1973 | Colorado | Mexico, United States |
| Recommendations for the solution to the border sanitation problems | 1979 | Rio Bravo/Rio Grande, Colorado | United States |
| Agreement between the United States and the United Mexican States on cooperation for the protection and improvement of the environment in the border area | 1983 | Frontier or shared waters | Mexico, United States |
| Agreement of cooperation between the United States and the United Mexican States regarding pollution of the environment along the inland international boundary by discharges of hazardous substances | 1985 | Frontier or shared waters | Mexico, United States |
| Agreement between the government of the United States and the government of the United Mexican States concerning the establishment of a Border Environment Cooperation Commission and a North American Development Bank | | Colorado, Tijuana, Rio Grande | Mexico, United States |

## TABLE 10.6 (continued)
## Chronology of Major Events in the Colorado River Basin

| Document Name | Date Signed | Treaty Basin | Country Name |
|---|---|---|---|
| Minute No. 291 of the International Boundary and Water Commission, the United States and Mexico, concerning improvements to the conveying capacity of the international boundary segment of the Colorado River | 1994 | Colorado | Mexico, United States |
| Minute No. 294: Facilities planning a program for the solution of border sanitation problems | | Frontier or shared waters | Mexico, United States |
| Agreement by Minute No. 218 of the International Boundary and Water Commission, the United States and Mexico | 1965 | Colorado | Mexico, United States |
| Convention between the United States and the United States of Mexico touching the international boundary line where it follows the bed of the Rio Colorado | | Colorado, Rio Bravo/ Rio Grande | Mexico, United States |
| Joint Communiqué (the United States and Mexico) | 1972 | Colorado | Mexico, United States |
| International Boundary and Water Commission—Minute No. 288—between the United States and Mexico regarding the long-term plan to address wastewater and water quality problems at the international boundary | 1992 | New, Alamo | Mexico, United States |

*Source:* Adapted from Program in water conflict management and transformation, available from: http://www.transboundarywaters.orst.edu/database/interfreshtreatdata.html.

### 10.2.4 NILE RIVER BASIN

The Nile River, with an estimated length of over 6800 km, is the longest river flowing from south to north over 35° of latitude (Figure 10.6). It encompasses roughly three million square kilometers with about 266 million people residing within its boundaries (ORNL, 2008). It is fed by two main river systems: the White Nile, with its sources in the Equatorial Lake Plateau (Burundi, Rwanda, Tanzania, Kenya,

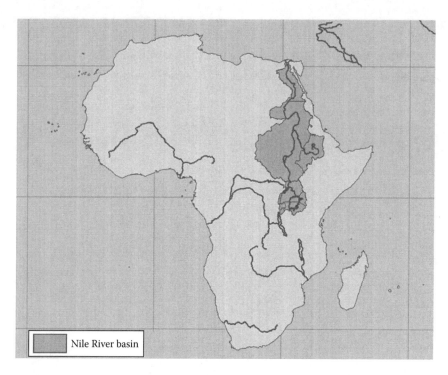

**FIGURE 10.6** Map of the Nile River basin.

Zaire, and Uganda), and the Blue Nile, with its sources in the Ethiopian highlands. The sources are located in humid regions, with an average rainfall of over 615 mm per year. The arid region starts in Sudan, the largest country in Africa, which can be divided into three rainfall zones: the extreme south of the country where rainfall ranges from 1200 to 1500 mm per year; the fertile clay plains where there is 400–800 mm of rainfall annually; and the desert northern third of the country where rainfall averages only 20 mm per year. Further north, in Egypt, precipitation falls to less than 20 mm per year. The total area of the Nile basin represents 10.3% of the area of the continent and spreads over 10 countries (Figure 10.6 and Table 10.7).

The potential impacts of climate change on the hydrology and water resources of the Nile River basin are assessed using a macro-scale hydrology model driven by twenty-first-century simulations of temperature and precipitation downscaled from runs of 11 GCMs and 2 global emissions scenarios (A2 and B1) archived for the 2007 IPCC report. The results show that, averaged across the multimodel ensembles, the entire Nile basin will experience increases in precipitation early in the century (period I, 2010–2039), followed by decreases later in the century (period II, 2040–2069, and period III, 2070–2099) with the exception of the eastern-most Ethiopian highlands, which is expected to experience increases in summer precipitation by 2080–2100. Efforts to integrate climate change into long-term planning and management of the Nile River basin have been limited, although recent efforts suggest that this may be slowly changing. Temperature predictions for all GCMs considered show increases throughout the twenty-first century, but the signature

## TABLE 10.7
### Country Areas in the Nile River Basin

| Country | Total Area of the Country (km²) | Area of the Country within the Basin (km²) | Percentage of Total Area of Basin (%) | Percentage of Total Area of Country (%) |
| --- | --- | --- | --- | --- |
| Burundi | 27,834 | 13,260 | 0.4 | 47.6 |
| Rwanda | 26,340 | 19,876 | 0.6 | 75.5 |
| Tanzania | 945,090 | 84,200 | 2.7 | 8.9 |
| Kenya | 580,370 | 46,229 | 1.5 | 8.0 |
| Zaire | 2,344,860 | 22,143 | 0.7 | 0.9 |
| Uganda | 235,880 | 231,366 | 7.4 | 98.1 |
| Ethiopia | 1,100,010 | 365,117 | 11.7 | 33.2 |
| Eritrea | 121,890 | 24,921 | 0.8 | 20.4 |
| Sudan | 2,505,810 | 1,978,506 | 63.6 | 79.0 |
| Egypt | 1,001,450 | 326,751 | 10.5 | 32.6 |
| Total | | 3,112,369 | 100.0 | |

varies substantially from sub-basin to sub-basin and from GCM to GCM. In the multimodel average over the entire Nile basin, warming increases to more than 3.5°C relative to the historical (1950–1999) average by the end of the 2070–2099 period, when the timing and magnitude of changes in temperature and precipitation will be critical to the hydrologic response of the Nile basin. The Nile River is expected to experience increases in streamflow early in the century at both gauging stations studied, the Blue Nile at El diem and the main stem Nile at HAD, mostly due to increased precipitation. Subsequently, streamflow is expected to decline during periods 2040–2069 and 2070–2099 as a result of both precipitation declines and enhanced evapotranspiration due to increased temperature (Beyene et al., 2010). Precipitation is to a large extent governed by the movement of the intertropical convergence zone (ITCZ) and its interaction with topography. In general, precipitation increases from north to south, and with elevation. Precipitation is virtually zero in the Sahara desert, and increases southward to about 1200–1600 mm/year on the Ethiopian and Equatorial Lake Plateaus (Mohamed et al., 2005).

The waters of the Nile basin offer great potential as a lever for development across a large part of the African continent. The fast demographic growth will accelerate demand for agricultural production and hydropower in this region, which is already vulnerable to drought. Realization of the economic potential of the basin requires targeted technical support in order to overcome barriers to joint management of the land and water resources of the basin. The basis for such cooperation is joint planning and equitable sharing of benefits flowing from the natural resource endowments of the basin. Current water management institutions, treaties, and infrastructure are the legacy of a long history of water management, but many were initially defined by colonialism and its dissolution (Table 10.8). In the early 1900s, a relative shortage of cotton in the world market put pressure on Egypt and Sudan, then under a British–Egyptian condominium, to turn to this summer crop, requiring perennial irrigation

## TABLE 10.8
## Chronology of Major Events in the Nile River Basin

| Document Name | Date Signed | Treaty Basin | Country Name |
|---|---|---|---|
| Protocol between Great Britain and Italy for the demarcation of their respective spheres of influence in Eastern Africa | | Nile | United Kingdom, Italy |
| Exchange of notes between Great Britain and Ethiopia | 1902 | Nile | Ethiopia, United Kingdom |
| Treaties between Great Britain and Ethiopia, relative to the frontiers between Anglo-Egyptian Sudan, Ethiopia, and Eritrea (railway to connect Sudan with Uganda) | 1902 | Nile, Sobat | Ethiopia, United Kingdom |
| Agreement between Great Britain and the Independent State of the Congo, modifying the agreement signed at Brussels on May 12, 1894, relating to the spheres of influence of Great Britain and the Independent State of the Congo in East and Central Africa | 1906 | Nile | United Kingdom, Democratic Republic of Congo (Kinshasa) |
| Agreement between Great Britain, France, and Italy respecting Abyssinia | 1906 | Nile | France, Italy, United Kingdom |
| Exchange of notes between the United Kingdom and Italy respecting concessions for a barrage at Lake Tsana and a railway across Abyssinia from Eritrea to Italian Somaliland | 1925 | Lake Tsana | United Kingdom, Italy |
| Exchange of notes between His Majesty's government in the United Kingdom and the Egyptian government regarding the use of the waters of the River Nile for irrigation purposes | 1929 | Nile | Egypt, United Kingdom |
| Jebel Awilya Compensation Agreement. 1932 | 1932 | NA | Egypt, Sudan |

## TABLE 10.8 (continued)
## Chronology of Major Events in the Nile River Basin

| Document Name | Date Signed | Treaty Basin | Country Name |
|---|---|---|---|
| Agreement between the United Kingdom and Belgium regarding water rights on the boundary between Tanganyika and Rwanda–Burundi | 1934 | Frontier or shared waters | Belgium, United Kingdom |
| Exchange of notes constituting an agreement between the United Kingdom of Great Britain and Northern Ireland and Egypt regarding the utilization of profits from the 1940 British government cotton-buying commission and the 1941 joint Anglo-Egyptian cotton buying commission | 1946 | Nile | Egypt, United Kingdom |
| Exchanges of notes constituting an agreement between the government of the United Kingdom of Great Britain and Northern Ireland and the government of Egypt regarding the construction of the Owen Falls Dam, Uganda | 1949 | Nile | Egypt, United Kingdom |
| Exchange of notes constituting an agreement between the government of the United Kingdom of Great Britain and Northern Ireland on behalf of the government of Uganda and the government of Egypt regarding cooperation in meteorological and hydrological surveys in certain parts of the Nile basin | 1950 | Nile | Egypt, United Kingdom |
| Exchange of notes constituting an agreement between the government of the United Kingdom of Great Britain and Northern Ireland and the government of Egypt regarding the construction of the Owen Falls Dam, Uganda | 1953 | Nile | Egypt, United Kingdom |

*(continued)*

## TABLE 10.8 (continued)
## Chronology of Major Events in the Nile River Basin

| Document Name | Date Signed | Treaty Basin | Country Name |
|---|---|---|---|
| Agreement between the government of the United Arab Republic and the government of Sudan for full utilization of the Nile waters | 1959 | Nile | Egypt |
| Agreement between the government of the United Arab Republic and the government of Sudan for full utilization of the Nile waters | 1959 | Nile | Sudan |
| Protocol (to the November 8, 1959 agreement) concerning the Establishment of the Permanent Joint Technical Committee, Cairo, on January 17, 1960 | 1960 | Nile | Egypt, Sudan |
| Agreement for the Hydrometeorological Survey of Lakes Victoria, Kyogo, and Albert (Mobutu SeseSeko) | 1967 | Lake Victoria, Lake Kyogo, Lake Albert (Mobutu SeseSeko) | Egypt, Kenya, Sudan, Tanzania, United Republic of Uganda |
| Agreement for the establishment of the organization for the management and development of the Kagera River basin (with attached map), concluded at Rusumo, Rwanda | 1977 | Kagera | Burundi, Rwanda, Tanzania, United Republic of Uganda |
| Amendment to the agreement for the establishment of an organization to manage and develop the Kagera River basin on May 19, 1978 | 1978 | Kagera | Burundi, Rwanda, Tanzania, United Republic of Uganda |
| Accession of Uganda to the agreement pertaining to the creation of the organization for the management and development of the Kagera River basin | 1981 | Kagera | Burundi, Rwanda, Tanzania, United Republic of Uganda |
| Framework for general cooperation between the Arab Republic of Egypt and Ethiopia | 1993 | Nile | Egypt, Ethiopia |

## TABLE 10.8 (continued)
## Chronology of Major Events in the Nile River Basin

| Document Name | Date Signed | Treaty Basin | Country Name |
|---|---|---|---|
| Convention for the establishment of the Lake Victoria Fisheries Organization with annex and final act | 1994 | Lake Victoria | Kenya, Tanzania, United Republic of Uganda |
| Agreement to initiate a program to strengthen regional coordination in the management of resources of Lake Victoria | 1994 | Lake Victoria | Kenya, Tanzania, United Republic of Uganda |
| Treaty for the establishment of the East African Community signed at Arusha | 1999 | Lake Victoria | Kenya, Tanzania, United Republic of Uganda |
| Protocol for Sustainable Development of Lake Victoria Basin, Arusha, November 29, 2003 | 2003 | Lake Victoria | Kenya, Tanzania, United Republic of Uganda |

*Source:* Adapted from Program in water conflict management and transformation, available from: http://www.transboundarywaters.orst.edu/database/interfreshtreatdata.html.

over the traditional flood-fed methods. The need for summer water and flood control drove an intensive period of water development along the Nile, and resulted in an agreement on allocation and infrastructure in 1929. With respect to hydrology and climate, there is already a high degree of variability in the flow of the Nile, with a standard deviation around 25%. Changes in variability are not consistent for any scenario or year. What is apparent is that some change occurs in every scenario and year, so a change in variability regime appears inevitable for some basins. Climate variability is already moderately high in upstream riparian systems such as Uganda and Kenya, and these countries may experience both moderate and high changes in variability under different scenarios. Downstream riparian and the primary users of the water, Egypt and Sudan, have high and medium variability, respectively, and experience moderate increases in one climate scenario each. Ethiopia's sensitivity to variability and climate may be the least, as its current variability is low and no climate scenarios result in a substantial increase in variability (Conway, 2005).

Examining the treaty information in Table 10.8 shows that since the end of colonialism, no basin-wide treaties have been signed; only the 1959 treaty between Sudan and Egypt has an allocation mechanism, excluding other riparian systems. Other treaties contain content related to dry season control or conflict resolution using arbitration and diplomatic channels (1994 establishment of the Lake Victoria

Fisheries Organization) that include other riparian systems, but not Egypt and Sudan. This is in part due to the region-specific nature of these bodies, but shows the lack of a unified international management structure that makes all countries party to addressing and managing variability. Water flowing from Ethiopia contributes the majority of runoff. Following Sudan and Ethiopia's contributions, Uganda, Tanzania, and Kenya all provide significant portions of the basin's runoff. In addition, of the 12 dams along the Nile, one third are in the upstream countries of Uganda and Kenya. These statistics point to an imbalance between the content of treaties, the countries signing them, and their hydrological relationships to the basin as a whole.

## 10.3 TRANSBOUNDARY WATER TREATIES

### 10.3.1 Climate Change and Transboundary River Basins

Climate change is expected to intensify security concerns within or between countries or within river basins (Gleick, 1992; Nordås and Gleditsch, 2007) and may have indirect negative effects on environmental resources that can undermine the legitimacy of governments, undermine economic livelihoods, and affect human health through food insecurity and increased exposure to new disease vectors (Barnett, 2003). IPCC has claimed that increased precipitation intensity and variability are projected to increase the risks of flooding and drought in many areas, which will affect food stability as well as water infrastructure and usage practices (IPCC, 2007). With changing climate, the resilience of social–ecological systems is expected to shift influenced by the existence and makeup of international treaties. International water treaties and river basin organizations can be particularly influential in managing or dealing with likely disputes among fellow riparian systems when faced with climatic change and water variability (Drieschova et al., 2008). Significant variability and changes in mean flows have already been observed as reported by many studies (Milly et al., 2007; Milliman et al., 2008; Dai et al., 2009; Xu et al., 2009). Drought-affected areas are expected to increase, impacting terrestrial and aquatic ecosystem. Increased flood risk will also pose challenges to society pertaining to physical infrastructure and water quality (IPCC, 2007).

Hydro–political balances are expected to shift in international river basins. Regions and basins not governed by treaties or water-related institutions and facing current and future variability may be more vulnerable to tension and conflict. In regions that are already governed by treaties, climate change and variability could affect the ability of basin states to meet their water treaty commitments and effectively manage transboundary waters, especially if such treaties are not suited to dealing with variability and new hydrological realities (Ansink and Ruijs, 2008; Goulden et al., 2009). Climatic variability and uncertainties also raise questions about the adequacy of many existing transboundary arrangements and may confer additional resilience to international treaties (Cooley et al., 2009; De Stefano et al., 2012). Given the links between climate change, water variability, conflict, and cooperation in international river basins, the existence of institutions and river basin organizations seems paramount.

## 10.3.2 INTERNATIONAL FRESHWATER AGREEMENTS/DECLARATIONS

The complex biophysical, sociopolitical, and human interactions within international river basins can make the management of these shared resources complicated and difficult. Decreasing water quantity, degrading water quality, unprecedented population growth, and increasing urban sprawl are known to be disruptive factors in co-riparian water relations. The combination of these factors has led academics and policy-makers alike to warn of impending conflict over shared water resources. Acknowledging the benefits of cooperative water management frameworks, policy-makers have been involved in institution-building efforts over the past century in a range of geographic scales. Globally, the international community has developed guiding principles and laws for international freshwater management. On a finer scale, regional bodies and individual understanding have advanced a goal of coordinated management within the world's international basins. Governments have developed protocols and treaties governing the management and protection of specific international water bodies. Together, these developments have encouraged greater interest between political and social communities to drive better management of these transboundary international water resources.

The Institute of International Law (IIL) published recommendations in its 1911 Madrid Declaration on the International Regulation regarding the Use of International Watercourses for Purposes other than Navigation. This agreement outlined general principles for cooperative water management, such as establishing joint technical committees and avoiding unilateral developments. In 1966, the Helsinki Rules on the Uses of Waters of International Rivers further elaborated these principles and outlined factors determining what constitutes equitable utilization of shared water resources (Caponera, 1985). The International Law Commission (ILC) was commissioned by the United Nations to codify the law on the non-navigational uses of international watercourses. The ILC's task with the United Nations General Assembly's adoption of the Convention on the Law of the Non-Navigational Uses of International Watercourses (UN Convention) was completed in 1970. This task regularized principles of "equitable and reasonable utilization" and the "obligation not to cause significant harm" and established a framework for the exchange of data and information, the protection and preservation of shared water bodies, the creation of joint management mechanisms, and the settlement of disputes (Wouters, 2000). Other international statements include the 1972 Declarations of the United Nations Conference on the Human Environment, the 1977 Declarations and Resolutions of the United Nations Water Conference, the 1992 Dublin Statement from the International Conference on Water and the Environment, and the 2000 Second World Water Forum's Ministerial Declaration.

## 10.3.3 REGIONAL ACCORDS

There are several multinational institutions such as the Organization for Economic Cooperation and Development (OECD), the European Union, and the Southern African Development Community (SADC) that have formulated agreements and protocols to support collaborative water resource initiatives at regional scales. The OECD Council issued a series of recommendations concerning the management and protection of transboundary resources relevant to international river basins. The Convention of Environmental Impact Assessment in a Transboundary Context (1991) and the Convention on the Protection and Use of Transboundary Watercourses and International Lakes (1992) are some examples of other regional agreements that address water quantity and quality issues. The SADC member states established the Protocol on Shared Watercourses in the Southern African Development Community (2000) based on the UN Convention.

### 10.3.4 River Basin Treaties

International water treaties have evolved over time. The history of international water treaties goes as far back as 2500 BC, when the two Sumerian city-states of Lagash and Umma agreed on ending a water dispute along the Tigris River (Wolf, 1998). More than 3600 international water treaties have been documented by the Food and Agricultural Organization of the United Nations that date back from AD 805. More than 400 water agreements have been signed since 1820, highlighting a number of positive trends in international river basin management over the past century. While individual sectors and countries may have exploited their riparian position or dominance at times throughout history, basin states have likewise demonstrated a remarkable ability to cooperatively capitalize upon their shared interests and to focus not only on the division of shared water resources themselves, but on the broader benefits from their use or control (Wolf et al., 1999). Here are some examples of such cooperation of shared water resources:

- As part of the 1957 Mekong River Agreement, Thailand agreed to provide financial support for a hydroelectric project in Laos in exchange for a proportion of the resultant power generation.
- According to the 1986 Lesotho Highlands Water Project Agreement, South Africa will support the financing of a hydroelectric/water diversion facility and in turn receive the rights to drinking water for its industrial heartland in Gauteng province.
- Uzbekistan and Kazakhstan will compensate in kind to the Kyrgyz Republic for the transfer of excess power generated during the growing season, under the 1998 Agreement on the Use of Water and Energy Resources of the Syr Darya Basin.
- A 1969 agreement between South Africa and Portugal on the Kunene River agreed on diversions of water entirely for human and animal requirements in Southwest Africa, as part of a larger project for hydropower.
- The 1994 Treaty of Peace allowed Jordan to store water in an Israeli lake while Israel leases Jordanian land and wells.
- India, under a 1966 agreement with Nepal, plants trees upstream in Nepal to protect its own, downstream, water supplies.
- In a 1964 agreement, Iraq "gives" water to Kuwait, "in brotherhood," without compensation.

- A 1957 agreement between Iran and the Union of Soviet Socialist Republics includes a clause that allows for cooperation in identifying corpses found in their shared rivers (Wolf, 1999).
- The 1987 Agreement on the Action Plan for the Environmentally Sound Management of the Common Zambezi River System allows for the future accession of additional riparian states to the treaty.
- The 1996 Treaty between India and Bangladesh on Sharing of the Ganga/Ganges Waters at Farakka, the 1986 Lesotho Highlands Water Project Agreement, and the 1992 Komati River Basin Treaty between South Africa and Swaziland are other examples of treaties with built-in flexibility including water allocation formulas that account for hydrologic fluctuations or changing needs and values.
- The treaties concluded in 1959 and 1966 between India and Nepal grouped projects related to irrigation, hydropower, navigation, fishing, and afforestation. The 1994 and 1995 agreements between Israel and Jordan and Israel and the Palestinian Authority, respectively, are other excellent examples where water resources are incorporated within a broader framework of peace in the region. Many agreements and treaties that were made in the twentieth century are notable in incorporating a broad framework of potential water use, beyond just a single benefit.

## 10.4 CONCLUSIONS

The examples presented in this chapter provide insight into the diversity of principles, policies, and institutions that guide the management of transboundary river basins. Transboundary agreements were developed based on a multitude of circumstances covering socioeconomic, political, climatological, and ecological aspects. Climate change will bring a wide array of challenges to freshwater resources, imposing complications in the global governance system. When those water resources cross borders and involve multiple political units, sustainable management of shared water resources in a changing climate will be especially difficult. While the shared water resources can be a source of conflict, they can also benefit many nations with appropriate negotiation and cooperation. Climate change increases the need for such cooperation, negotiations, and agreements to reduce the risk of potential future conflicts. Wide arrays of challenges such as water quantity, quality, and water transport; extreme events such as flood and drought management; and surveillance of aquatic habitat are commonly excluded in transboundary agreements. An integrated institutional framework that recognizes the interdependencies of all water uses to balance social, economic, and environmental objectives needs to be addressed through transboundary agreements. In addition to these factors, future climatic factors, extreme events, and their potential impacts on water quantity and quality also need to be part of future treaties and negotiations. Monitoring and evaluation of past treaties and agreements are often lacking. Joint monitoring programs can improve cooperation among river basins, states, and countries. Transboundary cooperation can broaden the knowledge base, enlarge the range of measures available for prevention, preparedness, and recovery, and so help identify better and more cost-effective solutions in this twenty-first century.

Different studies concerning climate change in all transboundary river basins are investigated by various researchers. The GCMs applied to these river basins are different, based on availability of GCMs' climate data for a different future time window. However, the GCMs' results represent the average hydrologic conditions for all river basins under the future climate. A changing climate affects all water-related sectors in different ways, both spatially and temporally. Therefore, disputes over the planning and utilization of suitable adaptation measures may increase. Additionally, adaptation measures in one sector may have retroactive, positive, or negative effects on one or more other sectors. To prevent possible conflicts and to foster common goals, cross- and interdisciplinary as well as integral approaches are necessary. Integral approaches also aim to enhance synergy effects, which should be sought. An example of a synergy effect is an increase in water retention areas that can lead to a higher groundwater recharge, a reduction of flood peaks, and positive effects for biodiversity. To improve the understanding of ongoing changes and their impacts, better observational data and data access are necessary. Quality assurance and the homogenization of data sets help to improve model projections and are a prerequisite for adaptive management required under conditions of climate change within transboundary regions or catchments. Shared waters can not only be a source of conflict but also a medium of cooperation, negotiations, and shared challenges.

## REFERENCES

Ahmad, Q. 2003. Towards poverty alleviation: The water sector perspectives. *International Journal of Water Resources Development* 19: 263–277.

Ansink, E. and Ruijs, A. 2008. Climate change and the stability of water allocation agreements. *Environmental and Resource Economics* 41: 249–266.

Arias, E., Albar, M., Becerra, M. et al. 2004. *Gulf of California/Colorado River Basin*. UNEP/GIWA Regional Assessment 27. Kalmar, Sweden: University of Kalmar, pp. 21.

Baillat, A. 2010. *International Trade in Water Rights: The Next Step*. London, U.K.: International Water Association.

Bakker, M.H. 2009. Transboundary river floods: Examining countries, international river basins and continents. *Water Policy* 11: 269–288.

Barnett, J. 2003. Security and climate change. *Global Environmental Change* 13: 7–17.

Beyene, T., Lettenmaier, D.P., and Kabat, P. 2010. Hydrologic impacts of climate change on the Nile River Basin: Implications of the 2007 IPCC scenarios. *Climatic Change* 100: 433–461.

Biswas, A.K. and Uitto, J.I. 2001. *Sustainable Development of the Ganges–Brahmaputra–Meghna Basins*. Tokyo, Japan: United Nations University.

Botterweg, T. and Rodda, D. 1999. Danube river basin: Progress with the environmental programme. *Water Science and Technology* 40: 1–8.

Burton, J.R. 1986. The total catchment concept and its application in New South Wales. *Hydrology and Water Resources Symposium*. Brisbane, Queensland, Australia: Griffith University, pp. 307–311.

Caponera, D.A. 1985. Patterns of cooperation in international water law: Principles and institutions. *Natural Resources Journal* 25: 563.

Cell, C.C. 2009. *Impact Assessment of Climate Change and Sea Level Rise on Monsoon Flooding*. Dhaka, Bangladesh: Climate Change Cell, Ministry of Environment and Forests.

Chakraborti, D., Rahman, M., Paul, K. et al. 2002. Arsenic calamity in the Indian subcontinent. What lessons have been learned? *Talanta* 58: 3–22.

Christensen, N.S. and Lettenmaier, D.P. 2007. A multimodel ensemble approach to assessment of climate change impacts on the hydrology and water resources of the Colorado River Basin. *Hydrology and Earth System Sciences Discussions* 11: 1417–1434.

Conway, D. 2005. From headwater tributaries to international river: Observing and adapting to climate variability and change in the Nile basin. *Global Environmental Change* 15: 99–114.

Cooley, H., Christian-Smith, J., Gleick, P.H., Allen, L., and Cohen, M. 2009. *Understanding and Reducing the Risks of Climate Change for Transboundary Waters*. Oakland, CA: Pacific Institute.

Cooley, H. and Gleick, P.H. 2011. Climate-proofing transboundary water agreements. *Hydrological Sciences Journal* 56: 711–718.

CWA. 1972. Clean Water Act (CWA). 33 U.S.C. §1251 et seq. 1972.

Dai, A., Qian, T., Trenberth, K.E., and Milliman, J.D. 2009. Changes in continental freshwater discharge from 1948 to 2004. *Journal of Climate* 22: 2773–2792.

De Stefano, L., Duncan, J., Dinar, S., Stahl, K., Strzepek, K.M., and Wolf, A.T. 2012. Climate change and the institutional resilience of international river basins. *Journal of Peace Research* 49: 193–209.

Döll, P. and Zhang, J. 2010. Impact of climate change on freshwater ecosystems: A global-scale analysis of ecologically relevant river flow alterations. *Hydrology and Earth System Sciences Discussions* 7: 1305–1342.

Drieschova, A., Giordano, M., and Fischhendler, I. 2008. Governance mechanisms to address flow variability in water treaties. *Global Environmental Change* 18: 285–295.

Faisal, I.M. 2002. Managing common waters in the Ganges–Brahmaputra–Meghna region: Looking ahead. *SAIS Review* 22: 309–327.

Fung, C.F. and Farquharson, F.A.K. 2004. Impacts of climate change and sea level rise on the Indian sub-continent. *Water, Department for International Development* 18: 4.

Giordano, M.A. 2003. Managing the quality of international rivers: Global principles and basin practice. *Natural Resources Journal* 43: 111–146.

Giordano, M.A. and Wolf, A.T. 2003. Sharing waters: Post-Rio international water management. In: *Natural Resources Forum*. Wiley Online Library, United Nations. Published by Blackwell Publishing, USA. pp. 163–171.

Gleick, P.H. 1992. Effects of climate change on shared fresh water resources. In: *Confronting Climate Change: Risks, Implications and Responses*, Cambridge University Press: Cambridge. pp. 127–140.

Goulden, M., Conway, D., and Persechino, A. 2009. Adaptation to climate change in international river basins in Africa: A review. *Hydrological Sciences Journal* 54: 805–828.

GRDC. 2013. Global Data and Map Products: Global Runoff Data Center (GRDC). Koblenz, Germany: Global Runoff Data Centre (GRDC) in the Federal Institute of Hydrology (BfG).

Harding, B.L., Sangoyomi, T.B., and Payton, E.A. 1995. Impacts of a Severe Sustained Drought on Colorado River Water Resources. Water Resources Bulletin, 31: 815–824. DOI: 10.1111/j.1752-1688.1995.tb03403.x.

Hoerling, M., Lettenmaier, D., Cayan, D., and Udall, B. 2009. Reconciling projections of Colorado River streamflow. *Southwest Hydrology* 8: 20–21.

Hutchins, W.A. 1977. *Water Rights Laws in the Nineteen Western States*. Washington, DC: Natural Resource Economics Division, Economic Research Service, United States Department of Agriculture.

ICPDR. 2012. ICPDR strategy on adaptation to climate change. International Commission for the Protection of the Danube River Vienna, Austria.

IPCC. 2007. Fourth assessment report, climate change 2007: Synthesis report, summary for policy makers. Geneva, Switzerland: Intergovernmental Panel on Climate Change.

Joint Rivers Commission Bangladesh. (JRCB). Treaty between the government of the People's Republic of Bangladesh and the government of the Republic of India on sharing of the Ganga/Ganges waters at Farakka. Available at: http://www.jrcb.gov.bd/ (accessed on May 23, 2013).

Lucarini, V., Danihlik, R., Kriegerova, I., and Speranza, A. 2007. Does the Danube exist? Versions of reality given by various regional climate models and climatological data sets. *Journal of Geophysical Research: Atmospheres (1984–2012)* 112: D13103.

Milliman, J., Farnsworth, K., Jones, P., Xu, K., and Smith, L. 2008. Climatic and anthropogenic factors affecting river discharge to the global ocean, 1951–2000. *Global and Planetary Change* 62: 187–194.

Milly, P., Betancourt, J., Falkenmark, M. et al. 2007. Stationarity is dead. *Ground Water News & Views* 4: 6–8.

Mirza, M., Warrick, R., Ericksen, N., and Kenny, G. 1998. Trends and persistence in precipitation in the Ganges, Brahmaputra and Meghna river basins. *Hydrological Sciences Journal* 43: 845–858.

Mohamed, Y., Hurk, B., Savenije, H., and Bastiaanssen, W. 2005. Hydroclimatology of the Nile: Results from a regional climate model. *Hydrology and Earth System Sciences Discussions* 2: 319–364.

Nordås, R. and Gleditsch, N.P. 2007. Climate change and conflict. *Political Geography* 26: 627–638.

Oak Ridge National Laboratory. (ORNL). 2008. LANDSCAN 2007 global population dataset. Oak Ridge, TN: Oak Ridge National Laboratory. Available at: http://www.ornl.gov/sci/landscan (accessed on May 24, 2013).

Pierce, D.W., Barnett, T.P., Santer, B.D., and Gleckler, P.J. 2009. Selecting global climate models for regional climate change studies. *Proceedings of the National Academy of Sciences* 106: 8441–8446.

Poff, N.L. and Zimmerman, J.K. 2010. Ecological responses to altered flow regimes: A literature review to inform the science and management of environmental flows. *Freshwater Biology* 55: 194–205.

Pradhanang, S.M., Anandhi, A., Mukundan, R. et al. 2011. Application of SWAT model to assess snowpack development and streamflow in the Cannonsville watershed, New York, USA. *Hydrological Processes* 25: 3268–3277. DOI: 10.1002/hyp.8171.

Pradhanang, S.M., Mukundan, R., Schneiderman, E.M. et al. 2013. Streamflow responses to climate change: Analysis of hydrologic indicators in a New York City water supply watershed. *Journal of the American Water Resources Association.* 49: 1308–1326. DOI: 10.1111/jawr.12086.

Ray, A., Barsugli, J., Averyt, K. et al. 2008. Climate change in Colorado: A synthesis to support water resources management and adaptation. Report for the Colorado Water Conservation Board. Boulder, CO: University of Colorado.

Salman, M. and Uprety, K. 2002. *Conflict and Cooperation on South Asia's International Rivers: A Legal Perspective*. Washington, DC: World Bank Publications.

Samal, N.R. and Mazumdar, A. 2005. Management of lake ecosystem. *Journal of Ekologia* 3(2): 123–130.

Samal, N.R., Rahman, R.V., Singh, A., Singh, V., and Singh, K.S.P. 2006. Water privatization: Pricing and conservation. In: Rao, V.B., Das, J.G., Sarala, C., and Giridhar, M.V.S.S. (eds.), *Second International Conference on Hydrology and Watershed Management, Hyderabad, India*. New Delhi, India: Center for Water Resource, Jawaharlal Nehru University, pp. 1197–1208.

Scheurer, K., Alewell, C., Bänninger, D., and Burkhardt-Holm, P. 2009. Climate and land-use changes affecting river sediment and brown trout in alpine countries—A review. *Environmental Science and Pollution Research* 16: 232–242.

Seager, R., Graham, N., Herweijer, C., Gordon, A.L., Kushnir, Y., and Cook, E. 2007. Blueprints for Medieval hydroclimate. *Quaternary Science Reviews* 26: 2322–2336.

Sommerwerk, N., Hein, T., Schneider-Jacoby, M., Baumgartner, C., Ostojic, A., Paunovic, M., Bloesch, J., Siber, R., and Tockner, K. 2009. The Danube river basin. In: Tockner, K., Robinson, C., and Uehlinger, U. (eds.), *Rivers of Europe*. Elsevier, London, pp. 59–112.

United Nations. 1978. Register of international rivers. *Water Supply Management* 2(1): 1–58. New York: Pergamon Press.

USGS. 2004. Climatic fluctuations, drought, and flow in the Colorado River. USGS Fact Sheet 3062-04.

Wolf, A. 1998. Conflict and cooperation along international waterways. *Water Policy* 1: 251–265.

Wolf, A.T. 1999. Criteria for equitable allocations: The heart of international water conflict. *Natural Resources Forum* 23: 3–30. Wiley Online Library United Nations. Published by Blackwell Publishing, USA.

Wolf, A.T., Natharius, J.A., Danielson, J.J., Ward, B.S., and Pender, J.K. 1999. International river basins of the world. *International Journal of Water Resources Development* 15: 387–427.

Wouters, P. 2000. *Codification and Progressive Development of International Law*. Dordrecht, the Netherlands: Kluwer.

Xu, J., Grumbine, R.E., Shrestha, A., Eriksson, M., Yang, X., Wang, Y., and Wilkes, A. 2009. The melting Himalayas: Cascading effects of climate change on water, biodiversity, and livelihoods. *Conservation Biology* 23: 520–530.

# 11 International Negotiations on Climate Change and Water

*Binaya Raj Shivakoti and Sangam Shrestha*

## CONTENTS

11.1 Introduction ........................................................................................................ 331
11.2 Climate Change Negotiations in General ...................................................... 333
11.3 International Environmental Negotiations Related to Water ....................... 336
11.4 Links between Climate Change and Water-Related Negotiations ............... 342
11.5 Bottlenecks of International Negotiations ..................................................... 344
11.6 Recent Developments and Emerging Paradigms
　　 in International Negotiations ......................................................................... 347
　　 11.6.1 COP Discussions and Post-Kyoto Regime ........................................ 347
　　 11.6.2 Rio+20 and Post-MDGs Development ............................................. 349
　　 11.6.3 New Paradigms in Water Management ............................................ 351
11.7 Conclusions ...................................................................................................... 353
References ................................................................................................................. 354

## 11.1 INTRODUCTION

Global climate change is an inconvenient environmental outcome of modern times. The Intergovernmental Panel on Climate Change (IPCC), through its Fourth Technical Assessment Report (AR4), makes a clear scientific warning that "warming of the global climate system is unequivocal," its impacts are observable, and time is running out to control the upward warming trajectory (IPCC 2007). The certainty of this warning has become even stronger in the latest Fifth AR (AR5), which maintains that it is mainly human action that is responsible for the climate change (IPCC 2013a). Excessive use of fossil-based energy has resulted in the accelerated accumulation of greenhouse gases (GHGs)—a key determinant for global warming and climate change—in the environment at a level of 400 ppm $CO_2$* at the time this chapter was drafted. Equivalent figures could be much bigger if all of the GHG types were added to that value. By 2010, the world had already exceeded 44 Gigatons (Gt)

---

* On May 10, 2013, daily measurements of $CO_2$ at a US government NOAA lab on Hawaii have topped 400 parts per million for the first time (BBC, 2013).

of $CO_2$ equivalent ($CO_2e$) emission, which is considered crucial for pegging global temperatures to 2°C or below over the twenty-first century, by a gap of 5 $GtCO_2e$ (UNEP 2012).

Water, which envelops the vast majority of the earth's surface, is one of the key mediums through which signals of climate change impacts will be felt by humans and the environment (Bates et al. 2008; Howard et al. 2010; Stern 2006; Xu et al. 2007). Water is involved in all components of the global climate system and climate change affects water through a number of mechanisms (Bates et al. 2008). Current and predictable trends point toward deteriorating water problems in the future and managing limited freshwater resources for over 7 billion people, while keeping ecosystem services intact and alive, will continue to become increasingly complex (Biswas 2008; WWAP 2012). The complexity is compounded by trade-offs arising from interlinks with development sectors such as agriculture, energy, industry, and transportation, as well as by multiple stressors such as urbanization and land use changes, pollution, sedimentation and degradation of natural water bodies, modification of watersheds, and water courses and wetlands. Climate change is a brand new stressor adding to the existing complexity of water management. Unlike other nonclimatic stressors, climate change is exceptional in that it adds high uncertainty to our prediction capacity about hydrologic conditions, and hence, about availability of water resources in space and time.

Various measures of slowing down global warming as well as finding effective ways to minimize potential negative impacts of climate change are being considered in almost every corner of the globe. In the meantime, water management has evolved to become multidimensional and diminishing water security is a major determinant for sustainable development. Alternative measures to improve access to water and avoid negative impacts of resource mismanagement are being employed. International negotiations have emerged as an effective means to handle common but delicate issues of climate change and water. Among others, international environmental negotiations serve as an important catalyst to increase awareness about the scale of the problem, to draw attention to threats and risks from the negative impacts, and to understand casual interrelationships of mutual actions. They boost the building of consensus and the forging of agreements on the responsibilities to pave paths for cooperation. Thus, they propose appropriate individual or collective actions that could assist in bringing major policy shifts at the regional and national levels. At this peak of globalization, international negotiations in tackling contentious issues of climate change, water resource management, and other environmental issues have become an absolute prerequisite.

For the past several years, there has been a surge in intensive debates and discussions on climate change and its potential impacts at various levels and in different forms. Those discussions have emerged out of concern for accumulated GHGs in the environment, which have a cascading impact, spreading to almost all areas and sectors that are sensitive to climate, directly or indirectly. In particular, climate change is feared to retard efforts to sustainable development, and in the worst case, could reverse the progress made so far in eradicating poverty and improving the livelihood of millions. Until now, climate change negotiations have concentrated mainly on two fronts—mitigation and adaptation. Mitigation deals with policies and actions for controlling global warming and climate change through an adjustment in the global carbon cycle by reducing GHG emissions, enhancing carbon sinks, and adopting other unconventional geo-engineering concepts to prevent further warming of the biosphere. Adaptation deals with initiatives or measures to reduce vulnerability of human and natural systems from the negative impacts of climate change or to exploit beneficial opportunities from climate change. Cobenefit is the third dimension, or by-product, of climate negotiations that could overlap with mitigation and/or adaptation but could be complementary to the mainstream socioeconomic developmental efforts.

Climate change is already a prominent issue during water-related negotiations or discussions such as water forums, conferences, or dialogues. But the same cannot be claimed in the case of climate change negotiations, where water issues are encountered intermittently or find their space in ex post discussions, directly or indirectly. However, despite close links between climate change and water, it is hard to single out any international climate change negotiation that was devoted specifically to water issues. This relatively unaddressed part of climate negotiations leads us to inquire about the need, relevancy, and issues to be addressed through water-focused climate change negotiations. The intent of this chapter is to present key negotiations that are considered turning points to highlight issues surrounding climate change and water, irrespective of their potential links. The chapter will further concentrate on exploring links, barriers, and recent developments on international negotiations on water and climate change. Finally, the chapter will conclude by discussing discourses of future negotiations on curtailing climate change and improving water management as a win-win strategy. While attempts have been made to grasp major developments that are essential in developing a comprehensive understanding by the readers on climate and water-related international negotiations, by no means should the contents be treated as exhaustive.

## 11.2 CLIMATE CHANGE NEGOTIATIONS IN GENERAL

Over the last century, there were several efforts to understand the mechanism of climate change—particularly the contribution of GHGs. Despite these efforts, formal discussions and resultant actions were out of mainstream focus amid other developmental priorities and new kinds of acute environmental problems that were physically apparent, such as pollution of air, freshwater, and sea; ozone layer depletion; deforestation; overfishing and habitat destruction; and so forth. In fact, climate change was hardly an agenda at the *UN Conference on the Human Environment (UNCHE)* held at Stockholm in 1972, which was a major international environmental negotiation that led to the formation of the United Nations Environment Program (UNEP). Major responses in climate change have only occurred at the latter quarter of the twentieth century when two major initiatives occurred under the United Nations (UN)—IPCC and the UN Framework Convention on Climate Change (UNFCCC) (see Table 11.1 for the major developments). The IPCC was established by the UNEP and the World Meteorological Organization (WMO) in 1988 to provide the world with a clear scientific view on the current state of knowledge in climate change and its potential environmental and socioeconomic impacts (IPCC 2013b).

## TABLE 11.1
## Timeline of Major Climate Change Negotiations

| Year | Major Developments and Outcomes | $CO_2$ (ppm)[a] |
|---|---|---|
| 1800–1870 | Beginning of Industrial Revolution. | 290 (around) (Weart 2013) |
| 1938 | G.S. Callendar claim about GHG warming (Weart 2013). | |
| 1958 | High-accuracy measurements of atmospheric $CO_2$ concentration initiated by Charles David Keeling at Mauna Loa Observatory in Hawaii (IPCC 2007). | 315.71 |
| 1967 | International Global Atmospheric Research Program established (Weart 2013). | 323.04 |
| 1979 | The first World Climate Conference (WCC). World Climate Research Program launched (UNFCCC 2013; Weart 2013). | 337.96 |
| 1988 | IPCC established (UNFCCC 2013). | 352.22 |
| 1990 | IPCC's First Assessment Report (AR) released. IPCC and 2nd WCC call for a global CC treaty (UNFCCC 2013). | 355.39 |
| 1991 | First meeting of the Intergovernmental Negotiating Committee (INC) (UNFCCC 2013). | 357.16 |
| 1992 | UNFCCC treaty agreed at Rio Earth Summit (UNFCCC 2013). | 357.81 |
| 1994 | UNFCCC enters into force (UNFCCC 2013). | 359.97 |
| 1995 | The first Conference of the Parties (COP1) in Berlin IPCC Second AR released (UNFCCC 2013). | 361.64 |
| 1997 | Kyoto Protocol (KP) adopted at *Conference of Parties 3 (COP3)* meeting held in Kyoto—a binding agreement among nations to curb GHG emission through a number of market and non-market-based mechanism (UNFCCC 2013). | 364.57 |
| 2001 | IPCC Third AR released. Marrakesh Accords adopted at COP7, detailing rules for implementation of the KP (UNFCCC 2013). | 372.12 |
| 2005 | Entry into force of the KP (UNFCCC 2013). | 380.38 |
| 2006 | Stern Review on the Economics of CC published—emphasis on early action in mitigation and adaptation, benefits of action outweigh cost of inaction (Stern 2006). | 382.56 |
| 2007 | IPCC fourth AR released; IPCC awarded 2007 Nobel Peace Prize at end of the year. At COP13, Parties agreed on the Bali Road Map, which charted the way toward a post-KP outcome in two work streams: Ad Hoc Working Group (AWP)-KP and AWG-Long-Term Cooperation Action under the Convention (UNFCCC 2013). | 384.34 |
| 2009 | Attended by close to 115 world leaders at the high-level segment, Copenhagen Accord drafted at COP15 recognizes scientific view on limiting warming below 2°C; countries later submitted emission reduction pledges or mitigation action pledges, all nonbinding developed countries agreed to support a goal of mobilizing US$100 billion a year by 2020 to address the needs of developing countries (UNFCCC 2013). | 388.52 |

## TABLE 11.1 (continued)
## Timeline of Major Climate Change Negotiations

| Year | Major Developments and Outcomes | $CO_2$ (ppm)[a] |
|---|---|---|
| 2010 | Cancun Agreements drafted and largely accepted by the COP (COP16)—comprehensive measures for mitigation, adaptation, financing, technology transfer, and capacity building (UNFCCC 2013). | 391.02 |
| 2011 | The Durban Platform for Enhanced Action at COP17: governments clearly recognized the need to draw up the blueprint for a fresh universal, legal agreement to deal with climate change beyond 2020 (UNFCCC 2013). | 392.45 |
| 2012 | COP18 sets out a timetable to adopt a universal climate agreement by 2015, to come into effect in 2020. The Doha Amendment to the KP adopted by the meeting of Parties to the KP (CMP) at CMP8 saw the launch of a second commitment period of the KP from January 1, 2013, to December 31, 2020 (UNFCCC 2013). | 394.37 |
| 2013 | On May 10, 2013, NOAA and Scripps Institution of Oceanography (SIO), for the first time detected daily $CO_2$ average concentration temporarily reaching 400 ppm (BBC 2013). | 399.89 (May average) |
| 2013 | Release of first volume, Working Group I of the IPCC Fifth Assessment Report (AR5) (IPCC 2013a). Subsequent volumes are expected to be published during 2013 and 2014. | |

[a] Mean atmospheric $CO_2$ concentration (ppm) for March 15 observed at Mauna Loa Observatory by Scripps Institution of Oceanography (http://scrippsco2.ucsd.edu/data/atmospheric_co2.html).

The UNFCCC is a major international treaty on climate change that took place at the Rio Earth Summit in 1992 when countries joined to cooperatively consider what they could do to limit the average global temperature increase and the resulting climate change, and to cope with whatever impacts were, by then, inevitable (UNFCCC 2013). The UNFCCC entered into force on March 21, 1994, and it has near-universal membership with 195 countries, that is, Parties to the Convention, which have ratified the convention.

The IPCC and the UNFCCC are closely linked, as shown in Figure 11.1. The IPCC has the status of an intergovernmental organization observer of the UNFCCC. Scientific evidence brought by the IPCC has a decisive role in contributing to the creation of the UNFCCC treaty (IPCC 2013b). The IPCC inputs are often used as a basis for negotiations and agreements that take place during the meeting of Conference of Parties (COPs), which is the supreme decision-making body of the convention (IPCC 2013b; UNFCCC 2013). Notable progress has been made in climate change negotiation after the inception of the IPCC and the UNFCCC. They include, among others, identification of issues that need collective decisions, establishing emission control targets and timetables, establishing a series of mechanisms to deal with climate change, monitoring of each country's activities, and establishing formal relationship with scientific communities and other actors (Gupta 2012).

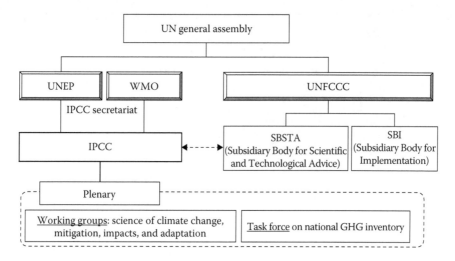

**FIGURE 11.1** Structures of the IPCC and the UNFCCC and their relationship.

By 2007, the IPCC had published its AR4—its flagship publication—and its first volume of AR5 (Working Group I) was unveiled in September 2013. Similarly, the UNFCCC has also matured by organizing annual COPs; so far, 18 COP meetings have been conducted. (KP, Nairobi Work Programs (NWP) on Impact, Vulnerability and Adaptation, Bali Action Plan, Cancun Agreement, and Doha Amendments are some of the important outcomes of the COP meetings. Besides that, both the IPCC and the UNFCCC have expanded their roles by producing key publications, acting as repositories of climate change knowledge, and facilitating postnegotiation and implementation of decisions. The IPCC and the UNFCCC have been instrumental in bringing key actors together such as technical experts, planners and policy makers, leaders and decision makers, civil society, private sectors and businesses, nongovernmental organizations (NGOs), and development organizations that are spread across the globe. For instance, over 3500 experts, coming from 130 countries, have contributed to the preparation of the IPCC AR4 in 2007 (IPCC 2013b). Similarly, the UNFCCC is composed of 43 industrialized developed country parties (Annex I, including the European Union, EU), 154 developing countries (non-Annex I, including 49 least-developed countries), and observer organization (1598 NGOs and 99 intergovernmental organizations) (UNFCCC 2013). This pool of human resources has contributed significantly in shaping the discourse of climate change negotiations and building consensus.

## 11.3 INTERNATIONAL ENVIRONMENTAL NEGOTIATIONS RELATED TO WATER

International negotiations on water are not so old, yet they have attracted global attention earlier than climate change issues. The chronology of water-related negotiation reveals three key priority shifts during this period: resource development (until the 1980s), resource management (1980–1990), and enhancing management through more integrated focus (since the 1990s) (Savenije and Van der Zaag 2008). Improving governance is a relatively recent policy focus as water management crisis is often related to the crisis of governance (WWAP 2006). Water, along with other environmental issues, gained significant highlight, particularly post–World War II, the period of rapid industrialization and population boom. It was also the period when negative externalities of rapid economic development became apparent in all three environmental dimensions—air, water, and land. The degradation of environmental commons across the globe has ignited discussion at multiple levels while the situation was seemingly slipping out of individual nation's control and capacity. Broader consensus at the international level to facilitate and cooperate on environmental issues was indispensible to handle the scale of the problem, minimize collateral damage from each other's activities, and safeguard environmental services for future human welfare. The creation of the UN and its specialized agencies in 1945 could be considered an early initiation of the modern international momentum to cooperate on environmental issues (Sands 2003). Table 11.2 highlights the important international processes that are related to water, directly or indirectly.

The UN has led several negotiations and resultant agreements and laws at the international level over the last few decades that have explicitly or implicitly mentioned water issues. Most of the responses were ad hoc and piecemeal types, while the progress has been gradual. Earlier attempts of intergovernmental environmental action could be followed after the Economic and Social Council (ECOSOC) resolution convening the 1949 UN Conference on the Conservation and Utilization of Resources (UNCCUR), in which water was one of the issues including integrated development of river basins (Sands 2003). Another notable international negotiation was the UNCHE at Stockholm, which could be regarded as a major step toward the international-level declaration devoted exclusively to environmental issues. A number of similar negotiations followed that underpin the seriousness of the environmental problems and need cooperation on an international level to deal with the problems. Among them, the Mar del Plata Action Plan in 1977 was the most comprehensive discussion dedicated to water, which formed the basis for future water-related negotiations (Biswas 1988). Later, the Brundtland Commission Report on "Our Common Future" highlighted the imminent environmental crisis, including water issues, and introduced "sustainable development," which became the unidirectional policy focus for international environmental diplomacy. "Our Common Future" eventually formed the basis for negotiations during the 1992 Earth Summit at Rio de Janeiro, which culminated in a comprehensive agreement "Rio Principles" and formulated a comprehensive plan of action "Agenda 21" on sustainable development. Just before the Rio Earth Summit, another *International Conference on Water and Environment (ICWE)* was held at Dublin, which resulted in four sets of principles that recognized, firstly, water as a finite and vulnerable resources, secondly underscored participatory management approaches and, thirdly, women's role at the center of that, and finally, valued water as an economic resource while recognizing the rights to drinking water and sanitation. Agenda 21, Chapter 18, which is dedicated explicitly to water-related issues and agendas, draws heavily on the Dublin principles.

## TABLE 11.2
## Major International Negotiations Related to Water

| Year | Major Development | Relevancy |
|---|---|---|
| 1949 | UNCCUR | Identified water, including integrated development of river basins, among the major issues to be covered under resource conservation and development. |
| 1972 | UNCHE, Stockholm/the Stockholm Declaration | First international gathering to address multiple global environmental problems. |
| 1972 | Establishment of UNEP | UN program on environment. |
| 1977 | Mar del Plata Action Plan | First of its kind, the most comprehensive discussion on water by UN at that time. Benchmark for future water negotiations. |
| 1980s | UN International Drinking Water Supply and Sanitation Decade | Established the target of universal coverage of a safe water supply and sanitation by 1990. |
| 1987 | Brundtland Commission Report on "Our Common Future" | Defined the Sustainable Development wherein water has also been highlighted. Basis for Rio Earth Summit in 1992. |
| 1991 | Establishment of Global Environmental Facility (GEF) | Financing mechanism to assist in the protection of the global environment and to promote environmental sustainable development. |
| 1992 | Dublin Statement on Water and Sustainable Development (ICWE) | Established four guiding principles for local, national, and international water resource management. |
| 1992 | United Nations Conference on Environment and Development (UNCED), alternatively Rio Earth Summit | In addition to agreeing on Rio principles on sustainable development and eradicating poverty, Agenda 21 (Chapter 18) brought various agendas related to water including Integrated Water Resource Management (IWRM) as a centerpiece idea. |
| 1995 | World Summit for Sustainable Development (Copenhagen Declaration) | Reduce poverty by providing water supply and sanitation. |
| 1996 | Established World Water Council (WWC) | High-level international platform on water; organizes World Water Forum (six forums by 2012 at the interval of 3 years), which among others also engages political leaders/decision makers such as Ministerial Declaration. |
| 1996 | Establishment of Global Water Partnership (GWP) | Global network to foster IWRM and to support the sustainable development and management of water resources at all levels. |
| 2000 | UN Millennium Declaration, later on known as Millennium Development Goals (MDGs) | Set eight specific goals and targets to meet those goals, including one target for water supply and sanitation, for 2015; water is connected with all eight goals, directly or indirectly. |

## TABLE 11.2 (continued)
## Major International Negotiations Related to Water

| Year | Major Development | Relevancy |
|---|---|---|
| 2001 | International Conference on Freshwater (Bonn) | Similar event, marking the 10th anniversary of ICWE that provided inputs to second World Summit on Sustainable Development (WSSD), alternatively Johannesburg Plan of Implementation (JPoI). |
| 2002 | UN Committee on Economic, Social and Cultural Rights adopted General Comment No. 15 on The Right to Water | States human right to water as indispensible for leading a life with human dignity and prerequisites for realizing other. For legal basis, it defined the right to water as the right of everyone to sufficient, safe, acceptable, and physically accessible and affordable water for personal and domestic uses. |
| 2002 | WSSD, alternatively JPoI (Rio+10) | Ask country to develop integrated water resource management (IWRM) and water efficiency plans by 2005; developing and implementing national and regional strategies, plans, and programs on integrated river basin, watershed, and groundwater management. |
| 2003 | Established UN-Water | UN interagency coordination mechanism for all freshwater-related issues. |
| 2003 | International Decade for Action "Water for Life" 2005–2015 started | By 2015, promote efforts to fulfill international commitments (MDGs, JPoI-WSSD, and Agenda 21) made to water and water-related issues by furthering cooperation at all levels. |
| 2010 | UN General Assembly adopts, through Resolution 64/292, The Human Right to Water and Sanitation | Build on earlier background efforts, including The Right to Water, to formally recognize the right to safe and clean drinking water and sanitation as a human right that is essential for the full enjoyment of life and all human rights. It calls upon states and international organizations to provide financial resources, and aid capacity-building and technology transfer to help countries to scale up efforts to provide safe, clean, accessible, and affordable drinking water and sanitation for all. |
| 2012 | UN Conference on Sustainable Development (Rio+20) | Issued outcome document "Future We Want" that, including others, highlights setting Sustainable Development Goals (SDGs) and transition to green economy (GE). |

Since the 1992 Rio Earth Summit, water issues have been hovering across three thematic issues, which are also closely tied with climate change impacts. The first issue is the lack of water supply and sanitation, which has remained as a lingering case since its formal admission in the Mar del Plata Action Plan (1977) and subsequently in the declaration of the 1980s as the International Drinking Water Supply and Sanitation Decade, Dublin Principle, and Rio Earth Summit. Continuous policy focus on water supply and sanitation benefitted a significant number of people, but it was not adequate to match with the rapid growth in population (WWAP 2003). After the Rio Earth Summit, water and sanitation agendas continued to remain in the limelight. For instance, the target of universal access was readopted for 2000, and, subsequently, the target was again reformulated by the Water Supply and Sanitation Collaborative Council (WSSCC) as part of the process to set targets for Vision 21: A Shared Vision for Hygiene, Sanitation, and Water Supply (ibid). The water supply and sanitation agenda regained its new height when the world leaders adopted the UN Millennium Declaration in 2000, which was later on known as the Millennium Development Goals (MDGs), of which, Target 7.C aimed to "halve, by 2015, the proportion of the population without sustainable access to safe drinking water and basic sanitation." In 2002, the UN Committee on Economic, Social, and Cultural Rights adopted General Comment No. 15 on The Right to Water that paved a way for the UN General Assembly, through Resolution 64/292, to adopt The Human Right to Water and Sanitation. The 10th anniversary of the Rio Earth Summit (Rio+10), also known as the Johannesburg Plan of Implementation (JPoI), affirms the UN's commitment to fully implement Agenda 21 alongside the MDGs, including the water and sanitation targets. Similarly, the International Decade of Action (2005–2015) "Water for Life" again emphasized on increased cooperation to implement international commitments. Impressive achievements have been made since the promulgation of the MDGs, as over 2 billion and 1.8 billion people gained access to improved drinking water and sanitation, respectively, between 1990 and 2010 (WHO/UNICEF 2012). Despite the progress, drinking water and sanitation are unfinished tasks because over 750 million people are still without access to improved drinking water sources and 2.5 billion lack improved sanitation facilities (ibid). There is growing concern that climate change could undermine the achievements in improving access to water supply and sanitation, and, therefore, needs proper attention during discussions.

The second issue is the implementation of the concept of Integrated Water Resources Management (IWRM), formally adopted for the first time at the 1992 Rio Earth Summit, into national water plans and policies that has been promoted as an undisputed means to address a range of water management problems in a sustainable way. The establishment of the GWP in 1996 is one of the concrete steps taken by international organizations to foster IWRM across the globe. The notion of IWRM, as its name implies, views water management problems and issues holistically as multidimensional, multisectoral, and multiregional, involving multi-interest, multi-agenda, and multicause purposes, which cannot be resolved by formulating strategies in silos but rather require appropriate multidisciplinary, multi-institutional, and multistakeholder coordination (Biswas 2008). According to the recent assessment across 130 countries, there has been widespread adoption of integrated approaches by 80% of countries that have embarked on reforms to improve the enabling environment as stated in Agenda 21 and affirmed in JPoI; 65% countries have already developed the IWRM plan and 34% are in an advanced stage of implementation as stated in JPoI (WWAP 2012). At present, the main challenge is in formulating and implementing IWRM principles in the real world in a timely, cost-effective, and socially acceptable manner (Biswas 2008), because without an operational framework, IWRM will continue to remain elusive and people will continue to derive its meaning differently. Despite disparities over operational implications, IWRM principles are considered effective in the planning and implementation of climate change impacts on water, especially adaptation (Bates et al. 2008; Cap-Net 2009).

The third issue is about managing shared, or transboundary, water resources that are not confined by political boarders but governed by natural geography. Transboundary freshwaters cover 45% of the world's land mass, connecting two or more countries/states in water resources above (surface) and below (groundwater) the earth's surface (INBO/GWP 2012). An estimated 148 states worldwide have international basins within their territory and 21 countries lie entirely within them, while to date 273 transboundary aquifers have been identified (WWAP 2012c). Transboundary issues are dictated by both geographical and political realities. So, any decisions on resource use and development could have short- and long-term implications to riparian nations or states. Negotiations on transboundary issues could be identified as far back as 1814; since then, over 300 bilateral and multilateral agreements on cooperative use and development of transboundary waters have been completed (Vollmer et al. 2009). Among them, the adoption of Helsinki Rules, on the Uses of the Waters of International Rivers in 1966, was an important step to identify the rights and duties in a comprehensive manner. Despite numerous negotiations in the past, international-scale formalization of transboundary negotiations materialized in 1997 during the UN Convention on the Law of the Non-navigational Uses of International Watercourses, nearly 27 years after the commencement of its development. For groundwater, such an agreement is even more recent as only in December 2008 the UN Assembly adopted a resolution on the Law of Transboundary Aquifer, which, among others, recommends the concerned states to make appropriate bilateral or regional arrangements for the proper management of their transboundary aquifers. Available bilateral or multilateral agreements, commissions, and laws on transboundary issues are intended, among others, to demarcate each other's rights and obligations to conserve the resource and protect the dependent environment during developments, minimize harm to other parties, settle disputes, ensure equitable utilization of benefits, provide exchange of information, and enhance cooperation. Yet, to date, transboundary waters remain an unresolved and critical issue in international negotiations as it is closely tied with the water security of individual countries. It also has similar management challenges as other pertinent water issues, that is, dealing with multidimensional issues in an integrated manner. Most importantly, multiple drivers, and increasing competition for water, such as for food, energy, and associated uncertainties, are likely to escalate existing transboundary issues to a new height (WWAP 2012). Climate change could further exacerbate the tension, especially on issues of sharing limited water. IWRM is again viewed as indispensible to enhance transboundary water resource management (INBO/GWP 2012) because coordination and cooperation could be enhanced only through proper integration.

Although these three issues are seemingly distinct, they scale closer on the issue of governance. Recently, water governance has gained wider policy attention on the backdrop of past policy failures, which were mainly dominated by resource development. Deficiency in the management focus and the resultant water crisis seen in many regions have led to rethinking water crisis seen in many regions as a crisis in governance, which is either in a state of confusion or is nonexistent in many regions (WWAP 2006). New facets such as a balance between top–down and bottom–up approaches, increasing the sphere of decision making by engaging relevant stakeholders, and strengthening institutional frameworks, equity, transparency, accountability, ethical concerns, and integration were identified as prerequisites for improving water governance. However, the big challenge posed by climate change is to design a governance framework that could skillfully tame the past management failures while making water resource management more adaptive to future changes.

## 11.4 LINKS BETWEEN CLIMATE CHANGE AND WATER-RELATED NEGOTIATIONS

The first UNCED Rio Earth Summit was a major international platform wherein both climate change (UNFCCC) and water issues (Chapter 18, Agenda 21) gained wider global attention than ever before. In fact, "an impact of climate change on water resources" was one of the proposed programs of Agenda 21 (Chapter 18, 18.5.G), which clearly underscores the importance of enhancing the information base and coordination in three areas of climate change and water: understanding and quantifying threats of climate change on freshwater, facilitating the implementation of national countermeasures against identified impacts, and studying the potential of climate change in areas prone to floods and droughts. Following this, discussions on climate change and water have progressed at different levels. The IPCC has also brought out a special report (Bates et al. 2008) and a specific chapter in AR4 (Ch3, WGII) that show a close link between water and climate change. To date, the majority of climate change negotiations are concentrated more around mitigation and relatively less on adaptation. Although water is connected to both realms, it probably weighs more on the adaptation side of discussions. For instance, water resources has been laid out in Article 4.1(e) of the UNFCCC as one of the focus areas where integrated approaches are necessary for preparing adaptation to climate change impacts. Water issues are mainly included in the context of adaptation activities under the National Water Academy (NWA), national communications, and national adaptation programs of action (NAPAs). In particular, the Subsidiary Body for Scientific and Technological Advice (SBSTA) has agreed to discuss the impact of climate change on water resources and IWRM under NWA. In the 34th session, SBSTA requested the secretariat to prepare a technical paper on water and climate change impacts and adaptation strategies under the NWP before the 35th session. The technical report points toward adopting IWRM principles for planning water adaptation to climate change by encouraging participatory and holistic processes, while transboundary cooperation is seen as both necessary and beneficial for climate change adaptation (UNFCCC 2011a). The UNFCCC has also published a synthesis of actions on climate change and freshwater by NWA partner organizations (UNFCCC 2011b).

# International Negotiations on Climate Change and Water 343

To date, 66 of the 209 NWP partner organizations are engaged in actions related to water resources and a total of 45 Action Pledges related to water resources have been made by partner organizations under the NWP (ibid). The importance of water adaptation to climate change has also been covered under NAPA, which serves as a direct channel to communicate the immediate and urgent adaptation needs of the LDCs. As of 2011, 74 priority NAPA projects of 485 projects submitted by LDCs were related to water resources (UNFCCC 2011a). Seemingly too much focus on adaptation does not imply that water and mitigation are incompatible. Energy is consumed, directly or indirectly, in almost all life cycle processes of water services such as uptake, transport, usage, and cleaning (pre- and post-uses). Similarly, dams, reservoirs, and polluted water bodies could emit highly potent GHGs like methane or nitrous oxide. Moreover, low carbon mitigation policies such as cultivation of biofuels, concentrating solar power (CSP), geothermal, hydropower, or REDD+ (Reducing Emissions from Deforestation and Forest Degradation) will also influence water balance through withdrawal and consumption. The KP, the only binding agreement on climate change, also identifies wastewater management (Annex A), though implicitly, based on the identified key categories of GHG emission sources/sinks by the IPCC, as one of the GHG sources/sinks. In fact, under Clean Development Mechanism (CDM), Joint Implementation (JI), or International Emission Trading (IET) of the KP, there are a significant number of projects aimed at reducing GHG emission through renewable energy production, which are directly dependent on the availability of water.

While the subject of water is not a central issue in international climate change discussions, there is an increasing tendency to discuss about climate change impacts in water-centric discussions, private and government organizations dealing with water, bilateral and multilateral donor agencies (such as the World Bank, Asian Development Bank, GEF), and global or regional institutes dealing with interdisciplinary areas such as United Nations Development Programme, Food and Agriculture Organization, and UN-Water. Nowadays, it is rare to find major water-related negotiations, events, or conferences lacking heated discussions on climate change impacts on water. For instance, the 5th World Water Forum in 2009 had "Adapting to Climate Change" as one of its key focuses. *World Water Weeks*, held annually at Stockholm, have also maintained water and climate discussions, delivering recommendations to deal with climate and water management. Major water publications such as the third (WWAP 2009) and fourth (WWAP 2012) reports by the UN World Water Assessment Program are more specific to climate change impacts. The rise in focused climate change discussions, however, has very limited impact on core climate change negotiations such as COP meetings. There has been only limited progress toward transferring outcomes of negotiations within the water community to key climate change discussions. In 2009, a group of stakeholders held a *Dialogue on Climate Change Adaptation for Land and Water Management* at Nairobi, cohosted by the UNEP and Denmark government, which agreed on five principles in the areas of sustainable development, resilience, governance, information, and economy/finance. These principles were aimed at feeding into COP15 negotiations. Similarly, in 2009, *World Water Weeks* at Stockholm outlined a message for COP15 specifically in the areas of water and adaptation, financing, and capacity building, among others.

A multistakeholder coalition, called Water and Climate Coalition, of 11 member partners are calling for a work program on water and climate under the UNFCCC that will have five specific functions: creating global policy discourse, establishing guiding policy principles, provide advice and guidance on financing, enhance implementation of priorities, and building coherence between global agreements and arrangements (WCC 2010). The GWP is also addressing the linkages between water security and climate-targeting COP events, especially since COP15 (GWP 2012a).

## 11.5 BOTTLENECKS OF INTERNATIONAL NEGOTIATIONS

Assessments done by the IPCC, and elsewhere, have provided a strong background about the science of climate change, thereby eliminating potential disagreements to a greater extent. Available scientific claims have been quite successful in drawing international attention, yet responses in orienting actions toward curbing the GHG emission trend are slow. Accumulation of GHGs is undesirably taking an upward trend, which is cultivating a pessimism among scientific and policy actors about the success of future negotiation challenges (Gupta 2012). Experiences till date have shown that global climate change negotiations are not a trivial affair as it involves long and curvy paths, confrontation with frequent deadlocks, and dealing with frozen-up relationships. Fundamentally, climate change is a global collective-action problem that demands collective efforts by various actors across the globe (Ostrom 2010). However, committing all to spend on, often costly, efforts to stop GHG emission is a practical dilemma because an individual may not see an incentive to do that without a strong enforceable mechanism or higher socioeconomic cobenefits such as through carbon trading. Surrounded with this reality, the core problem to negotiation, therefore, is anchored at building consensus on sharing roles and responsibility, "who should do what," in a justifiable manner (Gupta 2012). So far, the divide on sharing of roles and responsibility is heavily differentiated between developed and developing parties that have given rise to unwanted knee-jerk suspicion, defensiveness, and misunderstanding (Ghosh and Woods 2009). Parties are often involved in strategic bargaining such as maintaining a stringent position, strongly holding over the position, and allowing only few concessions that are intended to freeze the process and close opportunities for coordination and cooperation (DIIS 2011). On issues of equity, the developing South is taking a historical perspective on sharing responsibility and wants compensation for accumulated emission from the developed North, while the latter is arguing for meaningful participation of the former's increasing share on current and future GHG emissions (ibid.). Developing countries have gone even further by advocating per capita–based emission rights, which will allow comparative advantage over the developed North due to the high concentration of population in developing countries. Different stances maintained by each group of parties have therefore provided an excuse to escape from making stronger commitments on GHG emission reduction and maintain status quo. All of these have essentially resulted in little progress on negotiations.

At present, a source of major misunderstanding is not about the fact that climate is changing, and anthropogenic GHG emission is a major contributor to this, but it is the lack of preciseness about the timing and magnitude of consequences lying ahead, more at the local, subnational, or river basin scale, and less at the country, regional, or even global scale. No specific actions could be targeted other than retrofitting ourselves through experimenting and learning from a broad range of strategies. Lack of clarity on the course of action is not only limited to adaptation, it is also relevant to mitigation. In case of adaptation, there are debates about the effectiveness of adaptation and the recovering investments made in it. Such concerns have been addressed to some extent, such as by the Stern Review, which has shown that benefits from adaptation are immediate if well planned (Stern 2006). On the mitigation side, there is a problem of measuring the impacts of individual actions. First, GHG is an invisible entity without a competitive market value, but reliable monitoring and an incentive-based market mechanism could create market demand for it. Second, it is not straightforward to draw a fairly reliable time horizon required to stabilize global climate back to normal even if GHG reductions were brought back to preindustrial level; nor is it easy to precisely predict consequent risk, such as warming level, at different locations because of failure to control GHG. Finally, it is about the visibility of actions because individual contributions are often too minute to be weighed with massive-scale GHGs that are open dumped globally and that have accumulated into the atmosphere. These technicalities hamper the identification and scalability of the level of required actions in a particular setting, while they have created a fertile ground for chaotic debates and loggerheads by a certain group of actors who raise skepticism about the climate change phenomenon. Some, notably the United States and some emerging economies, are cautious over binding agreements or commitments on climate change mitigation as any such move could conflict with their economic progress, which is heavily dependent on carbon consumption. Likewise, others fear that indulging in climate change negotiations could be a threat to sovereign rights due to the resultant compromise over sharing of sensitive information.

Climate negotiations are a continuous process and experiences have shown that they involve preparations, several rounds of discussions, feedback processes, and subsequent cycles of processes. It is not unnatural to encounter shortcomings and challenges during the negotiations, especially where an increase in the number of issues on the climate agenda, a large number of actors, and group coalitions (such as G77/China, the EU, BASIC involving Brazil, South Africa, India, China, African Groups, LDCs, Umbrella, Alliance of Small Island States; Coalition of Rainforest Nations, Bolivarian Alliance of the Peoples of our Americas) with divergent and overlapping positions over an agenda may have helped or hindered in building compromises, finding common ground interests, and enabling smooth facilitation of negotiation (Betzold et al. 2012). In climate change negotiations, technical ambiguity, mistrust, lack of transparency, burden sharing, and parties' vested interests are some of the common limiting factors for the smooth transitioning of negotiations, the signing of agreements, or the implementation of decisions. Although there could be several nuts and bolts of negotiation processes that are invisible to the public, bottlenecks or challenges of climate change negotiations can be broadly divided into prenegotiations preparations, negotiations, and postnegotiation implementations. The first two are mainly caused by those factors or actors responsible for preventing an agreement or decision from happening. The final are those barriers causing ineffective implementation of the decisions such as meeting agreed targets on GHG emission reduction or smooth channeling of climate funds. The value of pledges for climate fund are increasing; for example, in COP15 in Copenhagen, developed countries promised to create an annual climate fund of US$100 billion by 2020, to help vulnerable developing countries to adapt to the negative impacts of climate change. Unfortunately, readily available climate funds have not been impressive so far. Where funds were made available, they could not be properly channeled and well utilized due to weaknesses such as the lack of sound planning of adaptation measures, developing countries' inability to administer such funds, and a lack of transparent mechanism to distribute funds to the local level.

The KP is an ideal case for showing ex ante and ex post complexities inherent within climate change negotiation because it has faced both welcoming and dubious responses, while its results so far have been mixed. The KP is a binding agreement, which is the first of its kind in the history of climate change negotiation. Although accepted by many, the United States, the then largest GHG emitter, refused to sign it. It was the first major blow to the KP takeoff. The United States' refusal to move onboard did not prevent the others from bring the KP into force. Complications on implementing the KP soon emerged as countries found it less meaningful to comply with their emission reduction (ER) targets as their ER was dampened by big emitters that were not under the KP, notably the United States, or were exempted due to their developing economy status such as China (current largest emitter), India, and Brazil. While differences were escalating, there were increasing pressures to reset new conditions under the principle of common but differentiated responsibility between the developed and developing countries. As a result, the countries could not comply with their commitments while enforcement was inherently difficult. Despite these hurdles, the KP was quite successful in establishing vibrant mechanisms such as CDM, JI, or IET. Within the KP, lack of sufficient trust among parties demanded more transparent and strict approval processes for the functioning of market-based mechanisms such as CDM. In due course, under the direct administration of the UNFCCC, CDM developed into more advanced processes and more refined methodologies, which are desirable for reliability, transparency, and maturity of the market. Despite the progress and the achievements, CDM is often criticized for its lower-than-anticipated performance both in GHG mitigation and in mobilizing sufficient funds (DIIS 2011). Potential project participants found it difficult to fit the eligibility for CDM due to strict criteria like additionality, lengthy bureaucratic process, and associated high transaction cost. Developed countries supporting CDM could not develop full trust with the recipient parties' claims for emission credits due to issues like double counting, false claims, and nonadditionality. In addition, there was the risk of failing CDM approval, for example, due to capacity gaps, strict validation, and verification, which could increase uncertainty over acquiring anticipated emission credits, the main financial attraction of CDM. The 2008 global financial crisis and uncertainty over the second CP of the KP was another setback for CDM that caused an oversupply of emission credits and the crash of credit price. All these obstacles resulted in fewer-than-expected climate change actions under the KP.

Including a water agenda into mainstream climate negotiations appears to be less likely at a time when progress on climate negotiations is slowing down and going through a tough and complicated period. A few factors that need mention include the

International Negotiations on Climate Change and Water            347

increasing number of actors, national self-interests, the growing number of international agreements, and prevalent misunderstanding on technicalities over key agreement issues. These facts reveal that barriers observed in climate negotiation could directly undermine any effort to incorporate water into mainstream climate agenda. Chances for linking water issues into climate negotiations are further diminished by the fact that there are no proper mechanisms, except more technical SBSTA processes, to table the water agenda in climate negotiations. For instance, the IPCC has already produced technical reports on water and climate change (Bates et al. 2008), but they could not be reflected in the policy agreements of the UNFCCC due to the lack of a corresponding mechanism (WCC 2010). Moreover, COP meetings are usually congested with regular agendas and fixed processes, so finding an entry point for water-centric discussion could be a problem.

## 11.6 RECENT DEVELOPMENTS AND EMERGING PARADIGMS IN INTERNATIONAL NEGOTIATIONS

Climate change negotiations are probably at the tipping point so no time could be further wasted on debating "whether to agree or disagree" over stabilizing GHGs within a safe boundary; more focus should be spent on action-oriented consensus building to limit the earth's biosphere from the overwarming trend. The first volume of the recently published IPCC AR5 (WGI) also indicates a similar urgency. As there are no other shortcuts to consensus building than through carefully planned and structured negotiations, ways should be found to orchestrate effective negotiation mechanisms. On the other hand, water resources have reached a stage of highly complex nexus such that isolated modus operandi is no more valid to fix the management puzzles and to establish a practice of good water governance. In order to coordinate the interrelated agendas, broader participation and enhanced cooperation are now viewed as both opportunity and challenge for future negotiations. Though late, stakeholders are gradually embracing the reality that a new level of integrated thinking is indispensible to understand and fix the complexities in a sustainable manner. The fact is also evidenced from recent developments in international environmental negotiations.

### 11.6.1 COP Discussions and Post-Kyoto Regime

COP meetings have grown both in size (participants, and workshops, side events) and areas covered (mitigation, adaptation, financing, technology transfer, and capacity building). With a few exceptions, COPs after the KP have delivered outcomes that are both unique and path-breaking such as the Bali Action Plan, Copenhagen Accord, Cancun Agreement, Durban Agreement, and Doha Amendment to the KP. For example, the Bali Action Plan came up with an idea of individual capacity-based Nationally Appropriate Mitigation Action (NAMA)* in a measurable, reportable,

---

* Excerpts of Bali Action Plan: "Nationally appropriate mitigation actions (NAMAs) by developing country Parties in the context of sustainable development, supported and enabled by technology, financing and capacity-building, in a measurable, reportable and verifiable manner."

and verifiable manner to facilitate countrywide mitigation actions. NAMA came as an alternative arrangement to expand mitigation action beyond the KP boundary and facilitate deep cuts in GHG emission. The Copenhagen COP15 raised climate change policy to the highest political level, with close to 115 world leaders attending the high-level segment (UNFCCC 2013). Despite its failure to strike an ambitious deal as anticipated, it did produce the Copenhagen Accord, supported by a majority of countries, which included the long-term goals of limiting the maximum global average temperature increase to no more than 2°C of pre-industrial levels and long-term finance by developed countries to support a goal of mobilizing US$100 billion a year by 2020 to address developing countries' needs. With near-universal agreement, the Cancun Agreement (COP16) forms the pillars of the largest collective effort the world has ever seen to reduce emissions, in a mutually accountable way, with national plans captured formally at an international level (Grubb 2011; UNFCCC 2013). The Cancun Agreement includes the most comprehensive package ever agreed by governments to help developing nations to adapt climate change through finance, technology, and capacity-building support and speed up their plans to adopt sustainable paths to low emission economies (UNFCCC 2013). Despite these developments, there was turmoil with the completion of the 5-year first commitment period (CP1) of the KP at the end of 2012. The fate of the second commitment period (CP2) was quite uncertain, especially after the unwillingness expressed by some big emitters to enter CP2 under status quo. As agreed at COP17 in Durban, the Doha Amendment (COP18) was successful in launching a CP2 of the KP from January 1, 2013, to December 31, 2020, which helped to clear the existing turmoil. All businesses under the CP1 will continue to function till 2020 after the amendment. COP18 also sets a timetable to adopt a universal climate agreement by 2015, coming into effect in 2020, which will bring all parties under a single umbrella. However, again the key question for us is whether 2020 is too far considering the accelerated accumulation of GHGs into the environment, warning by the IPCC for urgent action, as well as record climate incidents such as experiencing the warmest decade (2001–2010), below or above average precipitation and related disasters, record melting of ice caps and glaciers, rising sea levels, etc. (WMO 2013). Despite slow progress, the basis is that negotiations should continue and agreed actions should be implemented sincerely, and effective ways to accelerate the mitigation activities need to be explored.

In addition to the UNFCCC processes, there are also gestures by individual countries to cut their GHG footprint targets such as China, the United States, the EU, Japan, South Korea, New Zealand, and Australia. Recently, in a surprise move, the United States has introduced its climate change policy, which emphasized, among other issues, on improving its international presence by furthering its role in climate negotiations and cooperation with other countries. Based on the agreements since COP15, countries have disclosed pledges to reduce their emission levels by 2020. The United States and Canada have set the target of reducing GHG emissions by 17% below the 2005 level, and the EU, Japan, and Russia will reduce theirs by 20%–30%, 15%–25%, and 25% below the 1990 level, respectively, while China and India will reduce their emissions by 40%–45% and 20%–25% per GDP below the 2005 level, respectively. Another important development was the establishment of the domestic, bilateral, or international Emission Trading Scheme (ETS) by the EU, China, Japan,

South Korea, the United States (California), New Zealand, and Australia, which could eventually strengthen and spur the growth of the global carbon market (Höhne et al. 2012; Koakutsu et al. 2012). The EU ETS and EU Allowances (EUAs) accounted for the largest (84% trading value in 2011) carbon market by far. Japan, another large carbon market, has introduced its own mechanisms such as Japan-Voluntary Emission Reduction (J-VER) and Joint Credit Mechanism/Bilateral Offset Credit Mechanism (BOCM). Australia implemented a "Carbon Pricing Mechanism" on July 1, 2012, with a target of reducing emissions by 5% by 2020 and 80% by 2050. China has started a pilot emission trading program in selected provinces and cities, while South Korea will implement ETS starting in 2015. These domestic approaches offer relative advantages because each party will have a chance to innovate appropriate mitigation measures, reducing marginal abatement cost and increasing the magnitude of emission reduction. Countries will be able to set their own rules to administer GHG-related transactions to avoid losses through double counting or nonadditionality, to prevent outside interferences, and to minimize the level of mistrust. This will also allow them to explore business opportunities for transferring low carbon technologies, internally organizing and preparing themselves to deal with international opportunities and challenges. Similarly, when feasible, they could show their leadership in international negotiations by sharing successful mitigation measures developed at home.

### 11.6.2 Rio+20 and Post-MDGs Development

In retrospect, the outcome of the First Rio Earth Summit in 1992 (Our Common Future) was a major global realization about the unsustainable pattern of development and our shared responsibility to safeguard the future by pursuing sustainable development pathways. Since then, sustainable development has been a common slogan while undertaking developmental actions. Despite several efforts, there was uneven progress on sustainable development and poverty eradication during this period, while the world has witnessed major transformations in demographics, economic growth, environmental degradation, climate change, and loss of ecosystem services, which are responsible for retarding the progress made so far. After one decade, realizing unsatisfactory progress, world leaders came up with more focused goals and targets, known as the UN Millennium Development Goals (MDGs), to accelerate sustainable development within the 2015 target. Although results so far have been both satisfactory and unsatisfactory, the MDGs successfully created momentum to influence and reorient national and international development policies and resources to achieve measurable goals and targets. With the nearing timeline for MDGs, Rio+20 symbolized two decades of efforts toward attaining sustainable development. The outcome of Rio+20 (Future We Want) continued refocusing actions toward sustainable development but failed to generate comparable momentum to the level of the 1992 Rio Earth Summit and MDGs that the world community was enthusiastically looking for. Rather than digging up the weakness of Rio+20, it would be pragmatic to review key achievements that could help focus on setting future strategies and actions. Although Rio+20 emphasizes less on reinforcing a clear focus on water, it has emphasized on holistic and integrated approaches to sustainable development wherein water could be viewed as an implicit default. In particular, Rio+20 was quite open in acknowledging the positive momentum created by MDGs in fighting with poverty eradication and achieving sustainable development. As a result, Rio+20 also adopted a goal-, target-, and indicator-based approach for the formulation of holistic SDGs, keeping its coherence and integration with the UN post-2015 development agenda. It also outlined several guidelines for constructing SDGs. Among others, SDGs must be based on Agenda 21 and JPoI, respect the 1992 Rio Earth Summit principles, address and incorporate in a balanced way all three dimensions of sustainable development and their interlinkages, and avoid divergence from MDGs. Therefore, the formulation of SDGs will also provide a continuation to the MDGs, probably in a complementary manner, while caution will be exercised not to ignore existing commitments.

A set of 11 multistakeholder consultations on SDG development are ongoing in which water has been identified as one of the consultation themes. A recent final draft of the post-2015 Water Thematic Consultation has identified access to water as a fundamental right for all that cannot be taken away (UN-Water 2013). The report further proposes recommendations for Water and Sanitation, Hygiene (WASH), IWRM, and wastewater management and water quality. About setting goals and targets, it stresses on incorporating water into cross-cutting themes such as energy, food, and health. It also underscores addressing the unfinished and neglected MDGs related to water supply and sanitation by proper recognition of interlinkages and their proper integration. The IWRM principles are again viewed as integration mechanisms to manage nexus, climate change mitigation, adaptation, and building resilience. Whatever will be the final shape of goals, targets, and indicators, their effectiveness will depend on the clarity to address water governance complexities including climate change, easy communicability, and adopted processes to monitor and measure the quantity and quality of the progress.

The second notable theme of the Rio+20 Conference was "green economy (GE) in the context of sustainable development and poverty eradication." GE has recently become a spotlight of policy discussions, but the concept itself was coined nearly two decades earlier (Pearce et al. 1989). It was also included, though indirectly, during the 1992 UNCED declaration (through internalizing environmental cost and eliminating unsustainable production and consumption), but gained its prominence during the post-2008 global economic crisis period as a means to escape from the crisis and attain economic recovery by introducing green stimulus packages (Allen and Clouth 2012). In its simplest expression, GE is low carbon, resource-efficient, and socially inclusive (UNEP 2011a), which decouples economic growth from environmental externalities such as carbon emission, pollution, and resource exploitation, and promotes growth through the creation of new environment-friendly products, industries, and business models that also improve people's quality of life (ADB and ADBI 2013). GE is a major policy shift that the world is yet to experience as a transitioning global society toward the path of GE will require a massive overhaul in the way socioeconomic development is being defined and pursued till date.

Water is intricately linked with GE and is viewed as an engine for GE (MLTM-GRK/PCGG-GRK/K-Water/WWC 2012). In GE, water is the common denominator for agriculture, energy, and the industrial sector, and the role of water services in both maintaining biodiversity and ecosystem services, as "green infrastructure," is recognized, valued, and paid for, while the development and adoption of technologies for efficient use, recycling, and reuse are encouraged (UNEP 2011a; WB 2013). Similarly, the decoupling concept of GE in the context of water could be viewed in two ways that could be the guiding principles to achieve sustainability in water management (UNEP 2011b). The first one is resource decoupling, which aims to economize the use of water per unit of economic activities, while the second one is impact decoupling, which aims to delink economic growth from water resource degradation. The main challenge, however, lies in showing its operational pathways of decoupling concept such as improving water use efficiency, minimizing water footprints, promoting recycle and reuse, and maintaining the required flow of unpolluted water to sustain ecosystem services. Negotiations on water and GE are yet to gain full swing, but it could be a favorable means to comanage water and climate change. For instance, the success of GE also lies in the development, scale-up, and diffusion of low carbon technologies and other sustainable measures including real water savings and conservation of sources. A growing realization of the complex nexus between water and food, energy, health, ecosystem, and climate change will have a visible impact on the adoption of low carbon solutions, especially on renewable energy production. Accelerated investment in water-dependent ecosystems, in greening water infrastructures and in redesigning water governance could expedite a transition to GE.

### 11.6.3 NEW PARADIGMS IN WATER MANAGEMENT

Increasing agreement on recognizing water both as a human right and as an economic good on the backdrop of water security challenges has invigorated new patterns of policy discussions. Similarly, growing degradation of water resources has led to rethinking of restoring and sustaining ecosystem services, such as supporting, regulating, and provisioning, provided by water. Opposed to a fragmented and sector-wise discussion, there is growing attention to reinvent more integrated thinking based on a nexus approach—an approach that integrates management and governance across scales and sectors (Hoff 2011). Recognizing that water governance is no more confined to a sector but extends beyond that is absolutely crucial to develop a nexus understanding (WEF 2011; WWAP 2012). In particular, the nexus of water–energy–food has attracted wider policy attention as all three issues are becoming critical security concerns in the coming days, which demand coordinated management approaches. In fact, the nexus approach could be viewed as an opportunity to make significant progress on attaining sustainable water management by reducing trade-offs, exposing externalities to encourage efficiency gains, and finding management synergies (Hoff 2011). In this context, the nexus approach is also closely related with the concept of GE and better implementation of IWRM principles that aims to strike a balance among the three Es (economic efficiency, environmental sustainability, and equity) (GWP 2012b). Here, nexus tries to contextualize IWRM to the new requirements of various sectors that usually act in isolation. The nexus has been further extended to incorporate issues of climate change (water–food–energy–climate nexus) where water is a common thread because climate change policies could have feedback on water, food, and energy

(GWP 2012b; Hoff 2011; WEF 2011). For instance, mitigation via carbon sequestration, expansion of biofuels, or hydropower can create significant new water demands. However, a nexus approach could lead to a more water-smart strategy for mitigation (such as managing REDD+ beneficially to regulate the water cycle and conserve water) and adaptation (more water-efficient irrigated farming) (Hoff 2011). As water is placed at the nexus of so many global issues, the future boundary of this nexus could grow significantly to encompass more specific branches such as mitigation, adaptation, ecosystem services, land management, health, and industry (WEF 2011). Careful planning and management will be vital to maintain a focus and control the overbranching of the nexus in the coming days.

Closely coupled with the concept of GE and the nexus approach, "global commoditization of water" has gained huge policy discourse recently and has essentially brought the economic value of water back on the discussion table. It is a life cycle perspective of viewing a product or services that use (as green or blue water footprint) and discharge (gray water footprint) water. The metaphor of "virtual water" or "water footprint"—that is, viewing embedded water as a product or service in addition to real water content—has essentially helped to expose the hidden link between water and a commodity. The water footprint started to gain broad interest from about 2008, when the Water Footprint Network was established in order to ensure the establishment of one common language and a coherent and scientifically sound framework for Water Footprint Assessment (WFA), which serves different interests (Ercin and Hoekstra 2012). A growing interest in virtual water is also indicative of a growing realization of the decreasing share over accessible water due to multiple drivers and pressures on resources, which demands an alternative way to view and manage transactions involving water on a global scale. Meanwhile, it presents a way of understanding the transfers of water implicitly in global trade flows (Allan 2003). Its primary implication has been in raising awareness about the total water consumed by a product (irrespective of actual water content) as well as in globalizing water resources. On the one hand, it exposes externality risk to water resources in a producer country due to the external demand for water-intensive products; on the other hand, importing countries will increase their dependency on external countries for water-intensive commodities such as food (Chapagain and Hoekstra 2008). In addition to rising global awareness, there are at least two clear implications of this alternative viewpoint. First, virtual water-based discussion opens an opportunity for improving water use efficiency of a product during various stages of its life cycle, which is closely tied to the GE concept of resource decoupling. In this context, policy could be directed by tagging the import priority to water-efficient products rather than from areas where water is used less efficiently without considering negative environmental consequences. Second, it helps to shift production of water-intensive products from water-scarce regions to areas having a comparative advantage over water distribution and access so that pressures on domestic water resources could be minimized in water-scarce areas. While such considerations will provide a useful entry point into discussing inequities in international trading of virtual water, the concept could be beneficially utilized to establish a link with a similar concept of the carbon footprint, which is well established in the climate change arena (Ercin and Hoekstra 2012). In particular, the concept could be contextualized during discussions on water and energy nexus, in which energy productions are embedded with water footprint while water uses are embedded with carbon footprint. As mechanisms for reducing and offsetting carbon footprints are well developed and implemented, water footprint is yet to adopt a similar approach. Cross-fertilization of the concepts such as carbon–water efficiency, -offsetting, -neutral, -capping, -permits, or -labeling could open new opportunities for comanaging the footprints (Ercin and Hoekstra 2012).

## 11.7 CONCLUSIONS

There are no panaceas for complex problems such as global warming (Ostrom 2010); nor are negotiations alone capable of taming the climate change. It is evident that climate change and water crisis are likely to constrain our efforts toward sustainable development and poverty eradication unless ambitious goals are set and a corresponding scale of actions are multiplied. Well-orchestrated negotiations at the international level have the potential to trigger effective policy shifts and direct required actions down to the local level. Concurrently, the current stalemate in climate negotiations should come to an end as a broad objective of international negotiations is quite clear—control global warming. No spaces should be allowed during negotiations that could undermine the focus of discussion. Building upon recent progress, future negotiations are required to show creativity that will lead to both ambitious and actionable steps in addressing climate change and water side by side. Negotiations should also eliminate moldable administrative and organizational barriers, normalize broken relations, and show flexibility over individual positions. Similarly, it should avoid misunderstanding by retrofitting discussions with better preparations, sound facts, decision-supporting tools, and objective agendas.

This chapter has shown that discussions on the issue of climate change and water have intensified at different platforms and on various occasions in recent years although progress on both fronts is lagging behind. A strong and direct link between climate change and water negotiations is yet to be established, while this gap also represents a lost opportunity in dealing with the problem of climate change and threats to water security. Consequently, on the climate change side, a lack of space for discussion on water issues has partly influenced the implementation of water-dependent mitigation options; at the same time, progress on preparing a sound framework for investing in water adaptation is not so impressive. Similarly, there has been negligible progress in streamlining water-related mitigation and adaptation actions in parallel such as renewable energy production, REDD+. It was suggested elsewhere that connecting mitigation and adaptation is advantageous for minimizing trade-offs, identifying synergies, enhancing response capacity, and strengthening institutional coordination (Swart and Raes 2007).

On the water side, discussions have shifted more toward integrated approaches in which IWRM and improved governance are viewed as melting pots for addressing multiple issues and for achieving sustainability. A narrow view on water as a sector is no more valid as momentum is building up in support of the broader role that water has to play in all segments of economy. Complementing the past achievements, there is a growing understanding to view water as a "bloodstream" to link with multiple sectors, thereby involving stakeholders operating outside, but at the periphery, of the "water box." Unlike climate change negotiations, the water community seems to be well aware about the negative impacts of climate change on water, especially from the adaptation point of view. Only limited progress has been made in transferring the outcome of water negotiations into the mainstream UNFCCC discussion in particular toward establishing a work program on water and climate change under the NWA through SBSTA. In order to generate a more significant impact through water-centric discussions, operational implication of integrated approaches such as IWRM, nexus approach, GE, and footprints should be found that could be easily mainstreamed into future processes such as SDGs, post-2015 development agenda, adaptation (NWA, NAPA) and mitigation (post-Kyoto discussions, NAMA development), planning, and so forth.

It is extremely hard to find a way forward in the discourse of climate and water negotiation amid the existing situation of confusion stemming across various negotiation platforms operating independently. Similar existing integration gaps across sectors dealing with water issues make the situation more complex to propose an easy outlet. Yet, authors find it tempting to attempt the most obvious directions that could facilitate more coordinated discussions. Without a doubt, it could be said that consolidating climate change and water negotiation plays a pivotal role in opening win–win opportunities in terms of improved water governance and development of adaptive water management strategies that could enhance autolearning and adjustments, accommodation of cross-cutting issues and creation of a sitting environment for actors with different backgrounds, and effective implementation of mitigation. But materializing this symbiosis is also contingent upon the fulfillment of certain prerequisites that will support the negotiation process. Among them, relevant tools and approaches should be devised that could facilitate negotiations among actors with varied backgrounds. At the same time, available methods and tools should be customized so that they could be strategically targeted to raise general awareness, starting from the general public to decision makers, about the cost and benefit of linking water and climate issues, which will eventually create a pressure in favor of quick and fruitful negotiations. More importantly, there is a need to find fast-track mechanisms to channel available investment opportunities toward actions resulting in adaptation and mitigation synergies. Most importantly, negotiations should focus more effort in finding ways for enhancing the capacity of the relevant actors to enable them to identify, prioritize, and plan actions that result in better distribution and use of climate funds. Last but not the least, the framework for monitoring progress on the implementation of agreements and achievements of goals/targets should be established in order to measures the effectiveness of integrated approaches of water management.

## REFERENCES

ADB and ADBI. 2013. Low-carbon green growth in Asia: Policies and practices. A joint study. Tokyo, Japan: Asian Development Bank (ADB) and Asian Development Bank Institute (ADBI).

Allan, J.A. 2003. Virtual water—The water, food, and trade nexus: Useful concept or misleading metaphor? *Water International* 28(1): 106–113.

Allen, C. and Clouth, S. 2012. Green economy, green growth, and low-carbon development—History, definitions and a guide to recent publications. A guidebook to the Green Economy, Issue 1. New York: UN Division for Sustainable Development (UNDESA).
Bates, B.C., Kundzewicz, Z.W., Wu, S., and Palutikof, J.P. 2008. Climate change and water. 2008. Technical paper of the Intergovernmental Panel on Climate Change (IPCC), 210pp. Geneva, Switzerland: IPCC Secretariat.
BBC. 2013. *Carbon dioxide passes symbolic mark.* May 10, 2013. http://www.bbc.co.uk/news/science-environment-22486153 (accessed August 10, 2013).
Betzold, C., Castro, P., and Weiler, F. 2012. AOSIS in the UNFCCC negotiations: From unity to fragmentation? *Climate Policy* 12(5): 591–613.
Biswas, A.K. 1988. United Nations water conference action plan. *International Journal of Water Resources Development* 4(3): 148–159.
Biswas, A.K. 2008. Integrated water resources management: Is it working? *International Journal of Water Resources Development* 24(1): 5–22.
Cap-Net. 2009. IWRM as a tool for adaptation to climate change. Training Manual and Facilitator's Guide, Cap-Net/UNESCO-IHE. Pretoria, South Africa.
Chapagain, A.K. and Hoekstra, A.Y. 2008. The global component of freshwater demand and supply: An assessment of virtual water flows between nations as a result of trade in agricultural and industrial products. *Water International* 33(1): 19–32.
DIIS. 2011. Climate change negotiations and their implications for international development cooperation. DIIS report. Copenhagen, Denmark: Danish Institute for International Studies (DIIS).
Ercin, A.E. and Hoekstra, A.Y. 2012. *Carbon and Water Footprints: Concepts, Methodologies and Policy Responses.* Side Publication Series 4. Perugia, Italy: UN World Water Assessment Program (WWAP)/UNESCO.
Ghosh, A. and Woods, N. 2009. Governing climate change: Lessons from other governance regimes. In: D. Helm and C. Hepburn (eds.), *The Economics and Politics of Climate Change*, pp. 454–477. Oxford, U.K.: Oxford University Press.
Grubb, M. 2011. Cancun: The art of the possible. *Climate Policy* 11(2): 847–850.
Gupta, J. 2012. Negotiating challenges and climate change. *Climate Policy* 12(5): 630–644.
GWP. 2012a. GWP in action. Annual report. Stockholm, Sweden: Global Water Partnership (GWP).
GWP. 2012b. Water in the green economy. GWP perspectives paper. Stockholm, Sweden: Global Water Partnership (GWP).
Hoff, H. 2011. *Understanding the Nexus: Background Paper for the Bonn 2011 Nexus Conference. The Water, Energy and Food Security Nexus*, November 16–18, 2011. Bonn, Germany: Stockholm Environment Institute (SEI).
Höhne, N., Braun, N., Fekete, H. et al. 2012. *Greenhouse Gas Emission Reduction Proposals and National Climate Policies of Major Economies. Policy Brief.* Utrecht, the Netherlands: Ecofys/PBL Netherlands Environmental Assessment Agency.
Howard, G., Charles, K., Pond, K. et al. 2010. Securing 2020 vision for 2030: Climate change and ensuring resilience in water and sanitation services. *Journal of Water and Climate Change* 1(1): 2–16.
INBO/GWP. 2012. *Handbook for Integrated Water Resources Management in the Basins of Transboundary Rivers, Lakes and Aquifers.* Paris, France: International Network of Basin Organizations (INBO) and the Global Water Partnership (GWP).
IPCC. 2007. Climate change 2007. IPCC fourth assessment report. Geneva, Switzerland: The Intergovernmental Panel on Climate Change (IPCC).
IPCC. 2013a. Working Group I climate change 2013: The physical science basis summary for policymakers. IPCC fifth assessment report (AR5). Geneva, Switzerland: The Intergovernmental Panel on Climate Change (IPCC).

IPCC. 2013b. Intergovernmental panel on climate change. http://www.ipcc.ch/ (accessed June 2013).

Koakutsu, K., Usui, K., Fukui, A. et al. 2012. Towards the CDM 2.0: Lessons from the Capacity Building in Asia. IGES policy report 2012-07/IGES CDM Reform Series No. 3. Hayama, Japan: Institute for Global Environmental Strategies (IGES).

MLTM-GRK/PCGG-GRK/K-Water/WWC. 2012. Water and green growth. Thematic Publication, The Ministry of Land, Transport and Maritime Affairs-Government of the Republic of Korea (MLTM-GRK), Presidential Committee on Green Growth (PCGG-GRK), Korea Water Resources Corporation (K-water), World Water Council (WWC). Gyeonggi-Do, Seoul, and Daejeon, South Korea/Marseille, France.

Ostrom, E. 2010. Polycentric systems for coping with collective action and global environmental change. *Global Environmental Change* 20: 550–557.

Pearce, D.W., Markandya, A., and Barbier, E. 1989. *Blueprint for a Green Economy*. London, U.K.: Earthscan.

Sands, P. 2003. *Principles of International Environmental Law*. New York: Cambridge University Press.

Savenije, H.H.G. and Van der Zaag, P. 2008. Integrated water resources management: Concepts and issues. *Physics and Chemistry of the Earth* 33: 290–297.

Stern, S.N. 2006. *The Economics of Climate Change*. Cambridge, U.K.: Cambridge University Press.

Swart, R. and Raes, F. 2007. Making integration of adaptation and mitigation work: Mainstreaming into sustainable development policies? *Climate Policy* 7(4): 288–303.

UNEP. 2012. The emissions gap report 2012, Nairobi, Kenya: United Nations Environment Programme (UNEP).

UNEP. 2011a. Towards a green economy: Pathways to sustainable development and poverty eradication. Nairobi, Kenya: United Nations Environment Programme (UNEP).

UNEP. 2011b. Decoupling natural resource use and environmental impacts from economic growth. A report of the working group on decoupling to the international resource panel. Nairobi, Kenya: United Nations Environment Programme (UNEP).

UNFCCC. 2011a. Water and climate change impacts and adaptation strategies. Technical paper FCCC/TP/2011/5, UNFCCC, Bonn, Germany. http://unfccc.int/resource/docs/2011/tp/05.pdf.

UNFCCC. 2011b. Climate change and freshwater resources: A synthesis of adaptation actions undertaken by Nairobi work programme partner organizations. Nairobi Work Programme on Impacts, Vulnerability and Adaptation to Climate Change, United Nations Framework Convention on Climate Change (UNFCCC). Bonn, Germany.

UNFCCC. 2013. United Nations Framework Convention on Climate Change (UNFCCC). http://unfccc.int/ (accessed June 2013).

UN-Water. 2013. The post 2015 water thematic consultation report. The World We Want 2015 Water Thematic Consultation, UN. New York.

Vollmer, R., Ardakanian, R., Hare, M., Leentvaar, J., Van der Schaaf, C., and Wirkus, L. 2009. Institutional capacity development in transboundary water management. World Water Assessment Programme. UNESCO. Paris, France.

WB. 2013. At a glance: Water and green growth. Available at: http://water.worldbank.org/node/84118 (accessed September 2013).

WCC. 2010. Towards a work programme on water and climate under the UN framework convention on climate change. Discussion paper. Water and Climate Coalition (WCC).

Weart, S. 2013. *The Discovery of Global Warming*. Cambridge, MA: Harvard University Press.

WEF. 2011. *Water Security: The Water-Food-Energy-Climate Nexus World Economic Forum Water Initiative*. Washington, DC: The World Economic Forum (WEF), Island Press.

WHO/UNICEF. 2012. Progress on drinking water and sanitation: 2012 update. WHO/UNICEF joint monitoring programme for water supply and sanitation. New York: UNICEF and World Health Organization.

WMO. 2013. The global climate 2001–2020: A decade of climate extremes. Summary report. Geneva, Switzerland: World Meteorological Organization (WMO).

WWAP. 2003. UN world water assessment report 1: Water for people, water for life. UN world water assessment report. Paris, France: World Water Assessment Program (WWAP), UNESCO.

WWAP. 2006. UN world water development report 2: Water a shared responsibility. UN world water development report. Paris, France: World Water Assessment Program (WWAP), UNESCO.

WWAP. 2009. The United Nations world water development report 3: Water in a changing world. World water development report. Paris, France: UN World Water Assessment Program (WWAP), UNESCO.

WWAP. 2012. UN world water development report 4: Managing water under uncertainty and risk. UN world water development report. Paris, France: World Water Assessment Programme (WWAP), UNESCO.

Xu, J., Shrestha, A.B., Vaidya, R., Eriksson, M., and Hewitt, K. 2007. The melting Himalayas—Regional challenges and local impacts of climate change on mountain ecosystems and livelihoods. ICIMOD technical paper. Kathmandu, Nepal: International Centre for Integrated Mountain Development (ICIMOD).

# Index

## A

Adaptation
  climate change negotiations, 333
  framework, 189
  vulnerability assessment and, 188–189
  in water resources, 209–211
    barriers for mainstreaming adaptation, 229–231
    budgeting and allocation, 222–223
    CBWM, 222
    climate cobenefits and, 227–229
    decision support systems, 211–215
    differentiating adaptation actions, 215–217
    IWRM, 225–226
    managing uncertainty, 217
    multistakeholder engagement, 217
    nonstructural options and approaches, 217, 219, 221–225, 231
    pricing of, 223
    projected climate change impacts, 215
    restriction, 224
    scheduling, 223–224
    structural options/measures, 217–220, 230
    upstream and downstream integration, 226–227
    water trading, 225
Adaptive capacity, vulnerability assessment, 188, 195–196
Adaptive water management (AWM), 226
Aerosols
  climate change and, 70–71
  climate variability and change, 49–51
  human-induced increases to, 54, 56–58
Agenda 21, 337, 342
Agriculture
  climate change impacts, 129–132
  water pricing, 223
Anthropogenic climate change
  definition, 32–33
  human-induced increases
    to aerosols and other pollutants, 54, 56–58
    to greenhouse gases, 54–55
    land-use change, 58–59
    urbanization, 58–59
Asian brown cloud, 51
Asthenosphere, 16
Atmosphere, 3–4
  chemistry, 49–51
  Ferrell cell, 5
  general circulation, 4–8
  Hadley cells, 5
  intertropical convergence zone, 6
  role in climate, 7, 9
  stratosphere, 4
  troposphere, 3

## B

Bagmati River basin (BRB), climate change impacts, 135–138
  methodology, 138
  statistical downscaling model, 138
  streamflow, 140–141
  temperature and precipitation, 138–139
  water availability, 141
Bali Action Plan, 347–348
Bayesian model averaging (BMA) method, 87–88
Biosphere, 17

## C

Cancun Agreement, 348
Cascade of uncertainty, 241–242
Clean Development Mechanism (CDM), 343, 346
Climate
  definition, 1–2, 32
  nonuniform energy distribution, 2
Climate change
  adaptation, see Adaptation
  and aerosols, 70–71
  anthropogenic forcing, 71
  attribution of, 77–78
  Cancun Agreement, 348
  and $CO_2$, 69–71
  control mode, 72
  Copenhagen Accord, 348
  detection of, 71–77
  economics, see Economics of climate change
  global warming and, 331–332
  greenhouse gases (GHGs), 69–71
  impacts, 110–113
    global climate models (GCMs), 111
    nonclimatic drivers, 112
    regional climate models (RCMs), 113
    statistical downscaling, 113
    on water resources, see Water resources, climate change impacts
    on water use sectors, see Water use sectors, climate change impacts

359

international negotiations on water, 336
  Agenda 21, 337
  challenges, 344–347
  Conference of Parties and post-Kyoto regime, 347–349
  green economy, 350–351
  IWRM, 340–341
  major development, 338–339
  Mar del Plata Action Plan, 337
  "Our Common Future", 337
  Rio+20 and MDGs, 349–351
  Rio Earth Summit, 337, 340
  Sustainable Development Goals, 350
  UNCHE, 337
  water governance, 342
  water management, 351–353
Kyoto Protocol and, 343, 346–348
negotiations, 333–336
  adaptation, 333
  mitigation, 332–333
projected impacts, 215
rainfall observations, 73–76
sensitivity, 100–103
SST during boreal summer, 73–74
surface temperature record, 72–73
and transboundary river basins, 322
uncertainty in, see Uncertainty
vulnerability assessment, see Vulnerability assessment, climate change
and water-related negotiations, 342–344
Climate-informed decision analysis (CIDA), 257
Climate-sensitive decision making, 256–257
  science-first and decision-centric approaches, 258
Climate variability and change, 31–32
  aerosols, 49–51
  Asian brown cloud, 51
  atmospheric chemistry, 49–51
  definitions, 32–33
  Earth–Sun–Moon interactions, 46–49
  *El Niño/Southern Oscillation*, see *El Niño/Southern Oscillation* (ENSO)
  ENSO Modoki, 34
  geological drivers, 52–54
  Hadley circulations, 34, 37–38
  human-induced increases
    to aerosols and other pollutants, 54, 56–58
    to greenhouse gases, 54–55
    land-use change, 58–59
    urbanization, 58–59
  hydroclimate, 32
  Indian Ocean Dipole (IOD), 39–41
  Indian Ocean variability, 39–41
  Indonesian Throughflow (ITF), 41–42
  interdecadal Pacific oscillation (IPO), 32, 38
  jet stream, 42–43
  Kelvin waves, 38
  Northern Hemisphere climate modes, 43–44
  ocean–atmospheric circulations and interactions, 33–46
  Pacific decadal oscillation (PDO), 32, 38
  planetary wave, 43
  quasibiennial oscillation (QBO), 43
  Rossby waves, 38, 42–43
  SAM and NAM, 41
  solar cycle/Sunspot Cycle, 47–48
  stratospheric cooling, 58
  surface and deeper ocean circulations, 44–46
  thermohaline circulation (THC), 44–45
  trade winds, 34, 37–38
  volcanic eruptions, 49–51
  Walker circulation, 33–34, 37–38
  water management and planning processes, 31
  wave train, 43
Colorado River basin
  chronology of events in, 313–315
  country areas in, 311
  hydrology, 311
  reservoirs in, 312
  terrain, 310
Colorado River delta, 311–312
Community-Based Risk Screening Tool—Adaptation and Livelihoods (CRiSTAL) tool, 249, 255, 258
Community-based water resources management (CBWM), 222
Copenhagen Accord, 348
Cryosphere, 14–16
  albedo, 15
  frozen water, 14
  snow cover and sea ice, 14
  surface energy balance, 15

## D

Danube River basin, 299
  basin-wide approach, 309–310
  chronology of events in, 302–309
  country areas in, 300
  Danube River Protection Convention, 301
  Environmental Program for, 301
  WaterGAP Global Hydrology Model, 301
Danube River Protection Convention, 301
Droughts
  adaptation, 219
  climate change impacts, 127
  in Ethiopia, 156
  water rationing, 224
Dynamic and Interactive Vulnerability Assessment (DIVA) model, 172

# Index

## E

Earth–Sun–Moon interactions, 46–49
Economics of climate change, 153–154
   adaptation, 166–170
      assessing costs, 172
      DIVA model, 172
      pluvial flood risk assessment, 172–173
      top-down approach, 171–172
      tropical cyclones and storm surges in Bangladesh, 169
      water supply in Berg River basin, 168
   development, 154–155
   impacts, 155–159
      and adaptation costs, 175
      coastal storm in Dade County, 157–158
      drought in Ethiopia, 156
      flood in Japan, 156
      market/tangible, 159
      nonmarket/intangible, 159
      sea-level rise in coastal zones, 156–157
      warming, 158
      water resources in United States, 157
      water sector, 155–158
   methodological approaches, 170–173
      engineering rules of thumb, 162
      framework to assess local economic impacts, 164–165
      GRACE model, 161–162
      hydroeconomic modeling framework, 161–163
      IAMs, 159
      IIAWM, 161
      macroeconomic general equilibrium model, 161–162
      Mendelsohn model/global impact model, 159
      PAGE2002, 161
      "Tol" model, 160–161
      value of a statistical life (VSL), 165
      willingness to accept/avoid change (WTA), 165–166
      willingness to pay (WTP), 165–166
   wastewater treatment plants (WWTPs), 176
   water resources sector, *see* Water resources sector, climate change economics
*El Niño/Southern Oscillation* (ENSO), 33
   characteristics, 36–37
   interdecadal, 38–39
Expert elicitation, vulnerability assessment, 200

## F

Floods
   climate change impacts, 121–127
   impacts in Japan, 156

Food and water security integrated system (FAWSIM), 214
Freshwater, 22, 291
   cryosphere, 14
   hydrosphere, 7

## G

Ganges–Brahmaputra–Meghna (GBM) River basin, 292, 295
   chronology of events in, 298–299
   general circulation models, 296
   problems, 296–297
   regional climate models, 296
GE, *see* Green economy (GE)
General circulation models (GCMs), 296, 316–317
Global climate system components
   atmosphere, 3–4
      Ferrell cell, 5
      general circulation, 4–8
      Hadley cells, 5
      intertropical convergence zone, 6
      role in climate, 7, 9
      stratosphere, 4
      troposphere, 3
   biosphere, 17
   cryosphere, 14–16
      albedo, 15
      frozen water, 14
      snow cover and sea ice, 14
      surface energy balance, 15
   hydrosphere, 7, 9–10
      Antarctic bottom water, 12
      circumpolar deep water, 13
      Ekman transport, 11–13
      Hadal zone, 10
      layers of ocean, 9–10
      North Atlantic deep water, 12
      oceanic circulation, 11–13
      one atmosphere, 9
   lithosphere, 16–17
      asthenosphere, 16
   schematic view, 2
Global energy balance, 17–19
   Callendar effect, 20
   electromagnetic radiation, 17, 19
   greenhouse effect, 19–21
   greenhouse gases, 19
Global impact model (GIM), 159–160
Global Responses to Anthropogenic Changes in the Environment (GRACE) model, 161–162
Global warming, 69
   and climate change, 331–332
   irrigation, 114
Global Water Partnership (GWP) toolbox, 344
   for IWRM, 251

# Index

Green economy (GE), 350–352
Green growth, 210
Greenhouse effect, 2
   global energy balance, 19–21
Greenhouse gases (GHGs)
   accumulation, 331–332, 344, 348
   climate change, 69–71
   definition, 32–33
   emission, 345–346
   human-induced increases to, 54–55
Groundwater, 312
   availability, 117–120
   quality, 129–130

## H

Hadley circulations, 34, 37–38
Hydroeconomic model, 161
Hydrological cycle, 21–22
   flows/fluxes, 23–24
   reservoirs, 22–23
   residence time, 24
   saturation vapor pressure, 25
   water vapor, 24–26
Hydrological models
   CMIP3 and CMIP5 experiments, 97–100
   uncertainty in, 85, 89, 91–92
      bias correction, 90–91
      downscaling methods, 89–90
Hydropower, climate change impacts, 114–115, 132–134
Hydrosphere, 7, 9–10
   Antarctic bottom water, 12
   circumpolar deep water, 13
   Ekman transport, 11–13
   Hadal zone, 10
   layers of ocean, 9–10
   North Atlantic deep water, 12
   oceanic circulation, 11–13
   one atmosphere, 9

## I

Indian Ocean Dipole (IOD), 39–41
Integrated assessment models (IAMs), 159
Integrated irrigated agriculture water model (IIAWM), 161
Integrated water resources management (IWRM), 225–226, 245
   climate change international negotiations, 340–341
   GWP toolbox for, 251
   under NWA, 342
   principles, 350–351

Intergovernmental Panel on Climate Change (IPCC), 186, 210, 295
   climate change adaptation, 188
   uncertainty, 199
   and UNFCCC, 333, 335–336
International Commission for the Protection of the Danube River (ICPDR), 301, 309
International freshwater agreements, 323
Irrigation
   climate change impacts, 114
   scheduling, 223–224
IWRM, *see* Integrated water resources management (IWRM)

## K

Kelvin waves, 38
Kyoto Protocol (KP), 343, 346–348

## L

Lesotho Highlands Water Project Agreement, 324
Lithosphere, 16–17

## M

Mar del Plata Action Plan, 337, 340
Mekong River Agreement, 324
Mendelsohn model, 159–160
Millennium Development Goals (MDGs), 349–351
Mitigation
   climate change negotiations, 332
   vulnerability assessment, 188
Monte Carlo Method (MCM), 200

## N

Nationally Appropriate Mitigation Action (NAMA), 347–348
National Water Academy (NWA), 342
Natural climate variability/change, 32
Nile River basin, 317
   chronology of events in, 318–321
   country areas in, 317
   precipitation and, 316–317
Northeast Thailand, rice cultivation, 142–143
   climate projections, 144
   GCM scenario on, 145–146
   KDML105, 144–145
   methodology, 144
   temperature and $CO_2$ effect, 146–147
Northern annular mode (NAM), 41
Northern Hemisphere climate modes, 43–44

# Index

## O

Ocean-atmospheric circulations and interactions
  El Niño/Southern Oscillation, 33
    characteristics, 36–37
    interdecadal, 38–39
  ENSO Modoki, 34
  Hadley/Walker circulations, 34, 37–38
  Indian Ocean variability, 39–41
  Indonesian Throughflow, 41–42
  jet streams, 42–43
  Northern Hemisphere climate modes, 43–44
  oceanic Rossby and Kelvin waves, 38
  planetary wave, 43
  quasibiennial oscillation, 43
  SAM and NAM, 41
  surface and deeper ocean circulations, 44–46
  trade winds, 34, 37–38

## P

Planetary wave, 43
Policy Analysis of the Greenhouse Effect 2002 (PAGE2002) IAM, 161
Providing Regional Climates for Impacts Studies (PRECIS), 296

## Q

Quasibiennial oscillation (QBO), 43

## R

Rapid climate change adaptation assessment (RCAA), 249
Regional climate models (RCMs), 296
Reliability ensemble averaging (REA) method, 86–87
Reservoirs, hydrological cycle, 22–23
Residence time, hydrological cycle, 24
Rice cultivation, in Northeast Thailand, 142–143
  climate projections, 144
  GCM scenario on, 145–146
  KDML105, 144–145
  methodology, 144
  temperature and $CO_2$ effect, 146–147
Rio Earth Summit, 335, 337, 340, 342
  and MDGs, 349–351
River basin organizations (RBOs), 226
Robust decision making (RDM), 250, 257
Rossby waves, 38, 42–43

## S

Sanitation, water supply and, 134–135
Sea-level rise
  climate change impacts, 119–121
  in coastal zones, 156–157
SimCLIM tool, 250
Southern annular mode (SAM), 41
Streamflow
  Bagmati River basin, 140–141
  uncertainties, 100
Subsidiary Body for Scientific and Technological Advice (SBSTA), 342
Surface water
  availability, 115–116
  quality, 127–128

## T

Thermohaline circulation (THC), 44–45
"Tol" model, climate change economics, 160–161
Transboundary river basins
  climate change and, 322
  Colorado River basin, 310
    chronology of events, 313–315
    country areas in, 311
    reservoirs in, 312
  by continents, 293
  Danube River basin, 299
    basin-wide approach, 309–310
    chronology of events in, 302–309
    country areas in, 300
    Danube River Protection Convention, 301
    Environmental Program for, 301
    WaterGAP Global Hydrology Model, 301
  frameworks, 293
  Ganges–Brahmaputra–Meghna River basin, 295–299
  Institute of International Law (IIL), 323
  international freshwater agreements/declarations, 323
  International Law Commission (ILC), 323
  international water treaties, 322–325
  management, 291, 294
  Nile River basin, 315–316
    chronology of events in, 318–321
    country areas in, 317
    precipitation and, 316–317
  regional accords, 323–324
Tropical Pacific climate, characteristics, 36–37
  atmospheric feedbacks, 36
  Bjerknes feedback, 36
  delayed ocean adjustment, 36

intraseasonal variability, 37
small-scale features, 37
surface zonal current, 36
thermocline, 36
upwelling, 36

## U

Uncertainty, 81–83
  aleatory, 83–84
  approaches, 85–86
  bias correction, 90–91
  BMA method, 87–88, 91–92
  bottom-up approach, 82–83
  cascade of, 241–242
  CMIP3 and CMIP5 experiments, 97–100
  in decision making, 217
  definition, 198–199
  downscaling methods, 89–90
  global sensitivity analysis, 86
  GLUE method, 91
  in hydrological models, 85, 89, 91–92
  local sensitivity analysis, 86
  Monte Carlo simulation, 85
  Moy River basin, 92–94
    climate change sensitivity, 100–103
  Paint Rock River basin, 97–100
  probabilistic approach for impact assessment, 93–97
  quantifying, 86–88
  quantile mapping method, 90–91
  REA method, 86–87
  Republic of Ireland, 92–94
    climate change sensitivity, 100–103
  scenario-led approach, 92–93
  scenario-neutral approach, 83
  screening analysis, 86
  sources, 83–85
  Special Report on Emissions Scenarios, 84
  top-down approach, 82–83
  in vulnerability assessment
    assessment methodologies, 199–200
    definition, 198–199
    levels, 199
    multiple sources, 201
    scales and languages, 199–200
  for water sector, 241–244
UN Conference on the Conservation and Utilization of Resources (UNCCUR), 337
*UN Conference on the Human Environment (UNCHE),* 333
UN Framework Convention on Climate Change (UNFCCC), 333, 342
  IPCC and, 335–336
Urbanization, land-use change, 58–59

## V

Volcanic eruptions, 49–51
Vulnerability assessment, climate change
  adaptation planning and, 188–189
  components
    adaptive capacity, 188, 195–196
    assessing, 191–192
    exposure, 188, 192–193
    sensitivity, 188, 193–195
  evolution
    concepts and definitions, 184–186
    nomenclature of vulnerability, 187–188
    quantifying vulnerability, 186–187
  expert elicitation, 200
  gathering relevant data and expertise, 191
  goals and objectives, 190–191
  with hypothetical case, 202–206
  illustration, 202–206
  indicators, 194–195
  MCM, 200
  mitigation and resilience, 188
  scenario analysis, 200
  scientific use, 184
  steps, 190–192
  structuring, 190–191
  terminologies, 188
  uncertainties
    assessment methodologies, 199–200
    definition, 198–199
    levels, 199
    multiple sources, 201
    scales and languages, 199–200
Vulnerability index (VI), 187
  communication, 197–198
  computation, 196–197
  interpretation, 197–198

## W

Walker circulation, 33–34, 36–38
Wastewater treatment plants (WWTPs), 176
Water availability, climate change impacts
  Bagmati River basin, 141
  groundwater, 117–120
  surface water, 115–116
Water Evaluation and Planning System (WEAP), 250, 254
WaterGAP Global Hydrology Model (WGHM), 301
Water management, 332
  cooperative, 323
  new paradigms in, 351–353
  nexus approach and, 351–352

# Index

Water pricing, 223
Water quality
 groundwater, 129–130
 surface water, 127–128
Water resource management (WRM), 243
Water resources, 113–114
 adaptation in, *see* Adaptation, in water resources
 Bagmati River basin (BRB), 135–138
  methodology, 138
  statistical downscaling model, 138
  streamflow, 140–141
  temperature and precipitation, 138–139
  water availability, 141
 degradation of, 351
 droughts, 127
 floods, 121–127
 four dimensions, 195
 sea-level rise, 119–121
 shared, cooperation, 324
 water availability
  Bagmati River basin, 141
  groundwater, 117–120
  surface water, 115–116
 water quality
  groundwater, 129–130
  surface water, 127–128
Water resources sector, climate change economics
 climate and economic sensitivity matrix, 177
 impacts and adaptations, 178
 in New York State, 173
  approach, 174–176
  context, 174
  results, 176–178
Water safety planning (WSP), 249
Water sector, 239
 climate risks on, 240–241
  cascade of uncertainty, 241–242
  deep uncertainty and implications, 241–244
  quantifiable uncertainty, 241
 tools, 245, 261
  data and information provision tools, 246
  decision support, 244–245
  knowledge-sharing tools, 246
  in monsoon climates, 240
  process guidance tools, 246

  technologies for climate adaptation, 251
  tools
   for climate risk management, 245, 261–281
   and climate-sensitive support, 256–257
  CRiSTAL, 249
  emergence, 244–245
  functions, 246
  GWP toolbox on IWRM, 251
  *niches,* 251–253
  and robust decision support methods, 256–259
  SimCLIM, 250
  technologies for climate adaptation, 251
  types, 246–248
  user demand evidence for, 253–255
  WASH providers, 249
  Water Evaluation and Planning System, 250
  water safety planning, 249
  Water Security Index, 250
  weADAPT, 251
Water Security Index, 250
Water supply, and sanitation, 134–135
Water Supply and Sanitation Collaborative Council (WSSCC), 340
Water supply, sanitation, and hygiene (WASH), 243
 RCAA for, 249
Water trading, 225
Water use sectors, climate change impacts
 agriculture, 129–132
 CERES-rice model, 142, 144, 146
 hydropower, 114–115, 132–134
 industry, 132–134
 irrigation, 114
 rice cultivation in Northeast Thailand, 142–143
  climate projections, 144
  GCM scenario on, 145–146
  KDML105, 144–145
  methodology, 144
  temperature and $CO_2$ effect, 146–147
 water supply and sanitation, 134–135
Water vapor, hydrological cycle, 24–26
weADAPT tool, 251
World Water Development Report, 256